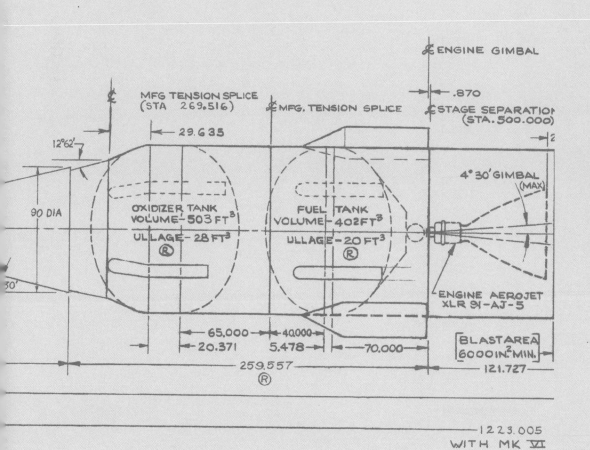

Titan II outboard configuration.
6 July 1960. The Martin Co., Denver.

TITAN II

TITAN II

*A History of a Cold War
Missile Program*

DAVID K. STUMPF

With a Foreword by
Lt. Gen. Jay W. Kelley, USAF, Retired

The University of Arkansas Press
Fayetteville
2000

22 21 20 19 18   9 8 7 6 5

Designed by Liz Lester

♾ The paper used in this publication meets the minimum requirements of
the American National Standard for Permanence of Paper for Printed Library
Materials Z39.48–1984.

LIBRARY OF CONGRESS CATALOGING-IN-PUBLICATION DATA

Stumpf, David K., 1953–
    Titan II : a history of a Cold War missile program / David K. Stumpf ;
with a foreword by Jay W. Kelley.
       p.   cm.
    Includes bibliographical references and index.
    ISBN 1-55728-601-9 (cloth)
    1. Titan (Missile)—History. I. Title.

UG1312.I2 S78 2000
358.1'754'0973—dc21   00-026881

*Funding provided by a Department of Defense Legacy Program grant to the
Arkansas Historic Preservation Program*

All photos are official U.S. Air Force photographs unless otherwise noted.

In Memory of Those Who Lost Their Lives
at the Operational Launch Complexes

308TH STRATEGIC MISSILE WING
Launch Complex 373–4   9 August 1965

Lucian J. Adams
William Q. Bailey
Kendall Belote
Bill E. Bennett
Jim S. Best
Willis L. Briley
Joe Cloud Sr.
Freddie Conway
Lowell H. Cook
Donald A. Dean
C. L. Dove
John P. Elkins
James F. Evans
Harry H. E. Fisher
Charles H. Fulson
Archie Hamilton
James C. Harvey
J. D. Headley

Henry H. Hegi
Raymond Helton
Sam C. Hicks
Thomas L. Hoggard
William C. Holden
William R. Kell
Hershell R. Linn
Archie A. Martin
Charles P. McMahan
Herbert E. Melton
William D. Merchant
Jack Milan
Junior E. Mobbs
Delphard H. Owens
Gillis R. Patterson
Foster W. Pemberton
Sidney S. Phillips
L. M. Phillips

Aubrey E. Reynolds
G. W. Richmond
Presley H. Sanson
Bobbie G. Scott
Marion A. Sewell
Charles H. Shaw
Harold Shelton
Kenneth W. Squires
James R. Stuckey
William H. Stuckey
Cecil F. Taylor
Echol Thurman
Odra M. Vaught
Herbert O. Wahrmund
James T. Wallace
Yonley A. Williams
Mervan H. Wood

308TH STRATEGIC MISSILE WING
Launch Complex 373–5   24 January 1968
MSgt Ronald E. Bugge

308TH STRATEGIC MISSILE WING
Launch Complex 373–4   8 October 1976
Sgt Larry G. South

308TH STRATEGIC MISSILE WING
Launch Complex 374–7   19 September 1980
SrA David L. Livingston

381ST STRATEGIC MISSILE WING
Launch Complex 533–7   24 August 1978
SSgt Robert J. Thomas

2 September 1978
A1C Erby B. Hepstall

CONTENTS

ILLUSTRATIONS

FIGURES

Following page 154

TABLES

FOREWORD

Definition: Titan—one gigantic in size, power, or influence.

This seems about right to me: Titan II was gigantic in size and in power. Gigantic in power certainly describes the role Titan II played in the cold war. Perhaps Norm Augustine, former chief executive officer of Martin Marietta, speaking at the inactivation ceremonies of the 390th Strategic Missile Wing (SMW), Davis-Monthan Air Force Base (AFB) on 30 July 1984, summed it up best: "The Titan II and the men and women of Titan II have, by any standard, done their job magnificently and with great dedication. Strategic deterrence has worked."

What follows in these pages by David K. Stumpf describes, better than anything else I have read, the story of the Titan II hardware and the people who made Titan so extraordinarily effective. I like Norm Augustine's words—"men and women of Titan"—because, while you might at first think this book is about a missile, technology in other words, it is really about people, the men and women of Titan. On the hardware side, the technology side, there are many names and labels that those of us who lived with Titan will never forget, such as LCCFC, CMG, PDC, MGACG, FPCB, and "CB-103, On . . . Set!" We remember launch duct, silo equipment area, prevalves (and leaks), P-1, 355-3, and MPHT in about that order! Right along with these abbreviations comes the human side, the people, with MCCC, DMCCC, BMAT, and MFT, all in white baggy coveralls. And don't forget the maintenance troops, the pad chief, and the topside guard, none of whom wore those flour bags! We were evaluated by Standboard, the dreaded SMES, and the SAC/IG. These, of course, were preceded by EWO training and MPT rides. I remember station wagons to the site and foil-pack meals.

Is it all coming back? Could you ever forget? I will never forget the first time I saw and touched a Titan II. Do you remember your first time in the launch complex? Impressed is hardly an adequate expression; from Level 8 up and Level 1 down, it was big! I clearly remember walking down the cableway to Level 2. Do you remember the smell as soon as you opened Blast Door 9? I will never forget it: sharp and piercing, distinctive beyond description. Better yet, do you remember the smell when you opened Blast Door 6 coming in early in the morning? Breakfast!

I started Titan II training right after commissioning at the Air Force Academy and was sent to the 390th SMW at Davis-Monthan AFB, Tucson, Arizona. I believed then, and am even more certain today, that I was indeed fortunate to start off with Titan II. Not just because of the hardware, but because of the hardware and the people. Titan II had a crew, and the crew was directly in touch with its missile, its sortie. We operated it, tested it, occasionally fixed it; it was a great system for a crew member because there was a very direct relationship with the weapon system. It was a great system for an officer because of the close contact with the enlisted members of the crew, a unique educational experience for all involved.

As a deputy missile combat crew commander, I believe that I was assigned to the best

Titan II crew commander in the business, Capt John W. Haley III. Haley was a great officer and professional, very demanding, but he took care of his people and our site, 571-8. He taught me more about hands-on leadership, discipline, and professionalism than I could ever put into words. John Haley was my right, great start for an interesting and rewarding career in the Air Force.

Nineteen years after my introduction to Titan II, I had the honor and privilege to engage it once again. This time as the wing commander of the 381st SMW at McConnell AFB, Wichita, Kansas. I was still very much impressed, just as before, but this time there were added feelings: fear and respect. In the intervening years there had been a number of fatalities and injuries; Titan II was not forgiving. Additionally, I had spent about seven years in the Joint Strategic Target Planning Staff and joint operations at Joint Chiefs of Staff J-3. I knew what Titan II could do, would do if deterrence ever failed and World War III started. I remember how much retargeting was necessary by other weapon systems when a Titan II was retargeted to another designated ground zero.

Of course, I also recall the tremendous, healthy rivalry that existed between the six Minuteman wings and the three Titan II wings. Inevitably, the argument would start with size and work its way to who had the biggest warhead. A former Commander in Chief, Strategic Air Command, Gen Russell Dougherty, described it this way: "There is as much difference in the size of warheads as there is between cannonballs and bbs. It takes a lot of bbs to make up for a cannonball." Those of us in the Titan II program were affectionately referred to as "plumbers," given the liquid propellants of Titan II, while we referred to the Minuteman troops as the "bottle-rocket bunch," given the rather small size of Minuteman compared to Titan II. Others referred to Minuteman as a "start cartridge" for a real missile —Titan II! While some would brag about large and thundering, others would praise small and discreet. I would recommend that we leave such arguments for the armchair strategists, because you and I know the truth, we were there. We know how ready Titan II was, how good Titan II was, and how ready we were!

But just in case you run into one of the other breed, you might remind them that it takes a lot of bbs to make up for a cannonball!

Jay W. Kelley
Lieutenant General, USAF (Ret.)
December 1998

PREFACE

Lynne Braddock-Zollner, Arkansas Historic Preservation Program, is the key to this entire project. During her visit to the Titan Missile Museum in Sahuarita, Arizona, in April 1995, we had the opportunity to discuss the scope of the book proposal that Lynne had submitted to the Department of Defense Legacy Program. I suggested that while the main thrust of the book was going to be on the 308th Strategic Missile Wing, Little Rock Air Force Base, Arkansas, a chapter or two on the research and development of the Titan II missile would help put the program in context. Slightly more than one year later, I signed a contract with the Arkansas Historic Preservation Program to write a comprehensive history of the Titan II intercontinental ballistic missile program. Lynne left the project after 18 months, but Mark Christ and Ken Grunewald of the Arkansas Historic Preservation Program provided excellent continuity and support. Ken has the dubious honor of being my first Titan II crew member to be interviewed. Julia Cantrell from Brooks Air Force Base, San Antonio, Texas, was the Air Force progam manager for this project and was key to helping me obtain the clearances necessary to see the classified wing histories.

All of the Lockheed Martin Denver and Vandenberg Launch Operations staff that I contacted were extremely helpful. Krisstie Kondrotis and Maggie Dane were more than patient in helping arrange access to retired and current employees at the Denver and Vandenberg facilities, respectively. Don Kundich, Robert Rhodus, and John Adamoli gave enthusiastic interviews with great stories about the early days of Titan I and Titan II. Jim Purkey provided a detailed Vandenberg Titan II launch list as well as details on the aborted launches. F. Charlie Radaz holds the marathon interview record with excellent stories and supporting documents. Andy Hall helped with the Titan I and Titan II program background. Dale Thompson set me straight on the "belly bands" and Operation Wrap Up. Ed Carson filled me in on refurbishing the sites at Vandenberg Air Force Base (AFB), California. Ed Patrick helped corroborate more than one of the more interesting stories. Elmer Dunn helped with a description of the uniquely named "Twang" test. Felix Scheffler set me straight on several early Titan II issues at Vandenberg AFB. Al Schaefle solved the mystery of the N-4 thrust chamber instability incident at Cape Canaveral during a marathon conference call and also helped with several photograph identifications. Robert Stahl provided the amazing thrust chamber photographs for the aborted launch of N-4. Jim Greichen was a key resource for the early days of Titan I at the Cape as well as at Vandenberg AFB. Jack Cozzens and Ron Hakanson helped me tremendously with the coded switch system description. Ron also served as an invaluable resource for airframe numbers, manufacturing dates, and manufacturing techniques. His collection of photographs from the work trying to refurbish 533–7 are greatly appreciated, as well as his generous donation of the majority of his Titan II memorabilia, soon to be translocated to the Titan Missile Museum. Don Picker and I spent a memorable day in the Lockheed Martin Denver Research Library with the assistance of Mel Coffin, librarian, where I found a treasure trove of early Titan II program summaries and reports. Ron Underwood and Frank

Nash provided rapid drawing and photograph reproduction, respectively, at a price that could not be beat. Vernon Selby provided an incredible tour of the Titan II space launch vehicle (SLV) refurbishment facilities at Lockheed Martin Denver before it all disappeared. He also helped clarify some discrepancies in the launch record. Ben Rizutto and Keith Wanklyn helped sort out the oxidizer spill details at 533–7. Roger Reiger provided up-to-date information on the Titan II SLV program as well as a memorable dinner during a visit to Tucson.

Site construction stories from John Carlson Sr., Sundt Corporation, and Don Boomhower, Ralph M. Parsons Company, as well as their photographs and memorabilia, made the construction phases come to life. Harry Christman from Fluor Corporation helped set the scene for the Davis-Monthan AFB construction program. Charles Terhune Jr., Parsons Infrastructure and Technology Group, Inc., made the Parsons' photographic archives available to me and generously covered the cost of reproducing a comprehensive set of photographs of early silo construction models as well as the full-scale equipment. David Fox was responsible for actually duplicating the negatives. Gen William Leonard, USAF (Ret.), clarified several construction questions for me from both the days before and after he joined Ralph M. Parsons Company. BrigGen Charles Terhune Sr., USAF (Ret.), helped me make sufficient sense of the early days of the ICBM program to permit me to interview Gen Bernard Schriever with a solid base of information. After hearing all the versions of what happened to the N-7 reentry vehicle after that fateful first launch, little did I know that Jack Easterbrook, Delco, a contact from my first book project, would prove critical in locating the divers, Leo Blickley, Sid Kuphal, and David Potter, involved in the search and recovery of the N-7 dummy warhead. Bob Popp, Bill Probert, Joe Koch, and Ed Stapp, from AC Spark Plug (Delco), were key factors in my guidance-system education. Aerojet-General staff and retirees in Sacramento, California, were of tremendous help. Gary Cook answered endless questions on the Titan I and Titan II engine design and identification of several photographs, as did Wally Dinsmore and Ken Collins. Roy Jones and Louis D. Wilson provided details on the Stage II hard-start problems that filled in several gaps. Norman Laux patiently described cavitating ventures. Rollo Pickford contacted me out of the blue one day and turned out to be just the person I was looking for to give me details about testing the feasibility of an in-silo launch at the Aerojet facilities in Azusa, California. Mary Abbott, the librarian at Aerojet, and William True Jr. helped me get through quite a bit of red tape with in-silo launch feasibility reports.

The staff at the Air Force Historic Research Agency, Maxwell Air Force Base, Alabama, are the key to the details of the operational history of the system. Archie Difante, Joe Caver, Esse Roberts, Melvin Watts, Ann Webb, TSgt Lee Morris (USAF), MSgt Jack Tant (USAF), and SrA Amy Stewart (USAF) provided excellent assistance during my two visits to Maxwell. Maj Neil Couch, USAF, provided long-distance research assistance at Maxwell on several occasions. Dr. Rick Sturdevant, History Office, Air Force Space Command, Peterson Air Force Base, Colorado, and Dr. Ray Puffer, former Ballistic Missile Office Historian, were extremely helpful with critical comments on the manuscript as well as in assisting with document location and copying. Mark Cleary, Chief, History Office, 45th Space Wing, Patrick Air Force Base, Florida, provided missing details for the Titan I and II launches from Cape Canaveral. Sgt Gary Johnson, USAF, the historian at McConnell Air Force Base, Kansas, was instrumental in providing rare photos of construction at the 381st SMW as well as a collection of wing histories on microfilm.

With over 200 interviews with former missile combat crew members, missile maintenance personnel, and guidance-system technicians, it is impossible to thank each one personally, yet several stand out and need to be thanked individually. SMSgt Bill Shaff, USAF (Ret.), provided photographs, details of an aborted launch, memorabilia, and a complete inventory of Titan II airframe locations as of 1987 that made my research much eas - ier. Bill's reviews of several drafts of the manuscript were insightful. LtCol Ted Suchecki, USAF (Ret.), provided not only a detailed review of the manuscript but also educated me on the alert status of the Titan II missiles at Vandenberg AFB, California, in the mid-1960s. Col John T. Moser, USAF (Ret.), and Col Richard A. Sandercock, USAF (Ret.), were invaluable in helping me reconstruct the acccident at Damascus, Arkansas. Only after interviewing them both did I locate an unclassified copy of the accident investigation through Col Lloyd Houchin, USAF (Ret.). BrigGen Ronald Gray, USAF (Ret.), gave me access to his collection of photographs and slides from the accident, as did Maj Mark Clark, USAF. The oxidizer spill at Rock, Kansas, was my most daunting challenge. A Freedom of Information Act request for records of the accidents involving Sgt Ronald Bugge, USAF, and Sgt Larry South, USAF, revealed that the records had been destroyed; further information was unavailable from the Air Force. Wayne Seals, records manager, Eighth Air Force, Barksdale Air Force Base, Louisiana, provided rapid turnaround time service on my last and successful attempt to find the official accident investigation report on the accident at Rock, Kansas. Bob Livingston provided me with copies of his extensive collection of newspaper clippings of the spill. Col Ben Scallorn, USAF (Ret.), helped tremendously with his recollections from the investigation. LtCol Craig Allen, USAF (Ret.), was the officer in charge for Project Pacer Down, the attempt at refurbishing the Rock, Kansas, launch complex. His recollections as well as his part in the Damascus explosion accident investigation were most helpful. Col Dan Jacobwitz, USAF (Ret.), provided me with critical information on the accident involving Sgt Ronald Bugge. Col Nathan Hartman, USAF (Ret.), filled in the remaining details. Capt Bill Howard, USAF (Ret.), and Jim McFadden enabled me to understand the accident involving Sgt Larry South. Bob Eagle provided helpful long-distance research in the Washington, D.C., area. SSgt Mark Hess, USAF (Ret.), and LtCol Ken Hollinga, USAF (Ret.), were critical resources for details on the Cooke spy case. Mark also provided me with the details behind the Jack Anderson "exposé" article. Maj Gregory Ogletree, USAF (Ret.), provided me with a comprehensive critique of the first rough draft, as did Nick Spark. Shivan Sivalingam from Stanford University and Desmond Ball from the Australian National University provided critical comments on the single integrated operation plans evolution during the Titan II program.

Becky Roberts, deputy director for the Titan Missile Museum, Arizona Aerospace Foundation, has been a most enthusiastic supporter of my research efforts from the start. LtCol Orville Doughty, USAF (Ret.), the resource of resources at the Titan Missile Museum, was an excellent mentor in learning how to research the technical orders and drawings so that I could conduct a concise and informed interview. Jim Austin from the Faculty Center for Instructional Innovation, University of Arizona, and his son Travis, provided expert assistance with digital rendering of many of the photographs and illustrations. The staff of the Faculty Center for Instructional Innovation, Jeff Imig, Maritza Martelle, Jose Noriega, and Casey Ontiveros, provided excellent assistance as well. Don Boelling worked hand-in-hand with me using his Titan II web site to reach many Titan II missile combat crews. Art Le Brun of the San Diego Aerospace Museum provided captions and photographs of both the Atlas and Titan II programs.

My thanks to Kevin Brock, Brian King, and Liz Lester from the University of Arkansas Press, as well as to Debbie Self, for their patience and expertise in making this book a reality.

My wife, Susan, said that after my first book, *Regulus: The Forgotten Weapon,* I could not write another book until I had a contract. Little did she know that two months later I would come to her with just such an opportunity. Now I have to have not only a contract, but also an advance! I think she is safe on that one. Thank you, Susan, for putting up with the constant trips and phone calls, the highs of finding that unbelievable source, and the lows of not being allowed access to classified contractors' flight reports.

Finally, a personal thanks to all of the Titan II contractor and military personnel, as well as their families, for their long hours, days, weeks, and years of attending to the overwhelming task of providing security for our nation. That Titan II was never used in anger stands as a fitting testimonial to your efforts.

INTRODUCTION

The story of the Titan II Intercontinental Ballistic Missile (ICBM) program has been over-looked for 35 years. This is due, in part, to historical hindsight which clearly views the liquid-propellant missiles of the 1960s with disdain. High maintenance costs and highly flammable liquid oxygen oxidizer, relatively small payloads, as well as the often-mentioned 15- to 30-minute response times for the Atlas and Titan I ICBMs were overshadowed by the simplicity, relative safety, and low price of the solid-propellant Minuteman ICBM with its almost instantaneous response time. For all the supposed disadvantages of liquid-propellant missiles, why was Titan II retained as a front-line weapon from 1963 to 1987?

Titan II was a technical marvel, a true second generation of liquid-propellant missiles, differing from the Atlas series D, E, and F and from Titan I in several ways. Titan II utilized storable propellants. The Atlas family, as well as Titan I, utilized a propellant combination of liquid oxygen and RP-1 (high purity kerosene). Liquid oxygen boils at -295 degrees Fahrenheit and could not be stored aboard the missile for prolonged periods. Loading the liquid oxygen took up to 15 minutes, and the oxidizer tank propellant level had to be maintained as the crew awaited the launch order. Titan II's propellants required a comparatively benign temperature of 60 degrees Fahrenheit within the launch duct so that both oxidizer and fuel were stored onboard the missile, ready for launch.

Titan II was housed in a protective silo built to withstand overpressures of 300 pounds per square inch. With the accuracy of Soviet nuclear weapons of that time period, this was deemed more than sufficient to ensure the ability to launch in retaliation. Both Atlas F and Titan I, while stored in silos, were not designed for launch from within the silo. Instead, after propellant transfer, the missiles were elevated to the surface and fired.

Perhaps the most important difference was the large increase in payload capacity with Titan II versus the Titan I and Atlas. Atlas E and F and Titan I carried the 4,000-pound Mark 4 reentry vehicle with a range of 5,500 nautical miles. Titan II's sheer size, with resulting payload capacity, permitted it to carry the 8,000-pound Mark 6 reentry vehicle 5,500 nautical miles. Housed within the Mark 6 was a W-53 warhead, the single largest warhead ever carried on an American ICBM, with an accuracy repeatedly demonstrated to be significantly less than one nautical mile.

Why, with all these advantages, was Titan II deactivated? The announcement of the decision to modernize our country's strategic deterrent forces was made in October 1981 with the concomitant announcement of the deactivation of Titan II. Even though a report commissioned by the Air Force and a parallel study by Congress concerning safety and utility of the Titan II had determined that it remained reliable and safe, Titan II was the oldest missile in the U.S. ICBM arsenal and a logical choice for retirement in the view of many.

Gen Bennie L. Davis, Commander in Chief, Strategic Air Command, in October 1981, made the fateful recommendation:

> In the discussions in the Department of Defense prior to the announcement of
> President Reagan's Strategic Modernization Program it was agreed that with the

hardness and improved accuracies in the Soviet third- and fourth-generation strategic systems the U.S. needed to improve accuracies, hardness and survivability of its current systems. Since Titan II has been in the Strategic Air Command inventory eight years beyond its predicted service life, was not survivable, had no hard target capability and was difficult to support logistically, I therefore recommend, as Commander-in-Chief, Strategic Air Command to the Secretary of the Air Force and the Secretary of Defense that we phase out Titan II.

Those interested in the liquid-propellant ICBM program have relatively few resources with which to understand the complexity and necessity of Titan II. This is the story of the civilian contractors who designed and fabricated the missile system and its silos and of the military personnel who manned the launch control centers and maintained the missiles for 24 years.

I

THE AIR FORCE
STRATEGIC MISSILE PROGRAM

The Air Force strategic missile program began on 31 October 1945, when the Air Force Air Technical Service Command released a request for a 10-year study focused on development of a family of four ballistic missiles with 10 to 20 times the 200-nautical-mile range of the V-2.[1] In April 1946, the Air Force awarded Vultee Field Division of Convair a $1.4 million study contract for the MX-774, the first strategic missile program.[2] Convair proposed studying a 5,000-nautical-mile range missile in two distinct forms: one subsonic, winged, and jet powered; the other supersonic and rocket powered. The MX-774 project manager was Karel J. Bossart. Several innovative features were quickly decided upon: thin-walled, pressure-stabilized propellant tanks; a nose cone that would separate from the main missile body so that only the nose cone had to withstand the rigorous reentry environment; and missile engines that gimbaled for control of the missile trajectory. Missile tank fabrication was begun in late 1946.

In July 1947, the program was canceled. MX-774 and ballistic missiles were competing with the more conventional winged-missile approach of the Navaho (MX-770) and Snark (MX-775) programs, and so, as the most expensive of the three, MX-774 lost. Convair was permitted to complete the test flight program of three launches on its own funding, the last flight taking place on 2 December 1948. While none of the launches was completely successful, valuable data had been gathered. All that remained was for the funding to resume, as the Convair engineering team continued to refine design concepts.[3]

Innumerable studies were conducted by the military services over the ensuing three years. A continuing stumbling block was interservice rivalry. The Air Force felt it should be the lead service to coordinate and develop what was logically an extension of bomber delivery of strategic—that is, nuclear—weapons. The Navy saw the forward deployment capability of aircraft carriers a natural fit with nuclear weapons. The Army did not want to depend on the Air Force or the Navy for tactical nuclear weapon support of its ground forces. Finally, two significant events occurred within 11 months of each other. The first took place in August 1949 when the Soviet Union detonated its first atomic bomb. Caught by surprise in that it had been felt such an event was still several years in the future, the Truman administration realized that the American monopoly on atomic weapons was now shattered. The second event was the beginning of the Korean War and, in a sense, the confirmation of a cold war between communism and democracy. In early 1951 the Air Force turned again to Convair and initiated the MX-1593 program, in effect picking up where MX-774 had left off.[4]

Continuing acrimony between the services, austere funding, and preoccupation with the Korean War led again to a flurry of studies, committees, and decisions. The decisive event that moved the Air Force into the missile age was an extensive review of defense research and development that was conducted at the beginning of the Eisenhower administration. In many cases the conflicting boards or agencies within the Department of Defense had been replaced with assistant secretaries of defense. As one of the highest priority needs based on campaign rhetoric, the Air Force was assigned the task of reviewing the missile development organizations with the Secretary of Defense special assistant for research and development Trevor Gardner leading the review process.

In June 1953, the Department of Defense Study Group on Guided Missiles was established to conduct a technical evaluation of all missile programs of the three services. The group realized that the intercontinental missile ballistic program (ICBM) component would be better analyzed by a separate group composed of the nation's leading scientists. In October 1953, Gardner appointed just such a select committee to evaluate the long-range missile program. The committee's charter was to evaluate the technical feasibility of the long-range weapon system, identify major constraints, and provide guidance in program direction. The chairman of the 11-member Committee on Strategic Missiles was Dr. John von Neumann, an eminent mathematician and head of the Institute for Advanced Study at Princeton.[5] Four months later the Teapot Committee, as the committee had been nicknamed, and the name by which it is best known, issued its report. The report was all-encompassing, containing a general overview of the technical aspects of long-range missiles followed by specific recommendations for the three major Air Force programs currently in progress, the air-breathing Snark and Navaho cruise missiles and the Atlas ICBM.

The general problems with all three systems in relation to the long-range program were that each system had out-of-date military specifications for the target circular error probable (CEP, defined as the radius of a circle within which half of the warheads would impact) and that missile basing configurations were insufficient to protect against attack and slow rates of launch in either single or multiple launch situations.[6]

There were six specific recommendations for the Atlas program. Crediting Convair with pioneering work on ICBMs, the committee urged that the design needed to take into account more modern approaches. Alternative approaches to several critical phases of the problem had to be explored. Atlas as it presently existed should not be accelerated into production. Rather, after reorganization of the entire missile program, acceleration could be considered. The committee felt that the initial operational capability would be possible in six to eight years if the recommended reorganization was adopted. The CEP for Atlas could be two to three miles instead of the present 1,500 feet now that the lightweight thermonuclear warheads would be available in time. Present basing plans were wholly inadequate. Design of the missile and basing configurations had to be adjusted to emphasize lower vulnerability and shorter launching time.

The final and most compelling finding, and the one that had a tremendous positive impact on the future of the program, was the clearly stated direction for how such reorganization might be undertaken:

> The nature of the task for this new agency requires that overall technical direction be in the hands of an unusually competent group of scientists and engineers capable of

making systems analyses, supervising research phases and completely controlling the experimental and hardware phases of the program.[7]

The result of the Teapot Committee's report and subsequent Air Force command staff meetings was the total reorganization of the strategic missile program. The only program goal left unaltered was to achieve the earliest possible deployment of a credible system. Additionally, the need to have each of the Atlas subsystems supported by at least one alternate contractor, as insurance against design failure, was fully realized and implemented. The alternate missile concept grew from this, giving birth to Titan I. On 14 May 1954, Gen Thomas D. White, Air Force Chief of Staff, assigned the Atlas program the top development priority in the Air Force. Simultaneously, White assigned overall management and development responsibility to a newly created Air Research and Development Command (ARDC) field office on the west coast, close to Convair's Atlas fabrication facilities.[8]

On 1 July 1954, Gen Thomas S. Power, ARDC commander, designated the new field office as the Western Development Division (WDD). In order to prevent unnecessary attention being drawn to the office, housed in Inglewood, California, military personnel were instructed to wear civilian clothes. BrigGen Bernard A. Schriever was assigned command of the WDD and the Atlas program. Schriever, the military advisor to the Teapot Committee, took command on 2 August 1954. With the advice of the Atlas Scientific Advisory Committee, again chaired by Dr. von Neumann, the WDD staff presented the framework for the new organizational effort. While three approaches had been studied, the one most favored was where the Air Force would continue with system responsibility and employ the consulting firm of Ramo-Wooldridge for systems engineering and technical advice. Convair would still be responsible for all fabrication of the missiles, but technical direction and decisions on the overall program would now be conducted by Ramo-Wooldridge. This format was accepted on 8 September 1954, and the Atlas program, indeed the strategic missile program as we now know it, came into being.[9]

Force Size—How Many, What Kind?

How were the actual numbers of missiles planned for the strategic deterrent forces decided upon? As originally envisioned within the Eisenhower administration in 1956, only 20 to 40 missiles were to be deployed. Cost was one factor because it was obvious from the postwar experiments that ballistic missile design was much more complicated than even the long-range jet aircraft weapon systems. An equally obvious point was that ballistic missiles were the ultimate deterrent since no method of intercepting such a weapon once en route or over the target could be envisioned.[10]

In response to a proposal by the Federal Civilian Defense Administration to spend $40 billion over several years to construct a nationwide civil defense infrastructure, President Eisenhower ordered a study of the concept by a select panel of civilians. This committee, chaired by H. Rowan Gaither, chairman of the RAND Corporation Board of Trustees, soon broadened its mandate to cover the entire scope of defense issues facing the nation as the cold war developed. On 7 November 1957 President Eisenhower received the Gaither Committee Report which recommended a nearly sevenfold increase in ICBMs and a fourfold increase in intermediate range ballistic missiles (IRBMs), as well as rushing the various weapons systems to completion as a national priority.[11] Greater protection and dispersal for Strategic Air Command (SAC) bases and aircraft were also urged in the report. Individual aircraft shelters hardened to 100 to 200 pounds per square inch overpressure were considered critical, and an increase in number of Atlas and Titan missiles for initial operational capability from 100 to 600 was recommended.[12]

The only force-specific decision on land-based missile deployment had been made by Secretary of Defense Charles Wilson on 5 October 1957 when he authorized an initial operating capacity and requisite funding for 4 Atlas squadrons and 4 Titan squadrons, totaling 80 missiles, by December 1962.[13] In April 1958 the strategic missile program was expanded to 9 Atlas squadrons; in January 1959 the Titan force was increased to 11 squadrons. One year later 4 more Atlas and 3 more Titan squadrons were added. By the end of the Eisenhower administration, a total of 255 Atlas and Titan missiles had been authorized. This was considerably more than originally planned, the numbers having increased under pressure from the Pentagon and Congress.[14]

Apparently the most influential study on strategic missile force requirements was that conducted by the Department of Defense Weapons Systems Evaluation Group (WSEG) beginning in September 1959. The report, "WSEG #50 Evaluation of Strategic Offensive Weapons Systems," was completed in 1960, but the incoming Kennedy administration was the first to see it outside of the Pentagon.[15] The main focus of the study was to determine the optimum mix of bombers, land-based ICBMs, and the Polaris fleet ballistic missile submarines. The report indicated that the first-generation ICBMs, the Atlas and Titan I, were too costly to continue to support. Retention of Titan II, then under development, was recommended due to its much larger yield warhead that could be useful against unforeseen future developments. The major question concerning the solid-propellant Minuteman was the ratio of fixed-base and mobile missiles.[16]

The Kennedy Administration

When the Kennedy administration took office in January 1961 only 12 Atlas D ICBMs were on strategic alert. The 576th Strategic Missile Squadron (SMS), Vandenberg Air Force Base (AFB), California, had 6 Atlas Ds deployed. Three of the 6 missiles were in vertical, exposed gantry (i.e., unhardened, or "soft") installations, and three were in above-ground "coffins," relatively thinly walled concrete structures which provided environmental shelter when the missile was stored in the horizontal position. These installations afforded minimal blast protection. The remaining 6 Atlas D launchers were part of the 564th Strategic Missile Squadron at Francis E. Warren AFB, Wyoming, and were also deployed in aboveground "coffin" installations.[17]

Within six months of President Kennedy's inauguration, six squadrons of Titan Is had been activated and construction was underway. Each squadron was composed of nine missiles, three per launch site. No further Titan I squadrons were funded. Titan II development was progressing well, with squadron activation set to begin in 1962. Titan II would be based as a single missile at each hardened launch site, nine sites per squadron, two squadrons per wing. The last two Titan squadrons, scheduled to be equipped with Titan II missiles, were canceled as a cost-saving measure in mid-March 1961.[18]

With Minuteman, the solid-propellant missile seen as a major competitor to all of the liquid-propellant missiles, the case was quite a bit different. The Eisenhower administration had originally proposed eight squadrons of 50 each in October 1958. By January 1961 when the 1962 defense budget was submitted, Minuteman planning called for nine squadrons, 50 missiles each, with 10 missiles per launch control center. Funding for three mobile Minuteman squadrons with 30 missiles each had been requested. The Kennedy administration used funding for the mobile Minuteman squadrons for an additional three silo-based squadrons. After deployment began in 1962, the point was soon reached where silo construction became a bottleneck since the missiles were being produced at the rate of 30 per month.[19]

By the beginning of 1963, 12 Atlas D, E, and F squadrons with 123 missiles and 6 Titan I squadrons with 54 missiles were fully operational.[20] In mid-1963 a review of missile system reliability by the Air Force recommended that the Atlas D and E, as well as Titan I, be removed from service. Atlas D would be first, beginning in 1964, Atlas E in 1966, and Titan I in 1967. Atlas F would be retained until 1969. Soon after the 1964 general elections, Secretary of Defense Robert S. McNamara amended his earlier recommendation to include Atlas F, rationalizing the retirement of all first-generation missiles as an economic decision, saving $117 million. On 31 March 1965 the last of the Atlas F missiles were removed from alert status. On 1 April 1965 the last of the Titan I missiles were removed from alert status.[21] The first complete Minuteman wing, composed of three 50-missile squadrons with Minuteman I missiles, became operational at Malstrom AFB, Montana, in October 1963.[22] Plans had been approved for a total of 16 Minuteman squadrons at 50 missiles apiece, for a total of 800 missiles. By 1 January 1964 the Titan II deployment was complete, with 6 squadrons for a total of 54 missiles. Funding was in place for an additional 200 Minuteman II missiles with greater range and payload capacity than Minuteman I.

At the time of the full deployment of Titan II a total of 581 ICBMs (Atlas, Titan I, Titan II, and Minuteman I) were available on alert status.[23] Four years later, in 1967, the Minuteman program was fully deployed with 1,000 missiles. Now the land-based forces totaled 1,054 ICBMs, a number that was never increased. By the end of the decade, modernization of the Minuteman force meant that 490 Minuteman Is (one reentry vehicle), 500 Minuteman IIs (three reentry vehicles), and 10 Minuteman IIIs (three multiple independently targetable reentry vehicles) were now deployed.[24]

How did this compare with the perceived threat from the Soviet Union? By mid-1963, the Defense Department estimated approximately 100 Soviet ICBMs were operational. Three years later the estimate was only 224 ICBMs deployed. This obvious deficiency did not last much longer, and by 1968 the Soviet Union had passed the 1,054 ICBMs deployed by the United States. By 1970 the Soviet Strategic Rocket Forces totaled 1,427 ICBMs.[25] By 1975, all Minuteman I missiles had been replaced by either Minuteman II or III. While the mix of Minuteman missiles changed, the total numbers did not. Titan II was retained with system upgrades but with no change in the numbers or types of warhead.[26]

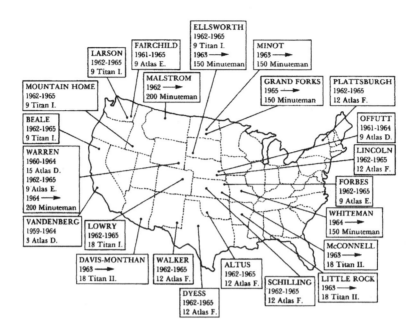

Figure 1. Locations of ICBM sites in the United States in 1965. *Courtesy of the Titan Missile Museum National Historic Landmark Archives, Sahuarita, Arizona.*

The 13 years from the inception of the Atlas program in 1954 to the completion of deployment of Minuteman saw the rise and fall of the first generation of liquid-propellant ICBMs: Atlas and Titan I. Atlas and Titan I paved the way for the Air Force ICBM program but were clearly a deadend with the relatively slow response time due to the need to load liquid oxygen just prior to launch. Titan II was the ultimate liquid-propellant missile, capable of the same instant response to a launch message as was Minuteman.

II

THE FIRST GENERATION:
ATLAS AND TITAN I

Atlas (SM-65/WS-107A-1)

In order to place the Titan II into proper perspective, a brief description of Atlas and Titan I, the nation's first-generation intercontinental ballistic missiles (ICBM)is necessary. From April 1946, when the Air Force accepted a Convair Corporation proposal to study a super-sonic ballistic missile design designated Program MX-774, to December 1954 when Atlas received top priority within the Air Force, Atlas underwent many convoluted program changes as world events, such as the Korean War, changed funding priorities.

On 28 February 1954, the first "dry" thermonuclear weapon was detonated in the Pacific, proving that a lightweight yet powerful warhead design would be available for the proposed ICBM. The first thermonuclear weapon test, Operation IVY MIKE on 1 November 1951, had used liquid deuterium, an isotope of hydrogen. The device weighed 60 tons, including the refrigeration system that kept the deuterium in the liquid state. Emergency capability bomb designs were created, but with the BRAVO and ROMEO shots of the Operation CASTLE test series of early 1954, the first to use lithium deuteride and lithium hydride salts, the heavy refrigeration systems were shown to be no longer necessary. Truly lightweight thermonuclear weapon designs were just around the corner, and now ICBMs began to show great potential.[1]

On 11 March 1954, the Air Force initiated another look at the feasibility of ICBMs. On 21 June 1954, the Air Force Air Research and Development Command received official noti-fication to accelerate Project Atlas into the nation's first strategic ballistic missile weapon system.[2]

In January 1955, the Air Force awarded the first Atlas ICBM contract for Atlas Weapon System 107A-1 to the Convair Division of the General Dynamics Corporation. Atlas was to have a range of 5,500 nautical miles, two booster engines of 150,000 pounds thrust each at sea level, one sustainer engine of 57,000 pounds thrust, and two smaller vernier engines of 1,000 pounds thrust each for roll control and final guidance corrections. Engine design and fabrication were the responsibility of the Rocketdyne Division of the North American Aviation Company. Concern about poor rocket engine ignition reliability in the early 1950s resulted in Atlas being designed as a one-and-a-half-stage missile. All engines ignited prior to liftoff. The booster engines were jettisoned after about two minutes of flight, and the sustainer and vernier engines, along with the entire propellant tankage, continued in flight until reentry vehicle separation. Since no propellant tankage was jettisoned with the booster engines, the booster stage was considered a half stage. The Atlas MA-series engines had regeneratively cooled thrust chambers, as had earlier rocket engine designs. Unlike the earlier thrust chambers which had double chamber walls, fuel was circulated through brazed tubes that formed the chamber bell, cooling the walls prior to flowing into the injector plate and mixing with the oxidizer. Another innovative design feature that the MA-series engine incorporated was gimbaling the entire engine bell to provide directional control and stability rather than using carbon vanes in the engine exhaust.[3]

A unique feature of the Atlas missile was the use of pressure stabilization for the propellant tank. This design was the brainchild of Karel J. Bossert, the project manager of MX-774. The extremely thin stainless-steel skin was lightweight but was not self-supporting. Less than the thickness of a dime, the single tank housed both the RP-1 (Rocket Propellant-1, a high-purity hydrocarbon fuel similar to kerosene) and liquid oxygen propellants separated by a common bulkhead. When the propellant tanks were not loaded with fluid, structural integrity was maintained using compressed helium. At first glance such a design might appear too delicate for a missile system. However, when inflated, one could hit the skin with a mallet and not damage the surface. The structural integrity of the propellant tanks was never a point of failure during the Atlas ICBM flight program.[4]

Atlas was designed in six configurations: Atlas A, B, and C were proof-of-concept configurations, while Atlas D, E, and F were the deployed ICBM configurations. Atlas D weighed 262,000 to 265,000 pounds at liftoff; Atlas E and F each weighed 260,000 pounds. Length varied slightly with the reentry vehicle payload dimensions, ranging from 75.1 feet to 82.5 feet. The principal diameter was 10 feet, tapering to a 5-foot diameter at the forward end of the liquid oxygen tank. The warhead was a single W-49 thermonuclear weapon, 1.44-megaton yield, housed in a Mark 2 heat-sink–type reentry vehicle in the first squadron; a Mark 3 ablative-type reentry vehicle for the remaining Atlas Ds and a W-38 thermonuclear weapon, 3- to 4-megaton yield, housed in a Mark 4 reentry vehicle for the E and F series. Thrust at liftoff was 360,000 pounds. While range is listed at 5,500 nautical miles, the missile was flown in one demonstration flight 7,860 nautical miles with a Mark 3 reentry vehicle, impacting within 1.8 miles of the intended target in the Indian Ocean.[5]

The first launch of an Atlas A missile on 11 June 1957 was unsuccessful, when, after gaining several thousand feet in altitude, one of the two booster engines failed and the range safety officer destroyed the missile as it tumbled end over end, out of control (Table 2.1). All was not lost, as the ability to gimbal the engines to try to control the missile was evident, as was the structural integrity of the thin propellant tank skin. On 17 December 1957, the first successful launch took place.[6] Twenty-one months later, on 9 September 1959, a crew from the Strategic Air Command's 576th Strategic Missile Squadron (SMS) launched the first Atlas D from Vandenberg AFB. That same day Gen Thomas S. Power, the Commander in Chief of the Strategic Air Command, declared Atlas D operational. On 31 October 1959, the first Atlas D was placed on strategic alert at Vandenberg AFB, beginning a new era of strategic deterrence for the United States.[7]

Figure 2.1. Atlas missiles in the General Dynamics factory in San Diego. Several unique features of the Atlas missile design are apparent. First, the frame used to support the missile skin when it was inflated with helium during storage can be seen on the aft end of the missile in the center of the photograph. The missile in the foreground shows the sustainer engine and the booster engine pair in the Stage I compartment at the lower left. Atlas was known as a stage and a half missile since all three engines ignited at liftoff but the two outer booster engines dropped off in flight. *Courtesy of the San Diego Aerospace Museum, San Diego, California.*

TABLE 2.1
ATLAS DEVELOPMENT LAUNCH CHRONOLOGY

DATE	DESCRIPTION
1957	
11 Jun	First launch partly successful but engine malfunction required missile destruction after less than one minute of flight.
25 Sep	Second launch partly successful but engine malfunction again required destruction of the missile.
17 Dec	First successful flight of Atlas A.
1958	
10 Jan	Successful flight of second Atlas A.
19 Jul	First launch of Atlas B. Flight terminated due to guidance system failure.
2 Aug	First successful flight of Atlas B with successful staging and sustainer operation.
28 Aug	Successful flight of Atlas B.
14 Sep	Successful flight of Atlas B.
18 Sep	Successful launch of Atlas B but missile was destroyed as range safety precaution.
28 Nov	First full range, 5,500 nautical miles, flight of Atlas B.
18 Dec	Entire missile 10B put into orbit to broadcast President Eisenhower's Christmas message of peace (Project SCORE).
23 Dec	First flight of Atlas C was successful. First use of General Electric Mod III radio-inertial guidance system.
1959	
4 Feb	Atlas B flight testing concluded with successful launch and flight of Atlas 11B.
14 Apr	First launch of Atlas D, the first operational configuration, was unsuccessful.
21 Jul	First full-scale test of ablation reentry vehicle using an Atlas C.
28 Jul	First successful flight test of Atlas D.
9 Sep	First SAC launch of an Atlas D from Vandenberg AFB, Complex 576A; Atlas D ICBM declared operational.
1960	
6 Jan	First flight test of a Mark 3 Mod 1 ablative reentry vehicle using Atlas D.
8 Mar	First Atlas D development flight with all inertial guidance successfully launched from Cape Canaveral.
22 Apr	First operational launch from Atlas complex 576B at Vandenberg AFB.
30 Aug	First operational Atlas D squadron, the 564th SMS at F. E. Warren AFB, Wyoming, transferred to SAC.
11 Oct	First launch of Atlas E with all-inertial guidance.
1961	
24 Feb	First successful flight test of Atlas E, 6,350 nautical miles downrange.
8 Aug	First successful flight test of Atlas F.

TABLE 2.1 (CONTINUED)
ATLAS DEVELOPMENT LAUNCH CHRONOLOGY

DATE	DESCRIPTION
1962	
28 Feb	First launch of Atlas E from a coffin-type launch facility at Vandenberg AFB.
1 Aug	First launch of Atlas F from a silo-lift type launch facility at Vandenberg AFB.
1963	
24 May	Air Force announces plans to phase out Atlas and Titan I ICBMs between 1965 and 1968.
1964	
27 Aug	Last Atlas E operational training flight was successful, part of the Nike-Zeus target program.
1965	
8 Jan	Last Atlas F operational training flight was successful.

Source: SAC Missile Chronology, 1938–1988, Office of the Historian, Headquarters SAC, Offutt AFB, Nebraska.

Deployment

During its deployment, Atlas was outfitted with two types of guidance systems. Atlas D had a radio-inertial guidance system where minor flight deviations were detected by ground-based radar, then guidance corrections were radioed to the missile guidance system during powered flight. This system was susceptible to jamming, and for the Atlas E and F configurations an all-inertial guidance system was utilized.[8]

Operational Atlas sites evolved through four designs. The first was referred to as PGM-16D (soft pad; ground launched; surface attack guided missile) and offered little protection from the environment, let alone a nearby nuclear explosion. The missile was in a vertical position and serviced by a gantry (i.e., mobile service tower), which was rolled back immediately before launch. Improved environmental protection came with the CGM-16D (coffin stored for a ground launch; ground launched; surface attack guided missile) where the missile was stored in the horizontal position in an above-ground coffin-shaped shelter with a roof that rolled back to expose the missile. The above-ground facilities were still exposed to the full effects of a nearby nuclear blast. A second version of the coffin design, referred to as CGM-16E, had the coffin buried with only the roof exposed. This bermed configuration improved the hardening against nuclear blast to 25 pounds per square inch (psi) overpressure.

The final and most secure deployment configuration was designated as HGM-16F (silo stored; surface launched; surface attack guided missile). This silo was 154 feet deep and 52 feet in diameter with reinforced walls that varied from two to nine feet in thickness and

Figure 2.2. An Atlas E is ready for launch from a below ground coffin launch complex. The ring of condensation just below the reentry vehicle is from liquid oxygen being vented during the countdown process. *Courtesy Dick Martin.*

was covered by a clamshell-type concrete door 2½ feet thick, hardened to withstand up to 100 psi overpressure.[9]

The PGM-16D and CGM-16D launcher systems were based as 3 x 3 configurations; that is, three missiles per launch control center complex and three complexes per squadron (the 564th SMS at F. E. Warren AFB had six missiles and one launch control center). The CGM-16E launcher system was based in a much more dispersed manner, with one missile and one launch control center at each site, 9 launchers per squadron. Dispersion was 18 nautical miles or more. The HGM-16F launcher system had one missile and one launch control center per site, but each squadron had 12 launchers. Dispersion was again 18 nautical miles or more. Two missiles in each squadron were ready for launch within 15 minutes of receiving a valid launch message; two more were ready within two hours and the rest within four hours. (See Table 2.2 for the Atlas deployment summary.) A total of 123 operational Atlas launch sites were built at 11 locations around the United States, with 72 in the silo-lift configuration that provided maximum protection.[10]

Figure 2.3. Schematic of Atlas F launch complex showing the missile sheltered (left) or raised (right) to the surface for launch. There was no capability for launching Atlas F in the silo since the flame deflector had to be above ground to function properly. *Courtesy of Dick Martin.*

Titan I (SM-68 /WS-107A-2, WS-107B)

On 21 July 1954 the Air Force Atlas Scientific Advisory Committee, composed primarily of members of the original Strategic Missiles Evaluation Committee and chaired by Dr. von Neumann, recommended that a second airframe, a true two-stage missile design, be developed as a contingency ICBM program to Atlas. While many subsystems would be interchangeable, the as yet unnamed ICBM was to represent an alternative design.[11] This new direction was based primarily on a report by Bruno W. Augenstein, an analyst with the RAND Corporation. Six months earlier, Augenstein's report had been a persuasive factor in revamping the Atlas program when the Strategic Missiles Evaluation Committee included many of his observations in their findings. A critical point was his view that a true two-stage rocket was now feasible. In addition to greater range and payload capacity, such a missile could be transported as individual stages, eliminating a possible logistics bottleneck on the nation's roadways with the large Atlas airframe.[12]

Even with Convair selected as contractor for Atlas, and with Atlas assigned a top priority, BrigGen Bernard A. Schriever, Commander, Western Development Division, felt that

TABLE 2.2
ATLAS DEPLOYMENT SUMMARY

SERIES	BASE	UNIT	CONFIGURATION	ON ALERT	FIRST OFF ALERT
D	Vandenberg	576 SMS	3x1 gantries 3x1 above-ground coffins	31 Oct 59	1 May 64
	F. E. Warren	564 SMS	3x2 above-ground coffins	2 Sep 60	15 May 64
		565 SMS	3x3 above-ground coffins	7 Mar 61	1 Jul 64
	Offutt	549 SMS[a]	3x3 above-ground coffins	30 Mar 61	1 Oct 64
E	Fairchild	567 SMS	1x9 buried coffin	28 Sep 61	17 Feb 65
	Forbes	548 SMS	1x9 buried coffin	10 Oct 61	4 Jan 65
	F. E. Warren	566 SMS[a]	1x9 buried coffin	20 Nov 61	4 Jan 65
F	Schilling	550 SMS	1x12 silo-lift	9 Sep 62	1 Feb 65
	Lincoln	551 SMS	1x12 silo-lift	15 Sep 62	10 Mar 65
	Altus	577 SMS	1x12 silo-lift	9 Oct 62	30 Dec 64
	Dyess	578 SMS	1x12 silo-lift	15 Nov 62	1 Dec 64
	Walker	579 SMS	1x12 silo-lift	30 Nov 62	5 Jan 65
	Plattsburgh	556 SMS	1x12 silo-lift	20 Dec 62	12 Mar 65

Source: Adapted from J. Neufeld, Ballistic Missiles in the United States Air Force *(Washington, D.C.: U.S. Government Printing Office, 1989), p. 234, Chart 7-4; SAC Missile Chronology, 1939–1988, pp. 24–50.*

[a]On 1 July 1961, the Atlas D squadron at Offutt and the Atlas E squadron at Warren exchanged designators.

[I]t is believed wise to sponsor an alternate configuration and staging approach with a second source. . . . It is possible that such an approach might provide a design substantially superior with the availability of future component development and thus would provide a chance for great advancement even with a late start. In line with this thinking, it is presently believed that the second design should be oriented around greater technical risks which might offer dramatic payoffs.[13]

In addition, Schriever realized that the alternative program would bring a sense of competition to the strategic missile program. He felt that Convair had grown somewhat complacent as the current sole ICBM contractor.[14] While Schriever's proposal had the endorsement of LtGen Thomas S. Power, commander of the Air Research and Development Command, Roger Lewis, the Assistant Secretary of the Air Force, was not as enthusiastic. Competition was not a critical factor in the program's success in his view, although the new capabilities described by Augenstein were significant enough to consider a second program.[15]

In early January 1955 a second report concurred with Augenstein's findings. A report by Dr. Louis Dunn of the Ramo-Wooldridge Corporation, an engineering consulting firm that worked closely with the Western Development Division, confirmed that configuration studies conducted by Convair, Lockheed, and the Glenn L. Martin Company (Martin

Company) had come to the same conclusion. A two-stage missile was definitely feasible. The result was that the Intercontinental Ballistic Missile Scientific Advisory Committee discussed Dunn's report in light of Augenstein's conclusions and recommended that an alternative staging approach become part of the strategic missile program and be given the same top priority as Atlas. On 12 January 1955 Schriever proposed the alternative program in detail. Interestingly enough, Schriever felt that only one of the two missiles would actually become operational because the resulting competition would result in a clearly superior system.[16]

A formal proposal was submitted to Air Force Headquarters in March 1955. On 28 April 1955 Secretary of the Air Force, Harold E. Talbott, approved the concept with the proviso that the facilities for design and development of the new missile be located in the central United States, not on the coast as in the case of Convair in San Diego. The Eisenhower administration dispersal policy was an attempt to overcome the overwhelming concentration of defense contractors on the seacoasts of the nation. New contracts were to be signed with companies that had facilities in the interior of the country. While originally stated as a way to move the fabrication facilities away from coastal areas that would be vulnerable to submarine missile attack, the very nature of the weapon system being produced—that is, an intercontinental ballistic missile—made this argument transparent.[17]

Slightly less than three weeks later, on 6 May 1955, the Air Materiel Command solicited proposals from Bell Aircraft, Douglas Aircraft Company, General Electric, Martin Company, and Lockheed Aircraft Company for WS 107A-2. On 12 July 1955, Air Force Headquarters issued General Operational Requirement Document 104 (SA-1C-1), stating that

> the weapon system be capable of launching missiles from bases within the continental United States carrying thermonuclear warheads with a desired weight of 3,000 pounds to ranges of 5,000 nautical miles with a circular error probable of five nautical miles or less . . . and must have the capability to strike a retaliatory blow against any attacking enemy in a minimum amount of time.[18]

Proposal evaluation began in early August and was completed by mid-September 1955. Three companies responded: Douglas Aircraft, Martin Company, and Lockheed Aircraft. Both the Douglas Aircraft and the Martin Company proposals were true two-stage designs, while Lockheed Aircraft proposed starting all three engines on the ground with two of the engines, each in its own airframe, being released at staging. Based on managerial plans and innovative engineering considerations, the Martin Company was selected.[19] On 27 October 1955, one and one-half years after Augenstein's initial report, Air Force letter contract AF 04(645)-56 was issued, authorizing the Martin Company to design, develop, and test a two-stage missile airframe designated as XSM-68. Additionally, Martin Company was to plan a program for the complete development of the weapon system, WS 107A-2, Titan. The final contract was not signed until 22 January 1957.[20]

Upon receipt of the contract, the Martin Company had one large and immediate concern, that of the exemption of missile manufacturers from the dispersal policy. Martin Company had kept to the original stipulation with which they had won the contract and narrowed down site selection to Denver and the surrounding area after looking at 94 cities

Figure 2.4. The Martin Company Titan factory in the Rocky Mountain foothills near Littleton, south-west of Denver, Colorado, in the early 1960s. The missile test stands were located in the hills to the right of the photograph. *Courtesy of Lockheed Martin Astronautics, Denver, Colorado.*

in 33 states. Site selection criteria included the need for a large labor pool, access to colleges and universities, and an excellent infrastructure such as roads and railway access. A key aspect of this site was easy access to remote canyons where fully assembled missiles and engines could be captive test fired in test stands. On 6 February 1956, a ground-breaking ceremony was held in the foothills of the Rocky Mountains west of the town of Waterton, Colorado. The fabrication facility was over 300,000 square feet in size along with another 103,000 square feet of engineering, administration, and cafeteria space. A peak employment of 5,000 personnel was anticipated by 1960.[21]

The Titan airframe design was completely different from that of Atlas. A thin-skinned design such as Atlas would be able to support only a limited amount of weight, severely limiting the second-stage performance. Since the whole point of developing this second missile was true second-stage operation, Martin Company engineers realized that unlike Atlas, where the tank walls were bands of stainless steel welded together and stored pressurized with helium, the solution for a lightweight but self-supporting airframe was to include the structural members in the propellant tank walls, a technique known as semi-monocoque construction. While this might at first appear to be a simple and obvious solution, this type of manufacturing for missile propellant tanks had not yet been attempted. This innovation was one of the keys to the success of the Titan program.

The Titan I airframe was fabricated from 2014 aluminum, a high-strength alloy consisting of copper and aluminum. Known to be extremely difficult to weld, and still considered unweldable to this day by the American Welding Society, the Baltimore Division of Martin Company had developed a tungsten inert gas welding process for use in fabrication with the 2014 alloy and by late 1957 had trained the Denver welding teams on the fine art of impossible welding. Vernon Selby, a supervisor in the welding shop in Denver, knew that they were working with special material when he found out that Martin Company had to buy the entire mill run of the 2014 alloy because no one else would buy the remainder.[22]

Fabrication of the Titan I airframe began with the chemical milling of the tank panels. Chemical milling permitted the propellant tanks to be fabricated for maximum strength yet minimum weight by the removal of aluminum in a complex pattern in specified areas. The process required that each component be masked with chemically resistant asphalt-like material in the desired pattern. Immersed in a sodium hydroxide bath, aluminum was removed at a rate of approximately 0.003 inches per minute of exposure. Those areas that had to be etched the most had no masking at the start of the process; those that were to be etched the least were masked until the last exposure process. Typically, three or four thicknesses had to be etched on each tank panel.

Once the flat panels had been etched and rinsed, they were moved to the horizontal weld fixture. The Stage I tank barrels consisted of 12 panels that were welded to form the tank cylinder, first into quarter panels, then the four quarter panels were welded to form the cylinder or barrel. The weld was made using a machine welding process and was performed by the weld torch traveling longitudinally over the weld joint. The tank barrels had to be supported by rings in the horizontal position until the domes were placed and welded. Every inch of each weld was x-rayed and hydro tested (the tanks were pressurized with water). No Titan Is were lost during flight due to tank weld failure.[23]

A feature unique to the Titan missile family airframes is an apparent slight discoloring of the exterior skin surface. This is the result of the application of Iridite, a chromium chemical conversion coating which is applied to the surface to prevent corrosion. The distinct coloring on the different panels showed how that particular batch of 2014 aluminum took the Iridite process.

Engines for Titan I were fabricated by Aerojet-General Corporation (Aerojet), Folsom, California. On 14 January 1955, Aerojet had begun research and development work on rocket engines for an as yet unnamed two-stage missile. The first stage would be powered by an engine with two thrust chambers, while the second stage would be powered by a single thrust chamber of similar design. Aerojet's design and development of these engines would serve as a backup to the North American Aviation team working on the Atlas engines, with the possible result of a better engine for use in Atlas. Contractual requirements were for two first-stage thrust chambers totaling 300,000 pounds thrust at sea level for 140 seconds of flight and a second stage of 80,000 pounds thrust with ignition at high altitude, for 155 seconds of flight. As with Atlas, these engines would use liquid oxygen and RP-1 as propellants. Basic engine components were a regeneratively cooled thrust chamber and a gas generator assembly to power the propellant turbopumps. Constant turbine speed, and thus constant propellant flow, was accomplished by metering main engine propellants to the gas generator that powered the propellant turbopumps.[24]

The first- and second-stage engines had two design configurations, the LR87-AJ-1 and LR87-AJ-3 (300,000 pounds thrust) and the LR91-AJ-1 and LR91-AJ-3 (80,000 pounds thrust), respectively. A key difference between the LR87 and LR91 engines was the use of an ablative thrust chamber skirt on LR91. The ablative skirt was necessary because of the LR91's larger expansion ratio at the high altitude where ignition would take place. The larger expansion ratio required a larger thrust chamber bell, which was difficult to effectively cool using fuel as in Stage I. Replacing part of the cooled chamber jacket with an asbestos-based ablative skirt greatly simplified engine operation, as well as saved weight. Stage I evolved to a gaseous nitrogen turbopump start that was then taken over by the propellant-supplied gas generator. Gaseous helium was used for the start up of the second engine turbopump. Engine tests began in mid-1956, with the first full duration firing taking place in March 1957. The first research and development engine was delivered to Martin Company for mating tests in November 1957. LR87-AJ-3 and LR91-AJ-3 series configuration work began in March 1959, and these engines were flight tested 13 May 1960. Production of the AJ-3 series ended after 506 thrust chambers were delivered.[25]

Due to the long lead times associated with guidance-system equipment purchases, the decision was made in April 1957 to use the Bell Telephone Laboratory radio guidance package. Development of the Bosch Arma Corporation inertial guidance system would continue as a research program. The radio guidance system was in actuality a radar guidance system. Ground-based radar tracked a missile transponder, deriving trajectory data for comparison with the computer-programmed sequence. Steering commands were then sent, via pulse-coded radar signals, to the three-axis reference system, which in turn signaled the autopilot for course corrections. In March 1958, a contract change was made to transfer the Bosch Arma inertial guidance system from Titan I to Atlas. Ten months later, the Titan program had a new inertial guidance system in development with the AC Spark Plug Division of General Motors Corporation for deployment in late 1962.[26]

Development

Titan I missiles were fabricated in eight lots for a total of 163 missiles; there were 67 flights with 20 conducted at Vandenberg AFB, California, and 47 at Cape Canaveral, Florida (see Table 2.3). The test philosophy for the Titan program had four major features. No special flight test vehicles were constructed other than the full-scale missile components. Testing started at the subsystem level at the point of manufacture and then continued into more complex testing at the Denver facilities as all the subsystems were mated into the missile airframe. The Denver facilities carried out captive fire tests for both Stage I and Stage II which featured all aspects of the flight profile except actual stage separation.

The 6555th Test Wing of the Air Force Ballistic Missile Division at the Atlantic Missile

Range, Patrick AFB, Cape Canaveral, Florida, was responsible for the Air Force side of flight test operations on the east coast. Hangar and shop facilities, propellant loading systems, tracking facilities, and launch scheduling were just a few of the Air Force responsibilities.

Each flight test required a detailed flight test plan. Space Technology Laboratories, the systems manager for all Air Force ballistic missile programs, prepared the intricate plan in coordination with Martin Company to ensure that each flight's performance built on the prior launch and covered all test objectives. Martin Company then coordinated with each subsystem contractor and the 6555th Test Wing in preparation for the actual launch. The flights from Cape Canaveral tested the missile system components, and while every effort was made to hold to proposed operational countdown procedures, et cetera, the operational testing details were left to be worked out at the Vandenberg AFB, California, test facilities.[27]

TABLE 2.3
TITAN I MISSILE FABRICATION LOTS

DESIGNATOR	LOT	# BUILT	DESCRIPTION
XSM-68	A	6	Simplified first stage; dummy second stage, limited range.
	B	7	Complete first and second stages with reduced second-stage engine duration; open and closed loop radio guidance.
	C	6	Complete first and second stages with reduced second-stage engine duration; radio guidance; separable scale-model reentry vehicle.
	D, E, F		Eliminated from the test program.
	G	10	Complete two-stage missile; closed loop radio guidance, separable reentry vehicle; range up to 4,000 nautical miles.
	H		Eliminated from the test program.
	J	22	Complete missile capable of flights up to 5,000 nautical miles; later missiles to carry operable reentry vehicle and warhead without reactive materials.
	K, L, S, T		Eliminated from the test program.
	V	4	Same as Lot J with exception of instrumentation and range safety equipment to be used as part of operational systems testing at Vandenberg AFB and the Pacific Missile Test Range. V1 & V4 were used in translation rocket tests, VS1 was launched from the silo launch facility and V2 was lost in the operational suitability test facility explosion.
XSM-68A	M	7	Same as Lot J except equipped with an inertial guidance system to serve as test bed for SM-68B (Titan II) guidance system.
	SM-68	101	Operational missiles.
	TOTAL	163	

Source: Titan Master Schedule, 31 July 1963, Air Force Historical Research Center, Maxwell AFB, Alabama.

Flight Test Program

Titan I missile A-1 was accepted by the Air Force on 17 June 1958, 28 months after ground was broken for the new fabrication facilities at Littleton, Colorado. Though the first launch of Titan I was contracted to take place in September 1958, slippages due to problems in the captive-fire testing program caused the tests to begin in December, finishing one month later.[28] A typical flight profile for the research and development missiles is listed in Table 2.4.

The Titan I test flight program began with four successful launches of Lot A missiles from Pad 15 at Cape Canaveral (Table 2.5). These missiles consisted of fully configured Stage I airframes and engines, an inert Stage II carrying water to simulate a fully loaded condition, and simple guidance equipment (see Appendix 1 for a detailed list of Titan I flights). Stage II operation awaited successful demonstration of the staging sequence using the dummy second stage. After first-stage burnout, two 5,000-pound-thrust solid-propellant rockets fired for approximately three seconds, permitting sufficient separation of Stage I and II to prevent Stage II engine ignition from fragmenting the Stage I tanks and possibly damaging Stage II. The results were encouraging as the radio guidance system functioned perfectly, the aerodynamic drag was found to be less than anticipated, and the stage separation system worked as planned.[29]

The Lot B missiles were used to test stage separation with successful firing of the Stage II engine and shortened Stage II flight. The first flight of the B-5 was delayed by a series of malfunctions and mishaps, not atypical of complex flight testing operations. On 14 August 1959 ignition was normal, but the missile hold-down bolts fired at T+1.3 seconds, 4.4 seconds early. At T+1.6 seconds the Stage I engine thrust was equal to the weight of the missile, and at T+1.8 seconds the missile lifted off the thrust mount. When one of the Stage I umbilicals was prematurely pulled free, 3.9 seconds earlier than programmed, a no-go signal was generated, which in turn caused a Stage I engine shutdown command to be automatically

TABLE 2.4
NOMINAL TITAN I TRAJECTORY

FLIGHT PROGRAM	STEP	TIME (sec)	ALTITUDE (nm)	RANGE (nm)	VELOCITY (ft/sec)
1.	Liftoff	0	0	0	0
2.	Stage I burnout	134	35	39	8,065
3.	Stage II burnout	290	150	345	22,554
4.	Vernier cutoff	340	199	515	22,371
5.	Apogee	1,061	541	not given	not given
6.	Reentry	1,920	50	5,400	23,540
7.	Impact	1,970	0	5,500	1,000

Source: Titan Master Schedule, 31 July 1963, Section X, pages 10–11, Air Force Historical Research Center, Maxwell AFB, Alabama.

TABLE 2.5

TITAN I DEVELOPMENT SUMMARY

DATE	DESCRIPTION
1958	
17 Jun	Air Force accepts first Titan I airframe from Martin Company.
1959	
6 Feb	First successful Titan I launch from Cape Canaveral.
1960	
10 Aug	First successful operationally configured Titan I launch from Cape Canaveral.
3 Dec	Missile explosion due to silo missile elevator failure destroys operational system test facility.
1961	
20 Jul	First Titan I ICBM equipped with decoys successfully launched to test radar capability to distinguish between decoys and reentry vehicle.
24 Jul	First full range test of Titan I with all-inertial guidance, 5,000 nautical miles, testing Titan II guidance system
23 Sep	First successful launch of fully configured operational Titan I using silo-lift facilities.
1962	
20 Jan	First SAC Titan I crew launch.
28 Jan	Forty-seventh and last launch of Titan I from Cape Canaveral. During this part of the program there were 34 successful, 9 partially successful, and 4 failed launches.[a]

Source: SAC Missile Chronology, 1939–1988, Office of the Historian, Headquarters SAC, Offutt AFB, Nebraska.

[a] *Martin Company records list 32 successful, 10 partially successful, and 5 failed launches.*

sent to the missile. The loss of thrust caused B-5 to drop back through the launch mount. The resultant explosion and fire damaged cabling and the umbilical tower at Pad 19.[30]

An additional series of problems at the Denver facilities caused the Air Force to suspend flight operations until Martin Company management was able to resolve production quality-control issues. On 28 August 1959 the suspension was removed, only to be followed by a helium line rupture on missile B-6, causing it to be returned to Denver, and on 10 October 1959 missile B-8, during airlift from Denver to Patrick AFB, was severely damaged when a pressure differential due to altitude changes during the flight was not noticed and the tanks collapsed.

On 12 December 1959, the first Lot C missile, C-3, was ready for launch. The Lot C missiles were identical to the Lot B missiles except for a separable scale-model Mark 4 reentry vehicle (see Chapter 3 for a complete discussion of the Mark 4 reentry vehicle). Ignition occurred normally and the hold-down bolts released as programmed, followed by the explosion of the missile as the range safety destruct package was unintentionally triggered. The cause of the explosion was innocent enough. Jim Greichen, a Martin Company engineer at the time, recalls that they had moved the range safety destruct relay approximately six inches as part of another modification and rotated it 90 degrees. Tests later confirmed that the vibration from the firing of the explosive bolts was much greater at the new location, causing the relay to close and trigger the range destruct package detonation.[31] This was the final straw for the Air Force managers, and a special team was sent to thoroughly review the Titan management team in Denver. In early January 1960, Air Force and Martin Company officials met again, with the outcome being Martin Company's president, George M. Bunker, deciding to personally take over Titan program oversight at Denver.[32]

On 2 February 1960, missile B-7A (composed of a first stage from B-7 and a second stage from B-6) was successfully launched on a 2,020-nautical-mile flight with nose-cone impact 2 nautical miles long and 1/10 nautical mile to the right of the target. This was the first successful attempt at staging. Over the next nine months, 18 launches of a mix of Lot C, G, and J missiles were conducted with 10 fully successful, 5 partially successful, and 3 failures. The Titan I program was well on its way as the Lot G and J flights demonstrated consistent, completely successful first- and second-stage engine operation and a high degree of target accuracy. The circular error probable for the Lot J and Lot G missiles was 0.8 nautical miles, 0.2 nautical miles below the design requirement. With an in-flight reliability of 78 percent, just above the contract stipulation of 77 percent, Titan I had met its requirements.[33]

All but one of this string of 18 launches had taken place at Cape Canaveral, while the operational system test facilities were being constructed at Vandenberg AFB, California. The test program at Cape Canaveral ended with the successful launch of missile M-7 on 29 January 1962. Forty-seven Titan I missiles had been launched with 32 fully successful, 10 partial successes, and 5 failures.[34]

Vandenberg AFB served as the operational base test launch site. Titan I launch facilities were built in two stages. The first was the Operational Suitability Test Facility (OSTF). Construction began in June 1958. Missile equipment installation began in July 1959, with the facility accepted for final testing on 6 January 1960. While not matching the proposed operational base facilities precisely, being a single-silo rather then a three-silo complex and having an above-ground launch control building, OSTF was used to test and confirm operating ground equipment compatibility. The silo was fully configured with blast doors, a propellant handling system, and a missile elevator system in the operational configuration so that operational launch procedures could be developed with the use of a battleship missile airframe. The battleship airframe was made of steel and could not be launched; it served only as a training aid for loading and unloading propellants and for operating the missile elevator system.

On 15 October 1960 Titan I missile V-2 was installed in the OSTF. The Lot V missiles were identical to the Lot J missiles except for the additional range safety instrumentation necessary for launch at Vandenberg AFB. Propellant transfer operations began soon thereafter. Titan I was stored with the fuel onboard. Liquid oxygen was transferred to the missile in the silo and then it was raised to the surface for launch.[35]

Figure 2.5. Aerial photograph looking southwest at Vandenberg AFB over Titan I Training Facility TF-1, also known as Launch Complex 395-A, -B, and -C. The trampoline-like structure over the launch control center in the center background of the photograph was a protective measure taken after the Operational Suitability Test Facility explosion in December 1960. The Titan I and Atlas F silo complexes share similar missile elevator and silo design. *Courtesy of Andy Hall.*

On Saturday evening, 3 December 1960, a full rehearsal short of actual launch was being conducted by Robert Rhodus, the Martin Company OSTF test conductor. This was the ninth attempt at completing this test; the earlier attempts failed because of minor equipment malfunctions and procedural difficulties. Missile V-2 was loaded with liquid oxygen within 15 minutes, the time frame required for operational launch conditions. The missile was raised to the surface and the countdown conducted to a point just prior to the ignition signal. All involved were relieved that this test had finally been successfully completed to the stage where the missile was on the surface, just short of the launch sequence. The standard operating procedure was to vent the propellant tanks and lower the missile back into the silo where the propellant probe reconnect crew was waiting in the blast lock area to offload the oxidizer.[36]

Rhodus was in the Training Facility Titan I Control Center, the control center for the operational base three-silo configuration complex built adjacent to OSTF, watching on a television monitor the missile's progress back down into the silo when it became apparent that the descent was too rapid. He watched in fascination as the elevator, carrying a missile fully loaded with propellants, plummeted to the bottom of the silo. The first explosion was a result of the first-stage oxidizer and fuel tanks rupturing and mixing and the propellants igniting. It was quickly followed by a second explosion as the second-stage tanks did likewise. As the propellants mixed and exploded, hurtling large chunks of concrete and structural steel into the sky, Rhodus realized with some trepidation that, unlike the buried operational control centers, the room he was in had only 12 inches of dirt on top, not much protection from the tons of concrete that were raining down all around. He also realized it was far too late to run.[37]

The propellant probe team that was on the far side of the blast lock was fully protected. Augie Chiarenza, a member of the probe team, heard the explosion and then the next thing he knew, they were all standing in the OSTF control center, out of breath, wondering what was going to happen next.[38] Chiarenza and the rest of the probe reconnect team were indeed fortunate. John Carlson, the Sundt Corporation project engineer for the OSTF, had received a change order from the Army Corps of Engineers to reinforce the blast lock area for the OSTF two months earlier. Being already overwhelmed with work to be done as the project was nearing completion, he said that he would get to it the next day. He specifically remembers being told to start on it immediately, and so they did. With the personnel tunnel already buried, his crews had to jack hammer out the door frames, add extra reinforcing steel to the frame, the hinge, and the door, and pack an extra-dense, high-strength concrete mix into the newly hung door. This change was the critical difference for the propellant team as the first of the two doors making up the blast lock partially failed while the second one held. Carlson still intuitively thinks that what happened was that an Army Corps of Engineers architect woke up in the middle of the night and told himself that the doors needed reinforcing *now*.[39]

Rhodus interviewed many of the witnesses of the blast. A young airman who was manning a searchlight on an adjacent embankment swore that he sprained his ankle getting underneath the searchlight frame, but when they went up to his searchlight station, they found tracks where he had run down the embankment, skidded to a stop when a piece of debris landed in front of him, reversed direction back up to the searchlight, and then dove underneath for protection.[40]

John Adamoli, Martin Company test conductor for the adjacent Silo Launch Test Facility (SLTF), barely 1,200 feet directly west of the OSTF, had a similar view to that of Rhodus. Adamoli and several of his staff were watching from the SLTF control room as the missile was being lowered back into the silo. When it started down they commented to each other that it was going down rather fast and continued to watch. They were awestruck as the entire elevator assembly, known as the crib, and missile launcher, a total of 160 tons of structural steel, came out of the silo, tumbling up out of the searchlight beams "in slow motion." This sight triggered a survival instinct when it dawned on them that the shock wave wasn't far behind. They ducked under a table as the shock wave hit, shattering all the

Figure 2.6. Aerial view of the aftermath of the 3 December 1960 explosion at the Operational Suitability Test Facility, Vandenberg AFB, California. The three operational training sites for Titan I can be seen in the upper right hand corner of the photograph. The large structure protruding out of the silo is the "crib" which supported the missile thrust mount and flame deflector. It was ejected straight up from the silo, cartwheeled 180 degrees, and slammed back down into the silo. *Courtesy of Fred Epler.*

windows and ripping all of the light fixtures from the ceiling. Amazingly, no one was seriously hurt.[41]

The OSTF was damaged beyond repair. The southeast half of the silo concrete cap was completely destroyed, while the northwest half was tossed into the air, rotated 180 degrees, and came to rest where the southeast half had been. The top 20 to 30 feet of the silo wall was shattered, and the remaining 90 feet, down to within 30 feet of the bottom of the silo, was severely cracked. For many years afterward, enormous chunks of concrete could be seen on the low hills near the approach road to the Titan I facilities. A reference system gyro was found on the base golf course, more than a mile away, as well as one of the large shock isolation springs used in the missile support system.[42]

On 31 January 1961 the official accident report was issued. The cause of the accident, a failed control valve in the elevator hydraulic system, was found in the debris field. The launch platform elevator brakes had failed, sending the missile and launch platform to the bottom of the silo. Since over 90 percent of the design information for the operational facilities had been gathered during the myriad of tests run prior to the explosion, the decision was made not to rebuild the OSTF. Appropriate modifications were incorporated into the nearly completed Titan I training facilities, (TF-I) and Launch Complex 395A.[43]

Construction on Launch Complex 395A, configured as an operational base with three missile silos connected to one launch control center, began on 13 May 1959. The construction lessons learned earlier on OSTF were used to streamline this prototype operational base. The three silos were referred to as 395A-1, -2, and -3. Construction was completed by late 1960, but the OSTF explosion pushed back initial testing nearly four months as modifications were made to the elevator hydraulic systems. Another modification, of which remnants can still be seen today, was the construction of a large netting structure over the above-ground launch control center to provide protection from large falling objects.[44]

The first launch of a Titan I from the TF-I complex took place on 23 September 1961 with the successful flight of SM-2, 5,300 nautical miles downrange. Flight operations continued through 5 March 1965 when Titan I SM-80 was successfully launched but fell short of the target area due to premature propellant depletion. In all, 19 Titan I missiles were launched from the training complex facilities with 11 successful, 7 partially successful, and one failure.[45]

Titan II actually evolved from the SM-68A concept, a Titan I that would be structurally modified for in-silo launch. The initial comprehensive studies on in-silo launch had been conducted in Britain as scientists of Dehavilland and Rolls Royce grappled with the in-silo launch process for their Blue Streak intermediate range ballistic missile (IRBM) which had a range of 2,400 nautical miles. The design of Blue Streak was based on the Atlas missile, employing the same thin-walled tank design. Two Rolls Royce RZII engines provided 135,000 pounds thrust each.[46] Developmental work on the various missile components had begun in 1955, and by 1959 sufficient data had been accumulated to complete the Blue Streak launcher for use in underground silos. Poised over a U-shaped tube that would deflect the blast and rocket exhaust away from the missile and back to the surface, Blue Streak was the free world's first in-silo launch weapon system concept. A prototype launcher was on the verge of being constructed in eastern England when the program was canceled in 1960.[47]

On 17 November 1958, Major General Schriever requested a detailed briefing on the conversion of the Titan program to in-silo launch concepts. The briefing was to include operational advantages, technical problems, and proposed methods to resolving any remaining technical problems.[48] Feasibility studies were conducted, including 1/6th scale model experiments conducted by Aerojet-General. On 17 August 1959, Charles P. Benedict, Acting Assistant Secretary of Defense, approved the 22 July 1959 proposal for the design and construction of a Silo Launch Test Facility. Benedict emphasized the need for economy:

> [I]t is essential that the conceptual designs of even this early test facility give every consideration to minimizing the size and scope of hardened underground construction

and to utilizing the most economical construction materials and techniques. This philosophy should be accentuated in any follow-on concepts for training and operational facilities, with particular care that requirements unique to test or training facilities are not carried over to the operational design.[49]

Knisely (K) Dreher was the Ralph M. Parsons Company (Parsons) project manager for the SLTF. Parsons was the lead engineering firm for the Titan I facilities at Vandenberg AFB. Dreher knew that the SLTF had fast-track priority and had been awarded to Parsons by the Army Corps of Engineers on a design-construct basis, an almost unheard of award at the time. The design and construct process took approximately 18 months. The design included the major features of an operational silo: the flame deflector, acoustical insulation, steel shells separating the launch and exhaust ducts, and folding service platforms. What was missing were the propellant transfer systems and an operational configuration silo closure door. Since the site was going to be for only one launch, considerable savings was realized by eliminating these two systems.

Of the many questions that needed to be resolved as construction started, two stood out in Dreher's mind. First was the design of the flame deflector. Would it be scoured by the flames and influence the launch? Would enough of the exhaust be deflected away from the missile? The second major area of concern was that of the acoustical energy. Initial estimates indicated that three feet of fiberglass insulation would be necessary. The fitting of hundreds of chicken wire–bound insulation cubes of fiberglass to both the launch and exhaust ducts was a task that was hard to forget.[50]

Because of the extremely tight schedule, the decision was made to slip-form the concrete lining of the silo. In this process the concrete emplacement is made as one continuous pour, with the form slowly moving up as the concrete was placed and quickly set. The continuous pour lasted 48 hours and proved to be the prototype for Titan II silo construction.[51]

Built at Vandenberg AFB, the SLTF was 145 feet deep and, unlike the British designs for the Blue Streak which utilized a J-shaped flame deflector, was equipped with a W-shaped flame deflector. With the increased thrust for the Titan II missile, the W-shaped deflector was chosen to improve the efficiency of exhaust gas deflection and removal. The SLTF was heavily instrumented to monitor the heat and vibration of the launch environment. While Titan II was to be the missile for in-silo launch, its test flight program had not begun, so a Titan I missile, designated as VS-1, was modified with a dummy second stage filled with water ballast and an inert reentry vehicle. Results of earlier acoustical energy studies required the strengthening of key areas of the VS-1 airframe (see Chapter 3 for further details on the in-silo launch design process).

Preparations for the launch, known as Operation SILVER SADDLE, began on 16 October 1960 when the first stage of VS-1 was lowered into the SLTF. The next day the second stage was attached and checkout began. The schedule for a static firing test slipped from late November into December and was further delayed by the explosion at the OSTF complex. A thorough review of procedures for the SLTF was conducted after the explosion. Originally the launch control center was going to be 75 feet from the SLTF silo. After the explosion they decided to use the OSTF control center for improved safety, located 1,200 feet east of the SLTF site. Instead of moving all of the launch control equipment from the

Figure 2.7. View down into the launch duct of the Silo Test Launch Facility at Vandenberg AFB. Titan I VS-1 is positioned in the launch duct. The silo closure door was for environmental protection only and was not a prototype of the operational facility door. The exhaust ducts are located above and below the launch duct, hidden by the door and partially visible at the center edge of the photograph. *Courtesy of Lockheed Martin Astronautics, Denver, Colorado.*

already completed control center, Martin Company engineers rigged a long surface cable to a table-top console where John Adamoli, the test conductor, would manually complete the firing circuit.

On 7 March 1961 a significant milestone in the Titan II program was reached with the successful captive firing of Titan I missile VS-1 in the SLTF. While highly successful, the need for additional sound proofing to absorb the tremendous acoustical energy of the rocket engine exhaust was clearly apparent when the data were reviewed. Much to everyone's relief, the remote console had worked perfectly, validating its use for the actual launch.[52]

On Wednesday, 3 May 1961, VS-1 was successfully launched. Strong winds with gusts up to 35 knots swept the ocean and swirled around an assemblage of journalists, base disaster team personnel, and others who gathered to witness the launch. A special readiness countdown had begun 24 hours before, which had included loading the propellants 9 hours earlier, and had proceeded smoothly up to the planned two-minute terminal count. A few seconds before 1400, Adamoli turned the "START FIRING SEQUENCE" switch to "ON" and then pressed the "FIRE" button, the only manual launch he ever conducted.

The Stage I engines ignited, sending clouds of steam and smoke into the air. Two seconds later the tip of the reentry vehicle appeared at the surface and then the massive cylinders of the first and second stage rose slowly and majestically before the eyes of the observers and the lens of whirring motion picture cameras. As the nozzles cleared ground level, the engine roar spread over the site. After six seconds, the autopilot began rolling the missile 50 degrees to fly due west as it pitched and climbed out over the Pacific. First-stage burnout came at 138 seconds into flight with the missile already many miles away. Several seconds later, a puff of smoke signaled the planned destruction by the range safety officer.[53]

The SLTF suffered relatively minor damage to the launch duct liner and exhaust duct walls. With in-silo launch now proven feasible for Titan II, the SLTF was to be a one-shot test site. Efforts by the Ballistic Systems Division (BSD) to have the SLTF converted into a fourth Titan II launch complex were rejected, and eventually the SLTF was converted to the Titan II Operations and Maintenance Missile Trainer (QMT) by removing the acoustical modules and concrete launch duct liner, allowing more ready access for training purposes.[54]

Deployment

The Titan I ICBM force was deployed in six squadrons in the HGM-25A configuration (silo stored; surface launched; surface attack guided missile). At least 18 nautical miles separated Titan I launch complexes of three missiles per launch control center (3 x 3), three launch control centers per squadron, hardened to withstand 100 psi overpressure. The silo-lift facilities were similar to those used in the Atlas F basing. Table 2.6 lists the deployment history for Titan I. Table 2.7 lists the general specifications for a Titan I Lot M missile.

On 24 May 1963, Headquarters Air Force approved the recommendation of an Air Force study that the Series D, E, and F Atlas and Titan I missiles be phased out of SAC between 1965 and 1968. On 20 November 1964, Secretary of Defense McNamara announced that all Atlas as well as Titan I missiles would be phased out by June 1965.[55] System complexity, due primarily to the difficulties of working with liquid oxygen and relatively slow

Figure 2.8. Titan I SM68-83 (61-4510), code-named Operation DAILY MAIL lifts off on 17 September 1963. This was the fourth flight of the Demonstration and Shakedown Operation program, a successful launch. *Courtesy of Andy Hall.*

TABLE 2.6
TITAN I DEPLOYMENT SUMMARY

BASE	UNIT	CONFIGURATION	FIRST ON ALERT	FIRST OFF ALERT
Lowry	724 SMS	3x3 silo-lift	18 Apr 62	17 Feb 65
	725 SMS	3x3 silo-lift	10 May 62	17 Feb 65
Mountain Home	569 SMS	3x3 silo-lift	16 Aug 62	17 Feb 65
Beale	851 SMS	3x3 silo-lift	8 Sep 62	4 Jan 65
Larson	568 SMS	3x3 silo-lift	26 Sep 62	4 Jan 65
Ellsworth	850 SMS	3x3 silo-lift	28 Sep 62	4 Jan 65

TOTAL TITAN I LAUNCHERS = 54[a]

Source: *Adapted from J. Neufeld,* Ballistic Missiles in the United States Air Force *(Washington, D.C.: U.S. Government Printing Office, 1989), 1990, p. 236, Chart 7-5.*

[a] *The Titan I sites at Vandenberg were training facilities. While they could have been placed on alert if necessary, the normal Titan I alert status did not include them in the force count. They were on strategic alert for a short period during the Cuban Missile Crisis.*

TABLE 2.7
TITAN I SPECIFICATIONS, LOT M

FULLY ASSEMBLED AIRFRAME	LENGTH	MAXIMUM DIAMETER
Stage I (including stage transition)	56.6 ft	10 ft
Stage II	25.4 ft	8 ft
Reentry Vehicle Adapter	4.62 ft	8 ft
Mark 4 Reentry Vehicle	10.79 ft	4 ft

ENGINES	THRUST
Stage I (LR87-AJ-3)	300,000 lbs at sea level
Stage II (LR91-AJ-3)	80,000 lbs at altitude

WEIGHT	STAGE I	STAGE II
Airframe Empty	7,741 lbs	4,484 lbs
Oxidizer	118,044 lbs	28,468 lbs
Fuel	51,682 lbs	12,441 lbs
Total Weight (including reentry vehicle)	223,211 lbs	

ACCURACY   CEP less than 1 nautical mile

Sources: *"Structural Description, SM-68," February 1961, Martin Company, p. 10; "General Arrangement, Lot M, Sheet 327-1000501" 18 July 1960; D. MacKenzie,* Inventing Accuracy: A Historical Sociology of Nuclear Missile Guidance, *Appendix A, MIT Press, 1990.*

response time, were but two of the many reasons for deactivation of these two first-generation ICBM systems.

The Titan I program served as a proving ground for two important aspects of the Titan II program. First were the Lot M missiles which served as the test bed for the Titan II inertial guidance system with a record of seven out of seven successful flights as far as the prototype guidance system was concerned. Second was the demonstration of the capability of a Titan I airframe to withstand the acoustical environment of an in-silo launch. The Titan II program was now poised to offer the best of both worlds. Robust enough to withstand in-silo launch, combined with storable noncryogenic fuel and oxidizer, the result was a second-generation liquid-propellant missile with a greatly decreased response time, increased payload capacity, and much greater protection against enemy attack.

III

THE SECOND GENERATION: TITAN II

Titan II (SM-68B, WS 107-C)

In July 1958 the Air Force reviewed several changes in the Titan I program, including changing from radio-inertial to all-inertial guidance; from the 3 x 3 configuration to 1 x 9 basing; to the use of storable propellants; and to in-silo launch instead of silo-lift.[1] The Martin Company was commissioned by the Air Force Western Development Division to perform a cost-reduction study on the Titan I system, and Robert Bolles, a Martin Company engineer working on the ground support equipment for Titan I, was reassigned to that task. At the time of these early missile systems, aircraft manufacturers were accustomed to the simple support equipment for aircraft such as fuel trucks, chocks for the wheels, ladders for cockpit entry, and the like. In contrast, missile systems had to be able to withstand a nearby nuclear blast, and then to launch; launch complex checkout equipment and launch control systems had to be monitored and repeatedly checked for what might be years before actual use, and with cryogenic liquid oxidizers, handling became a much more critical issue.

Bolles and others realized that in the next generation of the Titan ICBM, reduced complexity would mean reduced costs. By going through every system and subsystem in both ground and airborne equipment, 10 major points were revealed: (1) there should be an effort to change the Titan I checkout and launch philosophy by doing the end-to-end checks and only checking the internal sequences if an end-to-end check showed a problem; (2) there should be an elimination of the fast propellant loading system and problems with cryogenic oxidizer by storing propellants on board the missile; (3) there should be an elimination of the missile elevator by transitioning to in-silo launch, reducing the exposure time of the missile considerably as well as the size and complexity of the shock isolation system; (4) there should be an elimination of the radio guidance antenna system and its requisite hardening and shock isolation systems by going to an all-inertial guidance system; (5) there should be an elimination of the staging rockets by going to a "fire-in-the-hole" staging process (Stage II engine ignition would take place while the two stages were still attached); (6) the silo closure door design should be changed to reduce power needs; (7) engine igniters needed to be eliminated; (8) a new reentry vehicle design was needed to eliminate the uneven ablation (the reentry vehicle outer layer chars and sloughs off, dissipating the heat caused by reentry away from the interior of the reentry vehicle) on the Mark 4 reentry vehicle which resulted in inaccurate flight to target; (9) a change was needed to the basing mode to 1 x 9 (one missile per launch complex, nine launch complexes per squadron); (10) payload capacity should be increased by going to a 10-foot diameter second stage. As the review progressed, presentations were made to the Air Force

Western Development Division as well as to Ramo-Wooldridge on a weekly basis. The results formed the basis of the Titan II program.[2]

The development of storable propellants with sufficient power was a key factor in the birth of Titan II. Hypergolic (the propellants ignite on contact) propellant research had begun in earnest in 1945 when the Air Force and Navy both realized that tactical liquid-propellant missiles would need to have storable propellants in order to be useful weapons. Over the next six years a multitude of combinations were tried with varying degrees of success. Hydrazine was the fuel of choice, but it had a freezing point of 28 degrees Fahrenheit, far too high for a tactical or airborne missile that would routinely see much colder environments when carried at high cruise altitudes for prolonged periods. The military specification for liquid propellants required that they remain liquid at -65 degrees Fahrenheit.[3] In 1951 the Navy Bureau of Aeronautics Rocket Branch funded research at Metalelectro and Aerojet-General to develop hydrazine derivatives in an attempt to find a solution to the freezing-point problem.

Three derivatives were chosen for further research: symmetrical hydrazine, monomethyl hydrazine, and unsymmetrical dimethyl hydrazine. Symmetrical dimethyl hydrazine was of little value since its freezing point was 16 degrees Fahrenheit. Monomethyl hydrazine froze at -62 degrees Fahrenheit and unsymmetrical dimethyl hydrazine at -70 degrees Fahrenheit. Monomethyl hydrazine had an unfortunate tendency toward catalytic decomposition but this was not the case for unsymmetrical dimethyl hydrazine. Furthermore, mixtures of either compound with hydrazine lowered the freezing point to well below -65 degrees Fahrenheit.[4] Aerozine 50, Aerojet-General's name for a 50:50 (volume to volume) mixture of unsymmetrical dimethyl hydrazine and hydrazine, was selected for use in Titan II. Freezing-point depression was only one of the advantages of Aerozine 50. Another important feature was that Aerozine 50 was stable enough to use as a regenerative coolant for circulation in the thrust chamber walls. Hydrazine alone had a tendency to detonate when used as a regenerative coolant.[5]

The other half of the hypergolic propellant system was the oxidizer. Nitric acid was the first choice, but it had a major drawback: it was incredibly corrosive. An attractive alternative was nitrogen tetroxide. It was not as corrosive as nitric acid but had a high freezing point, well above -65 degrees Fahrenheit. Many alternatives were tried, but most ended up forming extremely sensitive explosive mixtures, a feature that was not compatible with use in a tactical or strategic missile. Nitric oxide was an obvious alternative but raised the vapor pressure of the mixture, generating vapor clouds when the liquid was spilled even more readily than nitrogen tetroxide. Still, since its addition did not cause explosive mixtures, further experimentation was warranted. If refrigeration was available, the nitrogen tetroxide/nitric oxide combination would be an optimum solution, compared to nitric acid.[6]

In September 1958, the Scientific Advisory Committee on Ballistic Missiles was informed of the feasibility of storable propellants by the Air Force. Earlier discussions between the Air Force and William M. Holaday, Director of Guided Missiles, had been extremely positive, with further feasibility studies indicated.[7] The success in movement toward storable propellants meant that an in-silo launch was now much more attractive.

On 17 November 1958, Major General Schriever requested a detailed briefing on the ongoing British studies of in-silo launch feasibility with an eye toward the Titan program. The briefing was to encompass operational advantages, technical problems, program scheduling, and cost comparisons.[8]

Col William E. Leonhard, deputy commander for Installations, Ballistic Missile Division, was assigned the task to further evaluate the British plans for the operational deployment of Blue Streak, an IRBM similar to the Air Force Thor IRBM. He traveled to London, reviewing the in-silo launch design that used a single U-shaped flame deflector.[9] Major General Schriever approved the conversion of future Titan facilities from silo-lift to in-silo launch on 19 January 1959.[10]

Between 1958 and late 1959, the fate of the Titan I program hung in the balance as economy of force size and structure was considered yet again due to budgetary constraints. With the solid-fueled Minuteman to be deployed in nearly the same time frame as Titan I, considerable effort was made to reduce or cancel the Titan program entirely. At one point, in April 1958, Titan I deployment was reduced to one squadron. The 1959 Eisenhower administration budget allowed for 4 Titan I squadrons. The Air Force had requested 11 Titan I squadrons.[11] By June 1958 planning had progressed to the point of site selection for the first 4 squadrons but two months later, the Office of the Secretary of Defense was still not convinced that Titan I should be continued. Holaday requested that a study be made of the effect of canceling Titan I completely and supplementing the Atlas force instead. In October 1959 Schriever not only pressed for continuation of the Titan program but also strongly repeated the requirement for an increase to 11 squadrons.[12]

Three months later, in January 1959, Holaday approved a force structure that included 11 Titan I squadrons. By April 1959 development plans had to be revised yet again as budgetary restraints required delay of Titan I by nine months for the first operational squadron. Restudy of the problems led to the first mention of in-silo launching or "fire from the hole." The Air Force Ballistic Missile Committee approved an 11-squadron Titan program with 6 squadrons of the 3 x 3 silo-lift configuration and 5 of the 1 x 9 with in-silo launch configuration, designated as Titan I SM-68A.[13] Permutations on Titan and Atlas configurations and basing modes continued for several months with Titan squadron numbers ranging from 4 to 11 to 14. Finally in December 1959 the Secretary of Defense Ballistic Missile Committee approved funding for an 11-squadron Titan I program.[14] Unfortunately, just when all seemed settled and done, the Titan program met with a series of flight test failures and again the call was made for cancellation.

After review on 31 March 1959 the Scientific Advisory Committee strongly supported efforts for the use of storable propellants in the fifth Titan squadron and urged that a high priority be placed on unsymmetrical dimethyl hydrazine/nitrogen tetroxide engines. Furthermore, in-silo launch was strongly endorsed. Interestingly enough, at this same meeting the committee members were skeptical of accelerating the Minuteman program but agreed on the condition that this did not detract from the Titan storable propellants program in any way. On 29 May 1959 a subcommittee reported to the full committee a recommendation that major modifications be made in the Titan program, beginning with the seventh squadron: all-inertial guidance, in-silo launch, and noncryogenic propellants.[15]

While liquid propellants were seen by the Minuteman proponents as an unnecessarily complex and awkard system when compared to the solid propellants of Minuteman, the fact that the improved Titan would be able to carry the Titan I payload much further served to overcome these concerns.

While official requirements for the Titan II weapon system were defined in Specific Operational Requirement 184, 10 April 1961 and revised on 5 September 1962, the first official Titan program development plan that included Titan II was published in April 1960.[16] This first plan showed the tentative specifications for an upgraded Titan with a fully loaded weight of 326,000 pounds, versus 221,500 pounds for Titan I; a length of 103 feet versus 97.4 feet for Titan I (from bottom of engine thrust chambers to tip of reentry vehicle), and both first and second stages were 10 feet in diameter, rather than the 8-foot diameter second stage for Titan I. Like Titan I, Titan II's first stage would have two engines, but they would develop 430,000 pounds of thrust at sea level, versus the 300,000 pounds thrust of the Titan I first stage; Titan II's second-stage single engine would ignite at 250,000 feet altitude with 100,000 pounds of thrust versus the 80,000 pounds of thrust for Titan I. Range with the Titan I Mark 4 reentry vehicle of 4,000 pounds was to be 8,400 nautical miles. The new Mark 6 reentry vehicle, weighing 8,000 pounds, was designated as the primary reentry vehicle at nearly the same time. While this reduced the range to 5,500 nautical miles, nearly identical to Titan I, the increased warhead weight was considered worth the tradeoff.[17] The Martin Company signed a letter contract with the Air Force for Titan II in May 1960, and development began one month later.[18] The Mark 6 reentry vehicle design contract was awarded to General Electric Company's Missile and Space Vehicle Department on 26 July 1960, indicating that Titan II would be carrying the Mark 6.[19]

All was not yet secure for Titan II. Within and outside of the Air Force, questions were raised about spending money to improve the liquid-propellant Titan when the solid-propellant Minuteman was much cheaper, costing approximately one-half of the proposed Titan II. MajGen. Osmond J. Ritland, commander of the Air Force Ballistic Missile Division, strongly supported Titan II since the first Titan II missiles would be true operational prototypes that built on the progress of Titan I. Titan II would carry a much larger warhead than Minuteman. Therefore, Minuteman and Titan II served a complimentary purpose in the SAC nuclear war strategy at least in terms of ICBMs. For those who sought to keep Titan I and delay Titan II, the fact that a squadron of nine Titan II missiles was estimated to cost $138 million versus $166 million for Titan I was yet another factor in favor of Titan II.

Funding for the North American B-70 "Valkyrie" supersonic bomber program was in jeopardy at this time. The manned bomber was the mainstay of the strategic nuclear deterrent forces, and LtGen Mark E. Bradley Jr., Air Force Deputy Chief of Staff for Materiel, argued that the proposed $400 million for Titan II development could be better spent on the B-70 supersonic heavy bomber program, whose budget had been slashed to $75 million due to questions about its ability to penetrate improved Soviet air defenses. The funding stayed with Titan II.[20]

Engine Design and Testing

Titan II engine development began on 9 October 1959 when Aerojet-General was given approval for the Titan Engine Storable Propellant Conversion Program.[21] The Stage I engine, with two thrust chambers, was designated XLR87-AJ-5 (Aerojet designation AJ23-132) and initially designed to provide 400,000 pounds of thrust at sea level. The operational Stage I engine generated 430,000 pounds of thrust. The Stage II engine, designated XLR91-AJ-5 (Aerojet designation was AJ23-133), was designed from the outset to provide 100,000 pounds of thrust at ignition altitude.[22] The objective of the program was to boost a payload of 4,500 pounds to a range of 8,500 nautical miles, or an 8,000-pound payload 5,500 nautical miles.

There was a superficial resemblance to the Titan I Stage I and II engines. Indeed, design of the Titan II engines began while the Titan I engine research was still taking place. While engine size and weight, as well as basic design, were very similar, significant differences existed for both the first and second stage. To improve reliability and maintainability, engine control parts were decreased from a total of 125 active control components in Titan I to 30 for Titan II. These changes are reflected in a similar decrease in power control operations, 107 to 21, respectively. Examples of the important changes were an autogenous pressurization system that used cooled gases from turbine exhaust to maintain propellant tank pressure rather than stored supplies of pressurized helium or nitrogen; use of solid-propellant start cartridges instead of stored pressurized gas to start turbopump operation; use of the Stage II turbopump exhaust stream as the power source for the Stage II roll nozzle, eliminating the need for an auxiliary power drive assembly for the vernier rockets, greatly increasing reliability; and perhaps the most innovative aspect of the Titan II engines, use of cavitating venturis and sonic nozzles to provide passive control to the gas generator and autogenous pressurization systems.[23]

Two key manufacturing differences were also important. In Titan I, the thrust chamber injector assemblies were milled from solid forgings, a time-consuming and costly process. With Titan II, the injector was formed from plates that were welded together. Titan I used both a fuel and oxidizer manifold, whereas Titan II used a fuel manifold and an oxidizer dome feed system.[24]

Louis D. Wilson, Titan I Stage II project engineer for Aerojet-General, was assigned to supervise configuration of engine hardware to demonstrate the feasibility of converting the Titan I engines to storable propellants with minimal hardware changes. Wilson and his team designed a Stage II thrust chamber injector and modified a turbopump for use with the storable propellants. This "bread board" engine was fired 15 to 18 times for proof of concept. More than sufficient data indicated that this idea had considerable merit for use in the design of the Titan II engines. With feasibility proven, development of the Titan II engines began in earnest. Engine test stands at Aerojet-General's Folsom Plant outside of Sacramento, California, were operating with both Titan I and II engines for several years. Money and people were thrown at the problem to maximize return in minimum time. The missile race was on, and the Air Force was not going to lose.

Aerojet's approach to the design of the Titan II engines was just like the previous development program for Titan I. Valves, pumps, and cooling jackets for the thrust chamber

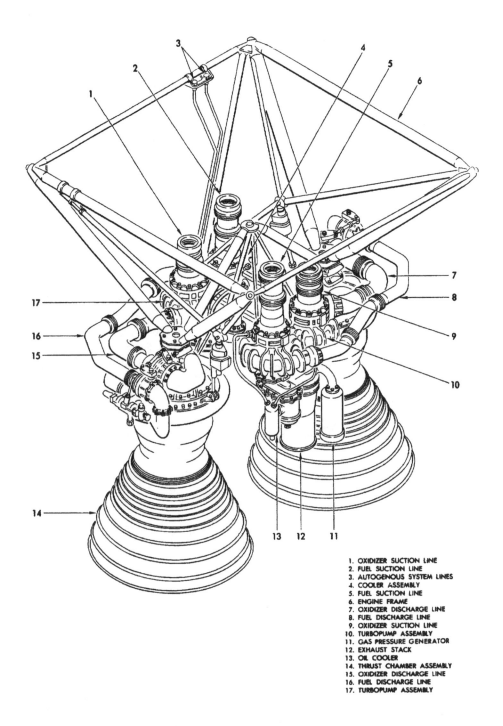

1. OXIDIZER SUCTION LINE
2. FUEL SUCTION LINE
3. AUTOGENOUS SYSTEM LINES
4. COOLER ASSEMBLY
5. FUEL SUCTION LINE
6. ENGINE FRAME
7. OXIDIZER DISCHARGE LINE
8. FUEL DISCHARGE LINE
9. OXIDIZER SUCTION LINE
10. TURBOPUMP ASSEMBLY
11. GAS PRESSURE GENERATOR
12. EXHAUST STACK
13. OIL COOLER
14. THRUST CHAMBER ASSEMBLY
15. OXIDIZER DISCHARGE LINE
16. FUEL DISCHARGE LINE
17. TURBOPUMP ASSEMBLY

Figure 3.1. LR87-AJ-5 Stage I engine set for Titan II. *Courtesy of the Titan Missile Museum National Historic Landmark Archives, Sahuarita, Arizona.*

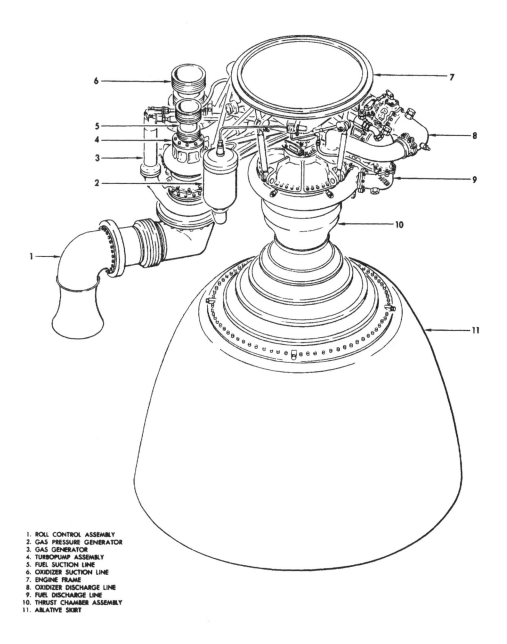

1. ROLL CONTROL ASSEMBLY
2. GAS PRESSURE GENERATOR
3. GAS GENERATOR
4. TURBOPUMP ASSEMBLY
5. FUEL SUCTION LINE
6. OXIDIZER SUCTION LINE
7. ENGINE FRAME
8. OXIDIZER DISCHARGE LINE
9. FUEL DISCHARGE LINE
10. THRUST CHAMBER ASSEMBLY
11. ABLATIVE SKIRT

Figure 3.2. LR-91-AJ-5 Stage II engine for Titan II. *Courtesy of the Titan Missile Museum National Historic Landmark Archives, Sahuarita, Arizona.*

were not seen as major hurdles. Workhorse steel injector patterns were fabricated, first in subscale and then full scale, to see how the propellants interacted in order to achieve maximum performance. These were hot-fire tested for limited duration using uncooled steel thrust chambers to determine design parameters such as combustion stability and chamber wall thermal loads, flow rate combinations, mixture ratios, and propellant temperatures. With determination of mixture ratios complete and initial injector plate patterns finalized, the timing of propellant movement through the engine cavities could be evaluated. Subsystems were being worked on simultaneously: for example, the turbopump team was designing the turbines, gearboxes, and impellers to move the propellants that the thrust chamber team needed for optimum operation; likewise, the gas generator team was developing the cavitating venturis concept; the autogenous pressurization team was working on the sonic nozzles, et cetera. Finally, the systems were placed together, and system integration began.[25]

For the Titan II Stage I engine this was all reasonably straightforward. Hundreds of tests were run around the clock to get the correct hydraulic balance or mass flow rate for the most efficient operation. Ken Collins, head of engine analysis for Titan II, had one engineer in particular who stood out with a critical contribution. Norman Laux, an engineer in the analytical design section at Aerojet, earned the nickname "Eyeball" due to his uncanny ability to evaluate the test data and recommend the change in orifice sizes for the various propellant lines almost as fast as the test data could be generated. Since each engine had slight differences in the internal cavities and restrictions, as each engine set was tested, thrust was measured and then the engine was shutdown and the orifice sizes changed, often with a call to "Eyeball" for a quick reading of the test results and decision as to the correct orifice size.[26] In March 1961 the first full duration firing of a Stage I engine was successfully accomplished, and in July and August 1961 the first production Stage I and Stage II engines, were accepted by the Air Force.[27]

One problem that was common to both the Stage I and Stage II engine valves and prevalves was the susceptibility of the 7075T6 aluminum alloy to stress corrosion (areas in these components under stress were more vulnerable to corrosion). This was only discovered once flight operations began at Cape Canaveral where the components were exposed to salt-laden air. Aluminum valve castings and stainless-steel bolts fractured due to the corrosive action of the salt air. Using 7073T6 aluminum alloy, though not quite as strong as 7075T6, solved the problem, and the modification was placed on both the ICBM and Gemini-Titan launch vehicle engines.[28]

Stage II "Hard Start"

The Titan II Stage II engine development was another matter. While rocket engine ignition at high altitude had been successfully demonstrated with Titan I, such was not to be the case with Stage II engine development for Titan II. Roy Jones, a development engineer for Stage II, was watching the television monitor during a Stage II engine test, when much to his surprise, he saw the thrust chamber drop away from the injector dome as if someone had taken a sharp knife and sliced it off. After several engines failed in this manner, review of the test data indicated that a combustion instability with a period of 25,000 cycles per second had swept around the injector face, cutting through the combustion chamber wall like an ultrasonic saw 1.5 inches below the attachment point. Thrust chamber pressure was cycling through ±200 pounds per square inch at 25,000 cycles per second.[29]

A combustion instability results from the combustion process at the injector plate pulsing instead of burning smoothly. This could happen in two ways. The first was the result of thrust chamber pressure increasing to the point where propellant injection is decreased. This caused, in turn, a sudden decrease in chamber pressure which would allow a surge of propellant to be injected which would in turn ignite, cause another pressure pulse, and result in a pressure oscillation. The second possibility was caused by the formation of detonation waves in the combustion chamber, analogous to the "pinging" in an automobile engine but with much more severe consequences. Instead of the propellant burning smoothly, a supersonic detonation wave moves through the unburned mixture and can literally tear an engine apart.[30] This did not happen each time an engine was tested and was in fact statistically almost insignificant for use in the ICBM program, occurring in just 2 percent of the ground tests. However, since Titan II had been selected by NASA as the Gemini Manned Spacecraft Program launch vehicle, even 2 percent was too much of a risk, and a solution had to be found.[31]

In September 1963, Aerojet-General began work on the Gemini Stability Improvement Program, also known as GEMSIP, to resolve the Stage II "hard start" problem. The direct cause of the problem was known. In Stage I, the propellants flowed into the engine cavities against sea-level air pressure, and engine bleed-in timing could be monitored and adjusted for. At the high altitude present for Stage II bleed-in prior to engine start, this process was very different from that at sea level since there was no air pressure to act as a barrier. The first real resistance "seen" by the fuel or oxidizer was the injector plate itself. This resistance was due to the small orifices that the fuel and oxidizer had to flow through to develop the spray pattern needed for efficient combustion. The physical shock was not a problem. The engine was robust enough, as was the airframe mounting, to take the impact. The problem was the resultant combustion instability at the injector plate face.

Wilson and Jones recall that Aerojet went through 20 to 30 Stage II thrust chambers trying to resolve the problem. The simple test of high-altitude bleed-in theory was to fill the thrust chamber wall tubes of the regenerative cooling system with water. When tested at 70,000 feet equivalent air pressure at the Aerojet facilities, the water provided enough hydraulic resistance to mimic that of the sea-level condition. Combustion stabilized significantly as the hydraulic shock was reduced to that found at sea level. However, the use of water was not an operational fix for an engine sitting in a launch duct for years, nor was it truly feasible for the Gemini Program. The water-filled thrust chamber tubes did, however, allow for continued engine system integration. The perfect fluid for use instead of water needed to have the vapor pressure of steel, so to speak, so that it would be stable in the tubes, yet still be fluid enough to move through the system and out the thrust chamber throat cleanly, not interfering with combustion. A number of fluids were tested. One was particularly promising but had an unfortunate tendency to detonate under oxidizing conditions, a tendency that was found out the hard way! The successful candidate was a fluid made by 3M called FC-43. Known throughout Aerojet as "distilled butterfly wings" due to both its exotic nature and price of $10,000 per 55-gallon drum, Jones remembers the visions of castles in the air in the eyes of the 3M sales representative as he imagined a fleet of 54 operational missiles, plus spares, requiring this 3M product.[32]

The primary solution, and the only one truly considered by both Aerojet and the Air Force, was a stable injector and a dry thrust chamber jacket start. Baffles were a logical control mechanism to break up the instability long enough for initiation of smooth combustion. The design evolved into a baffle that had oxidizer injection for thin film cooling. The final design was altitude tested in the Air Force Arnold Research Center Facilities at Tullahoma, Tennessee, and proved sound. The GEMSIP program took 18 months to complete and cost $13 million. The changes were incorporated into the ICBM program engines. Ironically, none of the research and development (R&D) missile failures were attributable to a Stage II hard start, and perhaps even more ironic, NASA launched the first six Gemini flights with the old-style injector plate.[33] Detailed Stage I and Stage II engine specifications are listed in Table 3.1.

Stage II Gas Generator

A second problem, and one that proved more troublesome, was that of Stage II gas generator failures in flight during high-altitude start-up. The gas generators utilized fuel and oxidizer to generate high-pressure gas for powering the turbopumps during flight. Solid-propellant start cartridges provided the initial high-pressure gas for spinning the turbines, and then the gas generators took over. The problem first occurred in the flight of N-1, the second launch of a Titan II. Telemetry indicated that the Stage II engine had reached only 50 percent thrust immediately after ignition, and the vehicle was destroyed

TABLE 3.1

TITAN II STAGE I AND STAGE II ENGINE SPECIFICATIONS

	STAGE I (LR87-AJ-5)		STAGE II (LR91-AJ-5)
	Subassembly 1	Subassembly 2	
Sea Level Thrust (lbs)	215,000	215,000	
Altitude Thrust (lbs)	236,900	236,900	100,000
Oxidizer Flow Rate (lbs/sec)	547.1	549.4	208.1
Fuel Flow Rate (lbs/sec)	283.5	284.7	115.6
Thrust Chamber Expansion Ratio	8:1	8:1	49.2:1
Turbine Speed (rpm)	24,000	24,200	23,800
Oxidizer Pump (rpm)	8, 360	8,420	8,450
Oxidizer Outlet Pressure (psi)	1, 210	1,210	1,150
Fuel Pump (rpm)	9,140	9,250	23,800
Fuel Outlet Pressure (psi)	1,380	1,380	1,220

Source: Titan II Propulsion Subsystem Handbook, *Aerojet Liquid Rocket Company, September 1981.*

Figure 3.3. Titan II engine test stand at the Aerojet-General facilities in Rancho Cordova, California. Built originally as a test structure for Titan I, this facility had three engine test bays: *from left to right,* G-1, G-2, and G-3. The tanks above the engines in G-1 were abandoned in place; the engines were fed from the G-2 tanks. The G-2 tanks were simply pressure vessels and did not mimic the Titan II tanks in any manner. G-2 was used to test the engines systems and not the tank pressurization system. G-3 was a full-up Stage I test stand with the engines evaluated for all functions. The tanks above G-3 were "battle-ship" tanks in that they were not flight hardware but were pressurized from the engines being tested in the G-3 stand. *Courtesy of the Aerojet-General Corporation, Sacramento, California.*

by the range safety officer. Unfortunately, the limited flight telemetry data provided insufficient information to the Stage II design team to solve this critical problem. The flight program continued with two partial failures in the next seven flights. Review of the accumulated telemetry data indicated that the small orifices at the injector plate for the gas generator were being partially plugged by particles on all the flights.

Careful review of the flight data indicated that backpressure was being developed due to the clogged orifices, decreasing propellant flow to the gas generator with subsequent loss of power. After trying to super-clean the gas generator components in a clean room prior to assembly, transporting the assembly to Cape Canaveral separately from the engine, and conducting a preflight nitrogen blowdown before each flight to verify the flight item cleanliness, the actual solution to the problem was found to be very simple and cost effective.[34]

At sea level the air trapped in the gas generator interior served as a cushion, preventing combustion gases and solid fuel particles produced by start cartridge ignition from reaching the injector plate of the gas generator on the Stage I engine. Due to the problems of vacuum testing large liquid-fueled rocket engines, the Aerojet facilities could reach only the equivalent of 70,000-foot altitude. This was assumed to be close enough to the Stage II start altitude vacuum at 250,000 feet, and the Stage II system was tested successfully.[35] However, even at 70,000 feet altitude, sufficient air was present to provide a barrier to the start cartridge combustion product particles. At 250,000 feet, the higher vacuum meant no such barrier existed and particles were being blown into the gas generator, clogging the oxidizer orifices. On many of the flights the result was not of sufficient magnitude to cause a problem, but on 3 of the first 20 flights it was significant. The solution to this problem was elegantly simple. A rupture disk was placed on the roll nozzle, the endpoint of the Stage II gas generator exhaust, entrapping the sea-level atmosphere (i.e., pressure) until start cartridge ignition took place. The cushion of air was retained at altitude, preventing combustion products from reaching and plugging the orifices.

After the N-22 Stage II gas generator failure, Louis D. Wilson and Ken Collins were at a meeting at the Ballistic Missile Office, Norton AFB, California, when this solution was presented to the Air Force and their advisors, TRW's Space Technology Laboratories. Arnie Hoffman from Space Technology Laboratories and Capt Clyde Smith, USAF, from the Ballistic Missile Office were present as Collins presented the recently developed solution to the gas generator problem. Hoffman grumbled that this was the third time that he had heard Aerojet say that the solution was at hand. Just what were they going to do if on the next flight the gas generator failed again? The room grew silent as Wilson looked at Hoffman and said, "I will quit my job at Aerojet, join the Government Accounting Office and come back to investigate the Ballistic Missile Office!"[36] There were no more losses due to gas generator failure after the Stage II roll nozzle rupture disk fix was implemented.[37]

Titan II Airframe Development

Testing of the in-silo launch concept began in April 1959 when Space Technology Laboratories, systems and technical engineers for the Titan program, contracted with Aerojet-General at the Azusa, California facilities, to build and test a 1/6th scale model of a proposed Titan II silo.[38] Rollo Pickford, the head of Aerojet's Applied Research Department at Azusa, was told that this was a "crash" program and that the development of this "ducted launcher," as it was then called, required only 60 days to build both the scale-model silo and scale-model Titan II airframe fitted with Nike-Ajax surface-to-air missile engines.

The scale-model silo was constructed completely above ground for easy access through hatches built in the silo and launch duct wall. The ground plane was simulated by a 35-foot diameter circular platform placed at the top of the silo. The entire silo, launch duct, and exhaust tubes were built by a steel fabricator in San Pedro, California, and trucked 40 miles to the Azusa facilities. The oversize nature of the load kept both Pickford and Robert Loya, the project engineer, up several nights plotting a route to avoid underpasses. As it was, telephone and power company crews still had to proceed ahead of the truck to disconnect or raise interfering wires.[39]

The first test firing took place on 6 June 1959, and by the time of the successful launch

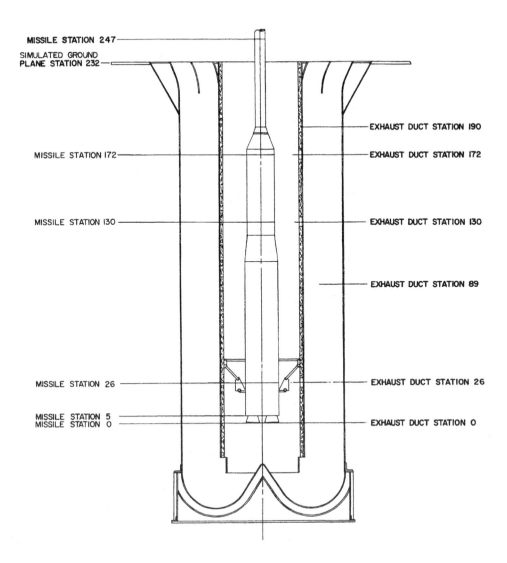

MISSILE STATION 247

SIMULATED GROUND
PLANE STATION 232

EXHAUST DUCT STATION 190

MISSILE STATION 172

EXHAUST DUCT STATION 172

EXHAUST DUCT STATION 130

MISSILE STATION 130

EXHAUST DUCT STATION 89

MISSILE STATION 26

EXHAUST DUCT STATION 26

MISSILE STATION 5
MISSILE STATION 0

EXHAUST DUCT STATION 0

Figure 3.4. Schematic drawing of the 1/6th in-silo launch scale model giving the general configuration of the launch duct, flame deflectors, and exhaust ducts. Previous work in Britain had used a J-shaped deflector. The W-shaped deflector demonstrated superior stability in airflow past the missile since it was symmetrical. Launch duct and exhaust duct acoustical lining position and thickness were also tested with this scale model. The model airframe was raised or lowered in the silo using the connection through the center of the airframe. *Courtesy of the Aerojet-General Corporation, Sacramento, California.*

of Titan I VS-1 from the SLTF on 3 May 1961, a total of 36 firings within the special silo test stand had been conducted. The first 23 were conducted using Aerojet Nike-Ajax production line engines. Originally designed for 2,500 pounds of thrust, two engines were modified to produce 4,200 pounds of thrust each.[40] These tests generated data on the general acoustic, aerodynamic, and thermal environments in a 1/6th scale-model W-tube type launcher. The feasibility of the concept was shown, but in late 1959 it was clear that the Titan I airframe

Figure 3.5. Construction of the 1/6th-scale silo model. Note the two workers at the center of the photo. The launch duct and exhaust duct were installed in one piece, and transportation through the streets of Los Angeles was quite an event. *Courtesy of Rollo Pickford.*

would have to be modified to withstand the in-silo launch environment. From February to September 1960, the test program concentrated on the specific design of the SLTF, developing and evaluating techniques for reducing potential damage to the missile systems.[41]

The last phase of the test program continued where the second phase had left off in September 1960 and was completed by February 1961. The final 13 tests were conducted using the same engine and a propellant supply package used in the first two phases, but modified for use with the Titan II propellants at a thrust of 6,000 pounds. Since engine start pressure pulse and exhaust products for the modified system were unknown, the acoustic, thermal, and aerodynamic environments were again thoroughly evaluated.[42] Combining the results of these tests provided a set of pressure pulse, temperature differentials, and acoustical energy profiles that permitted a launch duct acoustical liner concept to be developed.[43] The critical problem that had been addressed, modeled, and solved was that of sound-induced vibrations. A value of 148 decibels on the skin of the missile as it emerged from the silo had been predicted, and an actual value of 158 decibels was measured.[44]

The scale model provided insight on the design of the exhaust ducts. By positioning the scale model sequentially higher and higher in the launch duct, engineers discovered that by the time the guidance compartment of the missile emerged from the silo, an unacceptable 163-decibel acoustical energy level was present. This was a result of not only the acoustical energy in the launch duct itself but also the sound energy coming from the twin exhaust ducts. The solution was to line the exhaust ducts with acoustical panels, reducing the resultant decibel level and providing an adequate safety margin when combined with other design features. The pressure pulse generated by ignition of the engines was also a major design constraint. The scale model again proved invaluable as a water deluge directed into the engine exhaust plumes reduced the magnitude of the pulse to an acceptable level. The water deluge also reduced the exhaust plume temperature significantly.[45]

Concurrent with the research on engine and silo launch interactions was research into acoustical fatigue in the structural components of the Titan II airframe. Normally such work would have involved research of similar environments, followed by analysis and prediction, experimental testing, and then structural and component design. With the concurrency design concept in place, however, Martin Company had to conduct the first three aspects of the study while the basic structural design and fabrication of the missile were taking place. The Martin Acoustics Laboratory, Denver, used a multiple-disk siren and a high-pressure sinusoidal siren to produce the required acoustical energy. Compressed air to operate both noise sources was obtained using a compressor driven by two turboprop engines. Acoustical energy level information from the earlier captive fire test with Titan VS-1 in the SLTF at Vandenberg AFB was provided to the Martin laboratory for use in evaluating the robustness of the current airframe structure. Values varied from 156 decibels at Stage I engine level to 147 decibels at the launch duct opening.[46] A series of acoustic tests, using skin panels and missile structural elements, with and without compressive loads for simulating launch conditions, indicated where the Titan II airframe structure had to be strengthened. In many cases this meant increasing the tank skin thickness or the addition of internal ring frames to several of the larger panels.[47]

Airframe Fabrication

In an effort to reduce costs and reach deployment as rapidly as possible, the Titan II airframe was built using the same jigs and welding fixtures as Titan I. Where 12 extruded and chemically milled panels made up the Titan I tanks, with Titan II, only 4 integrally stiffened and mechanically machined panels were used. Chemical milling was again used to reduce weight in noncritical areas by 900 to 1,000 pounds. This was less than with Titan I due to the greater amount of machine milling.[48] The final design specification dimensions for Titan II are given in Table 3.2.

There were three major differences between the final Titan I and Titan II airframe design. The first and most obvious was that on Titan II the second-stage diameter was increased to 10 feet. The second difference was that overall missile length was increased from 98 to 103.4 feet (including reentry vehicle), mostly in the Stage II structure. Some structural modifications, mainly increasing skin thickness and adding ring frames, were necessary due to the in-silo launch environment as well as the increased density of the propellants. One source of problems in the Titan I airframe was the Stage I fuel tank longeron structures. The longerons served as the point of attachment for the missile to the launch mount. These were bolted onto the Stage I fuel tank skin and then sealed. Leakage had

Figure 3.6. A Titan II Stage I is being positioned for placement in the D-2 test stand at Littleton, Colorado. Stage II is already installed. In a full-duration firing test, Stage I would be fired for 165 seconds, followed by Stage II for 180 seconds. *Courtesy of Lockheed Martin Astronautics, Denver, Colorado.*

been a recurring problem in this area in the Titan I program. With Titan II, the longeron panel was welded directly to each quarter panel. After the quarter panels were welded together to form the fuel tank, a machined fitting was then riveted to the longeron panel, eliminating tank skin penetration.[49]

The third major difference was a change in the staging sequence. In Titan I, staging commenced with the detection of low liquid level in the Stage I tanks. After several seconds to allow for Stage I thrust decay, explosive bolts were fired and solid-fueled staging rockets on Stage II fired, pushing Stage II clear of the transtage structure, utilizing rails to prevent hitting the transtage area. The staging rockets also ensured that both Stage II propellants

TABLE 3.2
TITAN II ICBM FINAL DESIGN SPECIFICATIONS

FULLY ASSEMBLED MISSILE DIMENSIONS (FT)	LENGTH	
Stage I including interstage structure, Stage I engines	70.17	
Stage II (from interstage attachment to reentry vehicle adapter)	19.26	
Reentry Vehicle Adapter	3.77	
Mark 6 Reentry Vehicle	10.17	
TOTAL	103.4(a)	
Diameter (excluding conduits, air scoops)	10	

FULLY LOADED NOMINAL MISSILE WEIGHT (LBS)	STAGE I	STAGE II
Airframe empty (Stage I including engines; Stage II including engines, reentry vehicle, and adapter)	9,583	13,479
Oxidizer	160,637	37,206
Fuel	83,232	20,696
Ordnance, lubricants	1,543	743
TOTAL	327,119 (maximum 341,000)	

ENGINE THRUST (LBS)		
Stage I LR87-AJ-5 (sea level)	430,000	
Stage II LR91-AJ-5 (vacuum)	100,000	

RANGE (NM)		
Mark 6 Reentry Vehicle	5,500	
Circular Error Probable (nm)	less than 1.0	

Source: Detailed Design Specifications for Model SM-68B Missile (Including Addendum for XSM-68B), Lockheed Martin Astronautics Library, Denver, Colorado.

Note: The Mark 6 reentry vehicle weighed 8,380 pounds, including spacer with decoys; 7,575 pounds for just the reentry vehicle and warhead.

(a) The length of the Titan II ICBM is listed in official Air Force documents as anywhere from 108 feet to 114 feet. This is due to an error in defining Stage II length. The figure of 103.4 feet was verified by measuring missiles in storage.

Figure 3.7. A Titan II assembly line. The Stage I airframe is in the barrel welding machine. *Courtesy of Lockheed Martin Astronautics, Denver, Colorado.*

would be forced down to the turbopump intake lines. A timer ignited the Stage II engine once it was well clear of the Stage I structure.

In keeping with the concept of simplification that was a key in the Titan II program, a new staging sequence eliminated the need for the staging rockets and guide rails by igniting the Stage II engine while the stages were still attached and thrust was decaying from the Stage I engine. Large Stage II exhaust vent holes were a prominent feature of the Titan II airframe. These were positioned in the Stage I forward skirt assembly and the interstage assembly and facilitated venting of the Stage II exhaust during thrust buildup. The decaying thrust of the Stage I engines maintained sufficient acceleration to keep the Stage II propellants at the turbopump inlets prior to Stage II ignition.

Airframe Design Changes

The Titan II research and development missiles were nearly identical to the operational missiles, indeed this had been one of the selling points of the program. After fabrication of the first 10 Lot N missiles was well underway, a series of structural tests indicated that the skirt areas, the airframe portions that attached to the propellant tank dome structures, might fail. This had been discovered when bending loads were placed on an assembled Stage I and Stage II structure to simulate the most stressful portion of Stage I powered flight, the time just before staging when Stage I was under the highest acceleration environment. While the skirts were designed to withstand ultimate loads 1.25 times the expected flight loads, the tests, which used hydraulic jacks simulating flight conditions and a Stage II with propellant tanks full of lead shot, showed that failure did occur if the conditions were just right.

The production line fix was relatively simple and involved strengthening an area at the junction of the tankage and skirt. Missile airframes N-1 to N-9, already fabricated, were strengthened by rivetting reinforcing bands of alumium in the six problem areas. These bands were given the nickname "belly bands." Missile N-10 was not considered a flight airframe and was not retrofitted; N-11 through N-33 and all of the operational missiles had the areas strengthened internally during original manufacture.[50]

Contract AF04(647)-213, 15 May 1962, stated "[I]t shall be a design requirement that the allowable pressure decay with the propellant tanks loaded at flight pressures, shall be less than 2.0 p.s.i. in 30 days, except for Stage II fuel tank, which shall be less than 3.0 p.s.i. in 30 days. There shall be no visible leakage."[51] However, by mid-1963, early in the placement of Titan II missiles in operational silos, leaks began to appear in the oxidizer tank welds. Nitrogen tetroxide, leaking through holes too small to be detected by the original quality-control methods, was mixing with water vapor in the humid environment of the launch duct. The result was the formation of highly corrosive nitric acid, causing small leaks to turn into larger and more problematic leaks. The problem had not been detected earlier because none of the N-series flight test operations had necessitated the prolonged storage of propellants in the tanks.

At the operational bases, tank pressurization decays in excess of these requirements were observed, oxidizer vapor leaks sufficient to trigger the vapor detection system occurred, and finally, visible leaks were noted. Seventeen missiles of the 60 missiles deployed or awaiting deployment were recalled to Martin Marietta's Denver plant for inspection and rewelding. This recall program was given the name Operation Wrap Up.[52]

During production the tanks were checked via x-ray of each weld and then given hydrostatic and nitrogen pressure tests. The returned missiles had the oxidizer tank welds completely reexamined, and the problem areas were rewelded and then tested by pressurizing the tanks with helium and using a helium detector to evaluate each rewelded area. This new test equipment increased the leak detection sensitivity 10,000-fold. After hydrostatic testing, the tanks were baked to dry out all the water in the system, and the welds were painted with sodium silicate and then pressure-checked again prior to return to the field. A total of 15 production fabrication changes were made during Operation Wrap Up. Only three missiles built after October 1963 had to be returned to Denver for rewelding.[53]

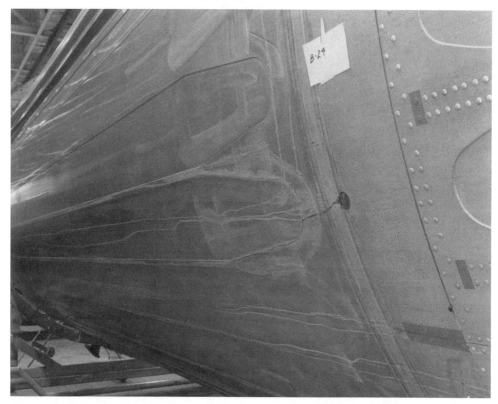

Figure 3.8. Missile B-24 (62-005) was returned to the Martin factory during Operation Wrap Up. Corrosion from minute oxidizer leaks can be seen at the tank weld points. *Courtesy of Ron Hakanson.*

Production Lots

A variety of designators were used in the Titan program. Originally Weapon System 107A-2 was used to designate the entire Titan program; LGM-25C was used to designate the operational missile which had been previously designated as the SM-68B missile. XLGM-25C designated the research and development missiles, also called the Lot N series. The former designation for the Titan II research and development missiles was XSM-68B. For clarity, the research and development missiles are referred to as N-series, and the operational missiles as B-series since these are the designators used within the production contracts.

No true production lots were assigned in the Titan II program. The N-series missiles were research and development program airframes, built under contract AF04(697)-576. Table 3.3 lists the major airframe modications of the N-series missiles. (A complete list of N-series airframes and serial numbers is given in Appendix 1.) The B-series were the operational missiles built under contracts AF04(694)-213 for 95 airframes, AF04(694)-589 for 7 airframes, and AF04(694)-785 for 6 airframes, a total of 108 operational missiles.[54] (A complete list of B-series airframes and serial numbers can be also be found in Appendix 1.) Unit production costs are listed in Table 3.4.

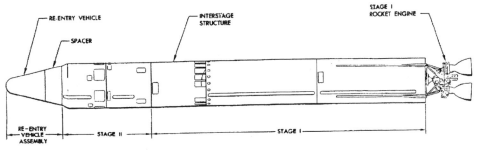

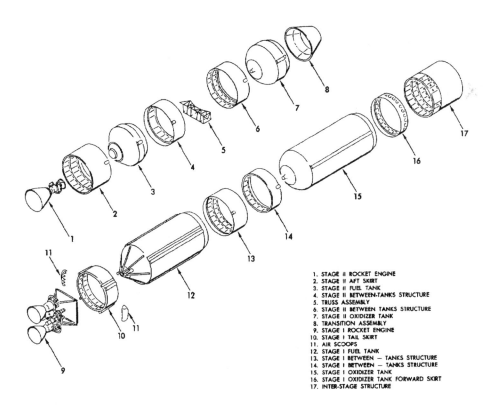

1. STAGE II ROCKET ENGINE
2. STAGE II AFT SKIRT
3. STAGE II FUEL TANK
4. STAGE II BETWEEN-TANKS STRUCTURE
5. TRUSS ASSEMBLY
6. STAGE II BETWEEN TANKS STRUCTURE
7. STAGE II OXIDIZER TANK
8. TRANSITION ASSEMBLY
9. STAGE I ROCKET ENGINE
10. STAGE I TAIL SKIRT
11. AIR SCOOPS
12. STAGE I FUEL TANK
13. STAGE I BETWEEN — TANKS STRUCTURE
14. STAGE I BETWEEN — TANKS STRUCTURE
15. STAGE I OXIDIZER TANK
16. STAGE I OXIDIZER TANK FORWARD SKIRT
17. INTER-STAGE STRUCTURE

1020L9A

Figure 3.9. Diagram of Titan II ICBM airframe showing assembled missile top and exploded view of major structural components below. *Courtesy of the Titan Missile Museum National Historic Landmark Archives, Sahuarita, Arizona.*

TABLE 3.3
MAJOR MODIFICATIONS TO TITAN II
RESEARCH AND DEVELOPMENT MISSILE AIRFRAMES

Change	1	2	3	4	5	6	7	8	9	
1. Belly Bands	▬	▬	▬							
2. Beefed Up Waffle & Skins										▬
3. One Piece Conduit Stage I	▬	▬	▬	▬						
4. Built-Up Conduit Stage I										▬
5. Built-Up Cone Fuel Tank Stage I	▬	▬	▬							
6. One Piece Cone Fuel Tank Stage I										
7. Oxidizer Dome Support Stage I Forward	▬	▬	▬	▬	▬	▬	▬			
8. Weld Land Area Increase		▬	▬	▬	▬	▬	▬	▬	▬	
9. Interstage Riveting										
10. R/V Adapter Martin (Mk-4)										▬
11. R/V Adapter G.E. (Mk-6)	▬	▬	▬	▬						
12. Translation Rockets										▬
13. Spectroradiometer										
14. Scientific Passenger Pod										
15. Malfunction Detection System										
16. 40-foot Staging Cable										
17. Air Duct Stage I Engine Compartment										
18. External Camera Pod										
19. Steel Feed Line (Suction)	▬	▬	▬	▬	▬	▬	▬	▬	▬	
20. Aluminum Feed Line (Suction)										▬
21. Beefed Up Transport Section										
22. Internal Camera Pod										

Improved Titan Feasibility Study

In 1964 Martin Marietta Company reported on a feasibility study funded by Air Force Systems Command aimed at increasing the performance of Titan II through the use of a gelled, aluminized fuel (Alumazine), increased propellant load, and uprated engines. The improvements under study would have enabled twice the payload to be carried yet would still use the same operational bases and equipment.[55]

An aluminized fuel/nitrogen tetroxide combination had been under study at Aerojet since 1962. New engines would need to be built since the new Stage I engine would pro-duce 630,000 pounds of thrust and Stage II had 163,000 pounds of thrust, resulting in an engine weight increase of 175 percent and 134 percent, respectively. Suspending the aluminum powder in the fuel would require the addition of Carbopol 904, a gelling agent. Alumazine composition would be 56.7 percent hydrazine, 43 percent powdered aluminum, and 0.3 percent Carbopol 904.[56] The result was a thixotropic gel which would present flow control problems that were next in the line of research efforts. The original gas generator concept would have to be modified to run on Aerozine 50 and not the aluminized fuel, requiring, in turn, separate valving and tanks to store the Aerozine 50 apart from the alu-minized fuel.

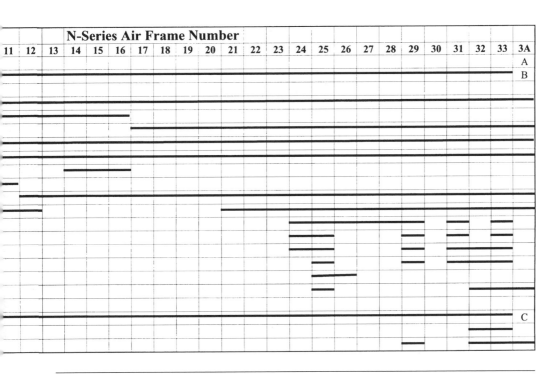

N-10 was never flown. N-3 was modified after the "Twang" test and flown as N-3A (See Chapter IV).

Note A: Belly bands Stage I oxidizer tank only.

Note B: Stage I fuel, Stage II fuel and oxidizer.

Note C: Suction line modified to Gemini Launch Vehicle standards.

Source: "GLV POGO Study Structural Configuration Changes of Titan II R & D Missiles, Addendum 1," February 1964 Lockheed Martin Astronautics Research Library, Denver, Colorado.

TABLE 3.4
TITAN II UNIT PRODUCTION COSTS, 1975

Airframe	$1,563,333	
Installed engines	$969,999	
Electric, guidance and control	$393,652	(original guidance system)
	$281,000	(universal space guidance system)
Other	$204,540	
TOTAL	$3,158,524	(original guidance system)
	$3,045,872	(universal space guidance system)

Source: Adapted from Fred Shaw, "Titan II Working Papers," p. 18, Reference Division, Air Force Historical Research Agency, Maxwell AFB, Alabama. This document is classified as SECRET. The information used is unclassified.

The improved Titan II would fit in the existing silos. Necessary modifications would include repositioning the Level 7 work platforms to accommodate the increased length of the missile; rebuilding the shock isolation system to handle the increased weight of the missile, now 473,400 pounds versus the current 330,000 pounds; modification of the engine water deluge system; and extension of the acoustical liner lower down in the launch duct. A new fuel transfer system would also be required.

The airframe would remain 10 feet in diameter but be extended to 111 feet in length, mainly through enlargement of the Stage I fuel tank. The external appearance would change slightly due to the presence of four cylindrical external tanks on Stage I to house the Aerozine 50 used in the gas generator system. Stage II tankage for Aerozine 50 would be carried internally. There is no evidence that this work went past the feasibility stage in 1964, though this did include some bench-scale production of Alumazine and limited engine test firings.[57]

Reentry Vehicle Development for Titan II

At a symposium at the University of California, Berkeley, in 1956, Theodore von Kármán, one of the fathers of ballistic missile development, stated that the "reentry problem," a body reentering the atmosphere at speeds of Mach 12 to 20, was "perhaps one of the most difficult problems one can imagine . . . a challenge to the best brains working in these domains of modern aerophysics."[58]

In order to reach targets 4,000 to 6,000 nautical miles away, ballistic missiles have to be accelerated to velocities in the region of 20 or more times the speed of sound (Mach 20). This is 10 times the speed of a high-powered rifle bullet. On encountering the atmosphere on reentry, a shock wave is formed at the front of the vehicle in which the air is heated to many thousands of degrees, in some cases exceeding the melting and boiling point of tungsten, the element with the highest known melting point, 6,170 degrees Fahrenheit. At this temperature the air plasma is also highly chemically reactive. There is a transport of heat by mass conduction from the air plasma to the vehicle surface which is dependent on both the temperature and density of the air in the plasma. At high altitudes where the air density is low, the mass transport of heat is low, in spite of the very high shock wave temperature. Conversely, at lower altitudes, the higher density plasma results in a higher heat flux for equal reentry vehicle velocities.

A key description of a reentry vehicle is its ballistic coefficient known as beta (ß). Beta is defined as $W/(C_d \times A)$, where W is the weight of the reentry vehicle, A is the cross-sectional area and C_d is the coefficient of drag. With reentry vehicle weight being held constant, reentry vehicles with a low ß (a high coefficient of drag and cross-sectional area), and thus more air resistance, would decelerate at a relatively high altitude where the density of the atmosphere is low and heat fluxes are lower but reentry times are longer. Alternatively, high ß reentry vehicles have shorter reentry times but encounter denser air at a higher speed, with the result that higher heat fluxes are developed but for relatively short exposure times. Obviously these considerations were critical to mission requirements but were constrained by both the materials and testing facilities available at the time. The result was that the first generation of ballistic missile reentry vehicles had relatively low ß values.

The early term for reentry vehicles was nose cone; the nose was a hemisphere, the afterbody a cone for aerodynamic stability (the Mark 3 had a cylindrical insert between the nose and cone portions to accommodate a specific warhead design). It was well known at the time that the severity of the temperature environment was a function of the radius of curvature of the nose. Earlier heat-sink designs used on the Thor IRBM had a large radius of curvature in order to reduce the heat flux to the point where the copper would carry the heat away from the interior of the reentry vehicle but not melt. The large weight of copper required for this type of reentry vehicle prohibited it from being of use in the ICBM program. The smaller the radius of the nose cone, the higher the temperature, and if an ICBM was to be developed in a timely manner, there was no other option but to go to a large radius, low ß reentry vehicle. It took five more years of research, development, and flight testing to successfully build a high ß operational reentry vehicle, the Mark 12, which was used in Minuteman.[59]

General Electric was awarded an Air Force contract in 1955 to design, develop, and manufacture a reentry vehicle for use on the Thor IRBM and the Atlas ICBM. The first deployed reentry vehicle, the Mark 2, utilized the heat-sink concept. In this design, the heat of reentry was conducted from the surface to a mass of high heat capacity material rapidly enough to keep the surface temperature below the melting point of the shield material. Copper was the material of choice although beryllium, cast iron, steel, and several other materials were tested. Known as a blunt conic sphere, the Mark 2 had a maximum diameter of nearly five feet and was five feet tall.[60] Size constraints to the heat-sink design, as well as the weight penalty of the heat-sink material, led to the desire for a second-generation reentry vehicle system. Many possibilities were reviewed at General Electric: ablation, transpirational cooling, re-radiating, and liquid metal cooling, amongst others. Ablation was chosen as simplest and most promising. Ablation depends on the mass transfer of heat to the shielding material which melts and vaporizes, carrying away the absorbed heat in the process. Don Schmidt, a researcher in thermal protection materials at the Air Force Materials Laboratory, Wright-Patterson AFB, Ohio, remembers that the Ballistic Missile Division advisors were not in favor of extensive ablation research. The heat-sink approach could be easily calculated and modeled, while ablation was an entirely new research direction.[61]

Successful development of an ablative material required laboratory test equipment that could generate, for short periods of time, the high temperatures associated with the heat of reentry. Many systems were used. An oxyacetylene welding torch could generate temperatures in the region of 6,330 degrees Fahrenheit and heat fluxes of 650 BTU/sec/ft^2. A solar furnace could generate 5,027 degrees Fahrenheit and heat fluxes up to 380 BTU/sec/ft^2 and had the added advantage of enabling testing to be done at pressures other than sea level. A new device, the water-stabilized arc, was created to reach temperatures of 26,000 degrees Fahrenheit, enabling a heat flux of 2,000 BTU/sec/ft^2 to be achieved.[62] Utilizing these tools, scientists tested a wide variety of organic and inorganic materials. General Electric's Aerosciences Laboratory selected a phenolic resin plastic as the key component for further testing.[63]

In December 1957, General Electric Company's Missile and Space Vehicle Department began participation in a three-part flight program, sponsored by the Air Force, to evaluate ablative materials for use in reentry vehicles: Thor-Able Phase I; Thor-Able Phase II (RVX-1); and Atlas (RVX-2). Thor-Able Phase I reentry vehicles were not recovered, while both the RVX-1 and RVX-2 reentry vehicles were recovered. Thor-Able Phase I demonstrated the successful reentry of an ablation-type reentry vehicle at ICBM ranges and velocities. Using a Thor IRBM for the first stage and the second stage of Vanguard, the "Able" flight vehicle was a biconic-sphere, approximately three feet long and two feet in diameter at the base, and weighing 600 pounds. Two successful flights proved the feasibility of ablative material-based reentry vehicles and paved the way for Thor-Able Phase II, which used the RVX-1 reentry vehicle configuration.[64]

The RVX-1, a flared-cylinder, conic-sphere design, had a maximum diameter of 2.5 feet and was approximately 5 feet long, weighing 700 pounds. A variety of ablative materials, those of both General Electric and AVCO, a second company involved in reentry vehicle design at the time, were flight tested using the RVX-1 shape. While still using the Thor-Able booster configuration, this program differed from the Phase I program in that the ß for the reentry vehicle was significantly higher. Four of six flights had successful test trajectories. Two flights, one for the AVCO product and one for the General Electric product, ended with the recovery of the vehicles for inspection and evaluation.[65] The results from the RVX-1 series led to the development of the General Electric Mark 3, the first ablative reentry vehicle used at intercontinental ranges, deployed on Atlas D.

The third and final phase of the program was the RVX-2, using a conic-sphere geometry with a base diameter of approximately 5 feet and a height of nearly 12 feet, weighing over a ton. The results of the RVX-1 program were utilized in selecting the ablative materials. The shape, similar to that of a badminton shuttlecock, also had the center of gravity at the front as did a shuttlecock. The heat shield and nose cap utilized a phenolic resin and phenolic resin–impregnated nylon, respectively. The Atlas program had progressed to the point that Atlas boosters were now available for the flight test program, and true ICBM flight conditions could now be examined. Three flights were attempted, but due to problems with

Figure 3.10a, b, c. Reentry vehicle evolution leading to the Mark 6. *Top:* (a) General arrangement of the General Electric Mark 2 heat-sink reentry vehicle. A heat sink of copper was used to protect the reentry vehicle payload. Maximum diameter of the reentry vehicle was approximately 5 feet, and it was approximately 5 feet long. The Mark 2 was used in the testing of the Atlas B and C missiles and deployed on Jupiter and Thor missiles.
Middle: (b) General arrangement of the RVX-1 experimental reentry vehicles with the first ablative heat protection system. The maximum diameter was approximately 2.5 feet, and it was approximately 5.5 feet long. The RVX-1 shapes were tested with both AVCO and General Electric ablative systems. The General Electric version led to the development of the Mark 3 reentry vehicle deployed on Atlas D.
Bottom: (c) General arrangement of the RVX-2 and RVX-2A experimental reentry vehicles. The RVX-2 had a maximum diameter of 5 feet and was approximately 12 feet long. Both General Electric and AVCO flew ablative systems during these tests. The AVCO system was used on the Mark 4 reentry vehicle, deployed on Atlas E and F and Titan I, similar in shape to the Mark 3 but larger. The General Electric system was used on the Titan II Mark 6 reentry vehicle. *Courtesy of General Electric Corporation, Philadelphia, Pennsylvania.*

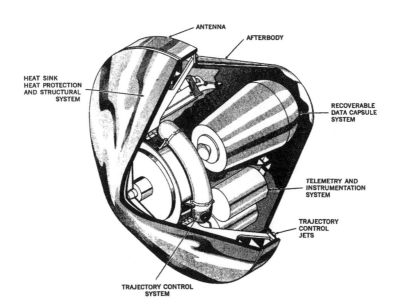

ANTENNA

AFTERBODY

HEAT SINK
HEAT PROTECTION
AND STRUCTURAL
SYSTEM

RECOVERABLE
DATA CAPSULE
SYSTEM

TELEMETRY AND
INSTRUMENTATION
SYSTEM

TRAJECTORY
CONTROL
JETS

TRAJECTORY CONTROL
SYSTEM

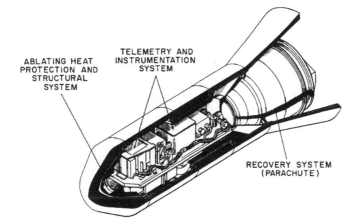

ABLATING HEAT
PROTECTION AND
STRUCTURAL
SYSTEM

TELEMETRY AND
INSTRUMENTATION
SYSTEM

RECOVERY SYSTEM
(PARACHUTE)

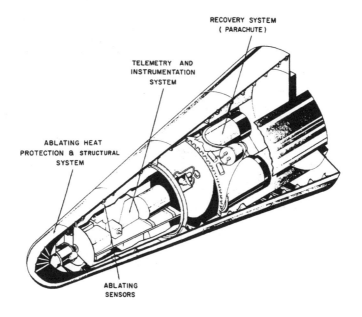

RECOVERY SYSTEM
(PARACHUTE)

TELEMETRY AND
INSTRUMENTATION
SYSTEM

ABLATING HEAT
PROTECTION & STRUCTURAL
SYSTEM

ABLATING
SENSORS

the Atlas booster, only the final flight was successful. The vehicle was recovered intact after a full-range ICBM flight in July 1959.

The results from the RVX-2 program led to the RVX-2A program that flew AVCO and General Electric heat-shield materials on separate flights in 1960. The first flight carried the General Electric Century Series plastic heat-shield and nose-cap components, the second carried AVCO components. While recovery was to take place on both flights, both vehicles were lost at sea. Fortunately, telemetry of over 100 measurement points was complete, covering the critical diagnostic and performance areas. On the third flight, the General Electric reentry vehicle was recovered. The results of these flights led to the AVCO Mark 4 reentry vehicle, which used an ablative Avcoite nose cap bonded to a metal support structure with reinforced plastic material in the cylindrical and flare sections. The Mark 4 shape was very similar to the Mark 3, only slightly longer and wider, allowing a larger warhead to be carried, and was deployed on Atlas E and F and Titan I, The General Electric design led to the Mark 6, deployed on Titan II.[66]

Detailed design documents for the Titan II ICBM list both the Mark 4 and Mark 6 reentry vehicles as possible payloads.[67] The reason for listing the Mark 4 may have been as

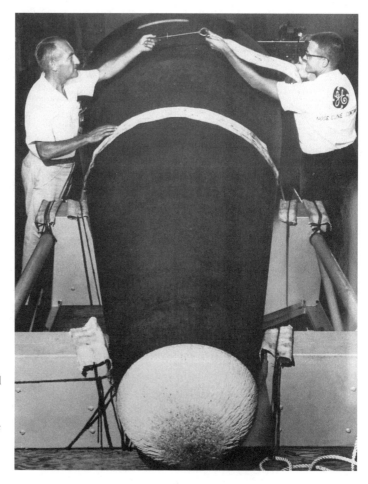

Figure 3.11. RVX-2 reentry vehicle being readied for transport after a successful recovery. Note the streaks in the ablative material on the nose cap section of the vehicle caused by reentry. *Courtesy of Don Schmidt.*

a fallback if the development of the Mark 6 was unsuccessful. Interestingly enough, a single and successful launch of a Titan II carrying a Mark 4 did take place on 6 December 1962 from Cape Canaveral; however, the flight was not successful (see Chapter 4). Using the Mark 4, the range of Titan II could have been extended to over 8,300 nautical miles, but the Mark 4 was not deployed with Titan II.[68]

The General Electric Mark 6 reentry vehicle deployed on Titan II utilized ablative materials for both the nose cap and the heat shield. The nose cap was composed of phenolic nylon (66-Nylon cloth impregnated with phenolic resin), chopped into 1/2-inch squares and pressure molded to the nose cone shape. This was the same material used in the Mark 3 nose cap. The heat shield was composed of the General Electric Century Series 100 plastic. The basic ingredients for the plastic were Dow Epoxy–Novolac 438; methylnadic anhydride, a curing agent; polypropylene glycol to increase flexibility to make fabrication easier; and n-butylphosphoric acid, a charring agent.[69] The Series 100 plastic was easily fabricated by casting the liquid epoxy into molds having the conical shape of the heat shield and hardening in an oven without pressure. The complete heat shield was assembled from three pieces: the nose cap and the two conical sections which comprised

Figure 3.12. General Electric Mark 6 reentry vehicle being readied for transport. The Mark 6 had a maximum diameter of 7.5 feet and was 10.17 feet long. The Mark 6 was painted white during the early Titan II ICBM flight test program but was deployed in the unpainted configuration. *Courtesy of Don Schmidt.*

the main body. The casting was so simple that the only machining required was to square off the top and bottom edges to the final length dimension.

The nose-cap ablative material had a maximum thickness of 2.3 inches, while the heat shield was 0.25 inches thick.[70] These dimensions seem thin when the material is to be exposed to thousands of degrees of heat. However, the shape of the reentry vehicle led to a relatively short time of exposure to the high temperatures. At this time, several porous char layers 1 to 2 millimeters thick are formed in sequence. The first one quickly plugs up, is sloughed off by aerodynamic forces, and is replaced instantly by the formation of a new char layer. Large amounts of pyrolysis gases that form as the material degrades serve to inhibit heat transfer from the very hot boundary layer to the ablating surface, greatly reducing the actual heating at the vehicle surface.[71]

The Mark 6 heat shield was bonded to the reentry vehicle airframe with neoprene rubber. The coefficient of expansion of the plastic was much higher than that of the aluminum airframe, and the rubber served both as an insulator and as an elastic interface which could stretch to accommodate heat-shield expansion.[72]

As one might expect, the detailed design of the Mark 6 reentry vehicle interior is not available for public release. Several reports from the Operational Test and Follow-On Test programs do shed some light on details of Mark 6 specifications.[73] The Mark 6, including decoys, reentry vehicle adaptor, and W-53 warhead, weighed 8,380 pounds. The W-53 warhead weighed 6,200 pounds and was the largest yield warhead used in the United States strategic missile forces.[74] The Air Force has not declassified the precise yield of the W-53 warhead. No official release of the warhead yield has been made, although published congressional reports estimate the yield at 9 megatons.[75] When launched from Vandenberg AFB, the Mark 6 carried either a denuclearized W-53 warhead that still contained the Grade II high-explosive components for air-burst tests, or a scoring kit utilized for surface-impact flight profiles. The Mark 6 Mod 3 could carry up to eight terminal decoys (Optically Enhanced, Model 1037J) and six mid-course decoys (Operational, Model 1026BP).[76] (See Chapter 4 for further information on the Mark 6 reentry vehicle.)

Improved Reentry Vehicle for Titan II

As early as 1960 Air Force Ballistic Systems Division began investigations into new multiple warhead payloads for the Atlas, Titan I, Titan II, and Minuteman missiles already under development. These were not independently targetable; rather, the separate warheads would land in a tight pattern that would increase the probability that at least one warhead would penetrate defensive systems. The Atlas and Titan I concept was named the Mark 14; the Titan II concept was named Mark 13; and the Minuteman portion of the study was named the Mark 12. Only the Mark 12 was funded.[77]

In May 1964 the Mark 12, under development, was joined by the Mark 17, a heavier payload version.[78] The Mark 12 and Mark 17 were now being designed as multiple independently targetable reentry vehicles (MIRVs). Each warhead was released from a "bus" that served to carry all of the reentry vehicles and release them at different points along the flight trajectory to targets separated by hundreds of miles.[79] Improvement in the yield-to-weight ratio (the pounds of explosive power divided by the weight of the thermonuclear material) of nuclear weapons made the MIRV, with its necessarily smaller individual weapons, feasible. Titan II with its large payload capacity of 8,380 pounds was an attractive candidate for the upgrade when compared to Minuteman I with a 1,400-pound payload and Minuteman II, with a payload of 1,625 pounds, both already deployed or Minuteman III destined to carry a payload of 1,975 pounds.[80]

In 1965 an internal study was conducted by the Ballistic Systems Division on the feasibility and costs of converting Titan II to an MIRV system. The new payload would consist of up to six Mark 17 reentry vehicles, each weighing approximately 1,120 pounds with yields estimated at 2 megatons. The system accuracy would be less than 0.3 nautical miles, with ranges from 5,740 nautical miles with six Mark 17 reentry vehicles to 7,600 nautical miles with four Mark 17 reentry vehicles. The only significant modifications to the missile would be a new payload interface between Stage II and the new 17.68-foot payload fairing in place of the Mark 6 ablative heat shield. The cost of developing and testing the MIRV system was estimated to be $278 million; retrofitting the Titan II fleet an additional $262 million. The total program, including retrofitting the fleet, was projected to take approximately 53 months.[81] The increased accuracy of the MIRVs, coupled with the ability to deliver as many as seven reentry vehicles per missile, would have converted the Titan II system into an even more formidable weapon with 378 targets covered instead of 54. The Air Force could now use Titan II against hardened ICBM silos as part of a counterforce doctrine—that is, targeting individual missile sites rather than the current wide-area, co-located targets.

Although originally intended for the Titan II and Minuteman III, by 1965 the Mark 17 was also being considered for use on the Poseidon C-3 fleet ballistic missile that was to replace the Polaris A-3. The Navy was not interested in the large warhead carried on the Mark 17, instead selecting the Mark 3 MIRV with a 40-kiloton-yield warhead. The C-3's guidance system allowed the much lower yield warhead to be targeted with more than sufficient accuracy to overcome the drawbacks of its much lower yield when compared to the Mark 17. The C-3 was programmed to carry 10 to 14 Mark 3s.[82]

In this same time frame the Air Force also realized that the total deployment of Minuteman was going to be limited to 1,000 missiles. One obvious solution to multiplying the capability of Minuteman was to move from the single Mark 17 reentry vehicle of megaton range yield to the Mark 12 reentry vehicle in a MIRVed configuration. Three reentry vehicles could now be used on Minuteman III with accuracy such that their 170-kiloton warhead would be effective against hardened targets. With this decision came the decision to cancel the Mark 17 program in 1968.[83] There is no evidence that a Mark 17 reentry vehicle was ever tested during the operational flight test program of Titan II.

Guidance

The Titan II missile utilized two inertial guidance systems during its 24-year existence. Inertial guidance is based on an inertial measurement unit and a computer. By measuring changes in inertia along the x, y, and z motion axis, the guidance computer can calculate its location in space and send appropriate corrective signals to the autopilot. The guidance system was installed between the Stage II propellant tanks. Guidance-generated steering signals and vehicle-sequencing discretes were routed to the autopilot (also located in Stage II), which sent the signals to the Stage I and II engines and staging and actuator circuits.

The original Titan II inertial measurement unit, the MX-6362/DJW-11E, nicknamed the "gold ball" because of the exterior gold resinate finish that provided thermal radiation protection, was built by the AC Spark Plug Division of General Motors, Milwaukee, Wisconsin. AC Spark Plug had a long history in inertial guidance and had teamed with Dr. C. S. Draper at the Massachusetts Institute of Technology to build the first inertial guidance system used in the Thor IRBM. The Navy Regulus II cruise missile also used an AC Spark Plug inertial guidance system during its research and development phase but had not been deployed. The Air Force Mace cruise missile was deployed with an AC Spark Plug inertial guidance system.

The original contract for the Titan II guidance system was awarded to AC Spark Plug on 14 April 1959.[84] AC Spark Plug contracted with IBM for the design, development, fabrication, and production of a rotating drum memory digital computer that interfaced with the inertial measurement unit. AC Spark Plug designers worked with Davidson Corporation and Perkin-Elmer Corporation in the development of the ground optical alignment system used to provide a precision prelaunch alignment reference. AC Spark Plug also designed, tested, and produced the associated aerospace ground equipment and operating ground equipment that was required to test, operate, and maintain the airborne components.

An intense effort to turn the laboratory version, created at the MIT Instrumentation Laboratory headed by Dr. C. S. Draper, into a device that could be mass produced finally came to fruition with the successful launch and flight of the first Titan II missile N-2 (60-6809) on 16 March 1962.[85] Over the next 16 years this first guidance system required only eight modifications, all of which were completed by 15 May 1965. The first modification related to air bubbles in the flotation fluid for the accelerometers observed during the flight of missile N-16 (61-2739) on 6 February 1963 and again during the flights of N-8 (61-2731) on 27 April 1963 and N-19 (61-2742) on 13 May 1963. Degassing the fluid and improving the gasket seals was the short-term answer, while the second modification, the addition of a bellows enclosure to maintain atmospheric pressure, was the permanent fix for the operational missiles.[86]

In the mid-1970s the Air Force faced a dilemma with the guidance system for the Titan II program. Nearly two decades after the original design of the system, advances in the electronics industry made the system difficult to support, as major suppliers, such as IBM, which made the drum memory unit for the on-board computer, were not interested in maintaining the capability of building obsolete equipment in small lot sizes. In some cases the older components simply did not exist, as suppliers had phased them out of their product line. Headquarters Strategic Air Command realized that at predicted failure rates, critical parts would no longer be procurable by December 1977.[87]

Fortunately, there was an existing and readily available state-of-the-art replacement: a modified Delco Electronics inertial guidance system called the Universal Space Guidance System (USGS). The USGS had been first flown in the Titan III-C program on 13 December

1973, with a total of six launches with one failure at the time of the decision to modify it for use with Titan II.

The Delco Electronics Carousel IV inertial navigation system was standard equipment in the Boeing 747 and had been retrofitted into Boeing 707 and McDonnell-Douglas DC-8 commercial aircraft.

Modification of the basic Carousel IV inertial reference unit for space applications was relatively simple, repackaging the instrumentation for the thermal environment as well as for vibrational stresses of a missile launch. The USGS hardware was composed of the Carousel IV inertial measurement unit and the Magic 352 computer (the commercial aircraft computer was the Magic III series). The Titan II autopilot was used with minor modifications, as was most of the airborne wiring. The umbilicals to the missile did not need to be replaced.[88]

Figure 3.13. In the upper photograph, the AC Spark Plug "gold ball" inertial measurement unit, *left*, and the missile guidance computer, *right*, are connected to an air conditioner during testing in the guidance shop. The lower photograph shows the Delco Universal Space Guidance System inertial measurement unit, *left*, and the missile guidance computer, *right*. The USGS did not require supplemental air conditioning. *Courtesy of Robert Popp.*

While the missile silo environment, as well as the missile flight profile, were obviously significantly different than that seen by the commercial aircraft Carousel IV and Magic III systems, the missile installation had a major difference: the guidance system would be turned on after installation, advanced to the "READY" mode, and, except for maintenance or repair requirements, remain in this steady-state operating environment. In the aircraft installation, the Carousel IV system was turned on and off several times a day depending on aircraft operations. This caused degradation in system accuracy and reliability due to the short-term operating times and the shock of heating and cooling. Once up and running, the USGS system self-calibration procedures continually fine tuned the system and was most stable if simply left on.[89]

The between-tanks truss that carried the guidance equipment needed to be rebuilt since the new system weighed slightly more than half that of the old system and occupied slightly less than half the space. To minimize developmental costs, consisting primarily of recalculation of target trajectories due to difference in guidance-system weight, the truss was reworked so that with the installed equipment the total weight and balance remained the same as the original installation. In addition, the new guidance computer software was written to emulate the signal timing of the old guidance computer. As far as the missile flight control system was concerned, it was as if no change had taken place. While the first-stage wiring harnesses were used without modification, the Stage II wiring had to be completely reworked.[90]

In hindsight this all seems straightforward. Robert Popp, an AC Spark Plug and Delco engineer involved with the installation of both guidance systems, completed the Project RIVET HAWK update, as the USGS modification was called, in record time. Evaluation of the "gold ball" system revealed that little if any of the system could be salvaged. It had been fabricated in a welded, encapsulated, module design configuration. Combined with the rapid advances in solid-state electronics, replacement of any module parts with new components would require additional flight qualification tests, a time-consuming and prohibitively expensive process since there were few spare Titan II missiles available.

Between 15 October 1975 to 27 June 1976, Delco engineers and technicians were able to modify two sets of flight systems from the already flight-proven USGS of Titan III. Included within this eight-month time frame was the design and fabrication of a new telemetry system for use during the qualifying process since the original telemetry system sets had been used up during the previous flight test program.[91]

Charlie Radaz was the Martin Marietta Company engineer who was in charge of interfacing the new system into Stage II at Vandenberg AFB, and Jim Greichen was the Martin Marietta Company, Denver, program manager. In November 1975 the USGS hardware modification program began. Working through the Air Force Logistics Center (AFLC), Hill AFB, Utah, Martin Marietta Company had the contract for modifications to install the USGS platform, designing all the adaptors and wiring changes, as well as an additional telemetry transmitter for the RIVET HAWK flight. All the modifications were done at Vandenberg AFB.

Denver did the initial design on the mechanical adaptions, working closely with Delco and TRW representatives. A 10-foot diameter wooden mock-up of the current truss and guidance section was built and sent to Radaz and his group at Vandenberg AFB. They then took out the guidance wiring harness from a current second stage, from the terminal points that interfaced with the first stage to the terminal connectors on the actual equipment housings, carefully measuring and photographing all of the attachment points. The harness was laid into the wooden mock-up and reattached. This gave the Denver group an exact mock-up to work from in designing the new harness and attachment points. The same autopilot was being used and the same thrust sensors; the only changes were to take the information from the sensors at the terminal board interface and rewire from that point forward to the new guidance equipment. Once the changes had been made, the new harness was removed from the mock-up and reinstalled on Stage II of of Titan II B-17 (62-2771).

With the successful flight of B-17, the wire harness tool was sent to Denver so that the fleet harnesses could be made more quickly. Radaz's group had modified the first mock-up as they went along and figured that Denver would use the modified tool to make a second one to speed up fabrication. Unfortunately, Denver used the original blueprints, and the first several harnesses for fleet installation were coming up inches short. The problem was quickly identified and corrected. The need for a new wiring harness derived from the fact that Delco had changed the guidance equipment housings so much for Titan III-A that the inputs were not in the same groups of wire harnesses as had been used for the original Titan II guidance set.[92]

Installation of the USGS equipment began at the 308th Strategic Missile Wing (SMW), Little Rock AFB, Arkansas, Launch Complex 374-6, on 5 January 1978, with the prototype installation to verify procedures for the operational sites. The first routine installation was at Launch Complex 373-4 on 24 April 1978, and all 18 sites were back on alert by 10 November 1978. Work at the 390th SMW, Davis-Monthan AFB, Arizona, began on 27 November 1978 at Launch Complex 571-9, and all 18 sites were back on alert by 4 June 1979. The 381st SMW, McConnell AFB, Wichita, Kansas, was the last wing for USGS installation, beginning at Launch Complex 533-1. The final USGS installation was completed at Launch Complex 532-5 on 13 December 1979.[93]

Coded Switch System

Unauthorized launch of a Titan II missile was prevented by the Launch Enable System (LES). By the early 1970s, a new system was needed to replace the aging and hard-to-support equipment that comprised the LES. Martin Company was contracted in 1971 to design, build, and field a new launch enable system. Named the Coded Switch System (CSS), the device had two locks. The first was a digital lock that was opened with a digital combination. The secondary lock was designed to lock the prevalve upon which it was mounted, preventing launch if a perpetrator tried to bypass the digital combination lock by opening the case.

Jack Cozzens worked on the digital lock, but was present when the first test of the explosive weld feature of the secondary lock was run. The inner and outer plates of the lock design had a steel cone through which the valve shaft could freely rotate. If the secondary lock was triggered, shaped charges on the steel cones would weld them into place on the shaft, inhibiting rotation and thus preventing the valve from opening. The test was conducted on a spare prevalve, with results that were spectacularly successful in the welding process but not so good in the fragmentation process. When they reentered the test room, the six high-strength stainless-steel bolts that held the inner and outer plates together had broken, and the outer plate was impaled in the cinder block wall, still climbing as it hit! With the initial design having backfired, so to speak, Cozzens and his team came back with the design that was initially deployed. A knife-edged pin was driven by a much smaller explosive charge into the prevalve shaft, preventing rotation until removed by a maintenance team.

The secondary lock firing circuit needed to be fast. One scenario was that a .357 magnum revolver round might be used to disable the lock. Cozzens traveled to three detonator vendors to test their claims of high-speed capability. He carried a small aluminum box with an ARM/SAFE/FIRE button on the top and the firing circuit inside to Stapleton International Airport and too late realized his mistake as the security guard asked him about the interesting box he had in his briefcase. Only fast talking and a few phone calls back to the plant convinced the guard that this was indeed just a humble piece of electronics.[94]

Ron Hakanson and John McDonnell were assigned the task of integrating the components of the CSS into a compact box that could fit onto the prevalve shaft without modi-fication of the Stage I engine. This was not a trivial task because the components had to fit together with extremely close tolerances to further prevent tampering with the system. Once the components were nested together and nearly ready for testing, a member of one of the review groups asked if one could drill through the housing. Back to the drawing board they went and located a company in Denver that produced a coating used on safes that was next to impossible to drill through. The inner and outer plates of the lock were coated with this material.

The CSS was composed of the Butterfly Valve Lock (BVL); the Butterfly Valve Lock Control; the Butterfly Valve Lock Status Encoder that transmitted the status of the BVL to the wing command post; the Butterfly Valve Lock Status Decoder and Display; and the Electronic Command Signals Programmer. The BVL, which was attached to the Stage I Subassembly A oxidizer butterfly valve, was composed of an electronics package, a drive mechanism, and an antipenetration system to prevent tampering with the lock equipment.

The basic principle of the BVL was that a six-character combination had to be entered from the Butterfly Valve Lock Control on Level 2 of the launch control center prior to the initiation of the terminal countdown for launch. The BVL code was transmitted in the Emergency War Order (EWO) message and decoded using the EWO documents housed at each launch complex. Each character was entered by turning one of six thumbwheels, each with 16 different letters, resulting in 16,777,216 possible combinations. To eliminate a trial-and-error method of entering the correct operational codes, on the seventh try at entering a combination, a signal was sent to the BVL logic, locking out the input signal and sending a signal to the wing command post indicating that the attempt counter had been exceeded. To reestablish BVL operation, the system had to be de-energized and reprogrammed by maintenance personnel.

If 28 volts DC power was removed from the BVL during an attempt to conduct an unauthorized launch, a 36-hour electronic timer began operation. At the end of 36 hours, power was removed from the penetration and secondary lock circuits, rendering the BVL safe for maintenance. Normal maintenance access was via a maintenance word, actually a six-letter combination, that rendered the BVL maintenance safe. The entire lock was protected from tampering by a multilayered cover composed of copper strands embedded in epoxy. Penetration of the cover, or even simple movement, triggered detonators that erased the coded word from the lock memory and locked the shaft of the butterfly valve in the closed position.[95]

On 10 September 1973 the first CSS-prototype installation began at Launch Complex 532-8 at the 381st SMW. Over a 30-day period, the prototype was installed, a combined systems test was run to ensure that the installation had not impaired launch capability, and then the prototype was removed for modifications prior to installation fleetwide.[96]

On 30 June 1974 full-scale CSS installation began at the 381st SMW, the first wing to be so modified. Two months later, BVL secondary locks fired unexpectedly at Launch Complexes 532-1 and 532-5. Installation was halted while the cause was investigated. The electronics had survived the effects of the detonation and were flown to Martin Company facilities in Denver for evaluation. Cozzens was in the middle of the debate concerning the cause. The major concern was sabotage. Careful inspection of the equipment indicated that no unauthorized entry had taken place. One of many possibilities was the quality of the launch complex DC power supply to the BVL. Bob Shuttle, one of the CSS team engineers for Martin Company, spent more than a month monitoring power at three launch complexes and clearly proved that power surges occurred that could easily have caused the premature firing of the secondary lock.[97]

Hakanson was involved with the redesign effort initiated after these two inadvertent firings of the BVL. The original BVL had a positive locking mechanism. The only way to repair the lock was to download the Stage I oxidizer, remove the valve from the missile, and take the valve apart to remove the lock mechanism. Three modifications were made during this redesign, the details of which remain classified. The first was mechanical with the result that the BVL could be replaced without conducting a propellant download. The second dealt with the detonators, and the third was a power filter contained on the BVL to condition the somewhat noisy launch complex power source. This redesign was successful in overcoming the random detonation problem, and only routine maintenance had to be performed on the BVL during the remainder of the program. Installation of the CSS was completed on 28 February 1975 with the final installation at the 390th SMW.[98]

Perhaps the best indication of the robust design of the Titan II airframe was the ability of the missile to far surpass the original design requirement of one year in the operational environment without major maintenance. Missile logs for the missiles remaining at the time of deactivation indicate that the record for remaining in a silo without a propellant download was 10 years. The average time a missile stayed in the same launch complex without removal for major maintenance was 15 years, with the longest time being 20 years.[99]

IV

TITAN II RESEARCH AND DEVELOPMENT FLIGHT TEST HISTORY

The Titan II research and development flight test program consisted of 32 flights of the N-series missiles, beginning on 16 March 1962 and ending on 9 April 1964. Twenty-three launches were conducted at the Eastern Test Range from Cape Canaveral Air Force Station, Florida, and 9 were conducted at the Western Test Range, Vandenberg AFB, California.[1]

The flight test program began at Cape Canaveral for two reasons. The first and most important was that the Titan I launch pads were in place and could be easily converted to accommodate the Titan II missile configuration. Furthermore, the Titan II airframe had been chosen as the launch vehicle for the Gemini-Titan Program, and these launches would be taking place at Cape Canaveral. Second, the Titan II silos at Vandenberg AFB were still under construction when flight testing began (see Tables 4.1a and 4.1b for a complete listing of research and development flights).

There were two categories for flight testing of the research and development missiles. Category I was design verification using contractor personnel and procedures. Category II was known as the weapon system effectiveness verification, which were still contractor operations but utilized proposed procedures for Air Force personnel.[2]

Eastern Test Range, Cape Canaveral

Two of the Titan I launch facilities at Cape Canaveral where modified for use in the Titan II test launch program. Launch Pads 15 and 16 had new propellant handling and transfer systems installed. Guidance-system alignment equipment was already in place due to the fact that the Titan I Lot M missiles had been used to test the Titan II inertial guidance-system operation. The missile erector equipment was modified to accommodate the 10-foot diameter second stage as well as the increased overall length of the missile. The liquid propellants used in Titan II, nitrogen tetroxide and Aerozine 50, had to be kept below 60 degrees Fahrenheit. With the missile erector in the vertical position, nearly surrounding the missile, this was not a problem, but once it was lowered in preparation for launch there was alimited launch time, depending on the local weather.

There were two primary test objectives for the Cape Canaveral flight test program. First was the need to establish missile flight characteristics without the interference from the still-unknown in-silo launch conditions. Second was verification of the operating ground equipment and maintenance equipment in support of missile flight preparations.

Titan II program launch operations started on 9 March 1962 with the successful static

TABLE 4.1A
TITAN II RESEARCH AND DEVELOPMENT FLIGHT OPERATIONS

FLIGHT #	DATE	AIRFRAME #	LAUNCH SITE	DESCRIPTION
	1962			
1	16 Mar	60-6809	Pad 16	N-2, the successful first Titan II launch operation was conducted on 16 March 1962, 28 months after the first storable propellant feasibility studies had begun. N-2 boosted the payload to the full 5,000-nautical-mile range and the R/V impacted in the target area.
2	7 Jun	60-6808	Pad 15	N-1 partial success, Stage II gas generator restricted, failed to develop full thrust.
3	11 Jul	61-2729	Pad 15	N-6, flight test objectives achieved.
4	25 Jul	60-6811	Pad 16	N-4, launch attempt #1 (27 Jun) aborted by combustion instability in Stage I engine, launch attempt #2 successful, flight was a partial success, Stage II fuel pump leak.
5	12 Sep	60-6812	Pad 15	N-5, first successful launch of a Titan II ICBM with decoys took place at Cape Canaveral.
6	12 Oct	61-2732	Pad 16	N-9, flight test objectives achieved.
7	26 Oct	61-2735	Pad 15	N-12, flight test objectives achieved.
8	6 Dec	61-2734	Pad 16	N-11, failure, Stage II bootstrap line severe vibration, thrust chamber pressure switch shut down.
9	19 Dec	61-2736	Pad 15	N-13, flight test objectives achieved.
	1963			
10	10 Jan	61-2738	Pad 16	N-15, partial success, Stage II gas generator restriction.
11	6 Feb	61-2739	Pad 15	N-16, flight test objectives achieved, first attempted launch of a Titan II by a SAC crew was successful on 6 February 1963 at Cape Canaveral. The missile traveled 5,800 nautical miles and the reentry vehicle impacted in the target area.
12	16 Feb	61-2730	Complex 395-C	N-7, first Titan II silo launch successful. Stage II umbilicals failed to disconnect properly, missile self-destructed at 18,800 feet. Successful in that missile cleared silo intact.
13	21 Mar	61-2741	Pad 15	N-18, flight test objectives achieved.
14	19 Apr	61-2744	Pad 15	N-21, partial success, Stage II bootstrap, premature engine shutdown.
15	27 Apr	61-2731	Complex 395-C	N-8, flight test objectives met.
16	9 May	61-2737	Pad 16	N-14, partial success, Stage II oxidizer leak, premature shutdown.
17	13 May	61-2742	Complex 395-D	N-19, successful, flight test objectives achieved.
18	24 May	61-2740	Pad 15	N-17, successful, flight test objectives achieved.
19	29 May	61-2743	Pad 16	N-20, failure, thrust chamber fuel valve failure, leak and fire in Stage I engine compartment.

Source: "Martin Marietta Titan II Chronology," undated. Kundich Collection.

Note: Pads 15 and 16 were located at Cape Canaveral; Launch Complexes 395-B, -C, and -D were located at Vandenberg AFB.

TABLE 4.1B
TITAN II RESEARCH AND DEVELOPMENT FLIGHT OPERATIONS

FLIGHT #	DATE	AIRFRAME #	LAUNCH SITE	DESCRIPTION
	1963			
20	20 Jun	61-2745	Complex 395-C	N-22, partial success, Stage II gas generator failure.
21	21 Aug	61-2747	Pad 15	N-24, successful, flight test objectives achieved.
22	23 Sep	61-2746	Complex 395-D	N-23, successful, flight test objectives achieved.
23	1 Nov	61-2748	Pad 15	N-25, successful, flight test objectives achieved.
24	9 Nov	61-2750	Complex 395-C	N-27, successful, flight test objectives achieved.
25	12 Dec	61-2752	Pad 15	N-29, successful, flight test objectives achieved.
26	16 Dec	61-2751	Complex 395-D	N-28, successful, flight test objectives achieved.
	1964			
27	15 Jan	61-2754	Pad 15	N-31, successful, flight test objectives achieved.
28	23 Jan	61-2749	Complex 395-C	N-26, successful flight test objectives achieved.
29	17 Feb	61-2769	Complex 395-B	B-15, successful, flight test objectives achieved, first technical data launch, first launch of operationally configured missile.
30	26 Feb	62-1867	Pad 15	N-32, successful, flight test objectives achieved.
31	13 Mar	61-2753	Complex 395-C	N-30, successful, flight test objectives achieved, last flight of N- series missiles from VAFB.
32	23 Mar	62-1868	Pad 15	N-33, successful, flight test objectives achieved.
33	9 Apr	60-6810	Pad 15	N-3A, successful, flight test objectives achieved, last Titan II ICBM launch from the Cape.[a]

Note: Pads 15 and 16 were located at Cape Canaveral; Launch Complexes 395-B, -C, and -D were located at Vandenberg AFB.

[a] *N-3 was used in the twang test. Often listed as N-34, N-3 was modified after the test: Stage I oxidizer tank had a belly band , other tanks had strengthened waffle section; aluminum feed line (suction) Gemini Launch Vehicle flanged joint replacing the Alfin Joint configuration; Stage I oxidizer tank salvaged from N-3 and updated to N-28 configuration; transportation section salvaged from N-3 and updated to N-28 configuration; Stage II fuel tank is scrapped Gemini Launch Vehicle 2 tank; Stage II oxidizer, forward and aft domes are operational domes chem-milled to R&D configuration, skirts are operational configuration. The modified airframe was designated as N-3A.*

firing of the Stage I engine of missile N-2 (60-6809) on Pad 16. On 16 March 1962, 22 months after the initiation of program development, N-2 was successfully launched from Pad 16.[3] Unlike the earlier flight test program for Titan I where the second stage was ballasted with water, N-2 was configured for full Stage I and Stage II performance. As far as the Air Force was concerned, both stages operated perfectly as the reentry vehicle landed in the target area 5,000 nautical miles downrange.[4] NASA, on the other hand, was deeply concerned about a vibration anomaly which had appeared 90 seconds into the flight. The missile had vibrated lengthwise at 11 cycles per second, for 30 seconds. Missile acceleration at this point in the trajectory was 2.5g under normal conditions; this unforeseen oscillation added ±2.5g. Under these conditions an astronaut would be hard-pressed to respond quickly to an emergency

Figure 4.1. Titan II N-2 (60-6809) lifts off on 16 March 1962 on a successful flight from Cape Canaveral, Florida. Unlike the first Titan I launch, both stages of N-2 were fully powered. The white finish on the reentry vehicle was for thermal control but was discontinued once it was found to be of little value. *Courtesy of the History Office, 45th Space Wing, Patrick Air Force Base, Florida.*

situation. These longitudinal oscillations quickly became known as "pogo stick" or Pogo for short.[5]

The next flight was N-1 (60-6808), launched from Pad 15 at Cape Canaveral on 7 June 1962. After successful Stage II ignition, full thrust was not achieved, and the range safety officer sent a manual fuel shutoff signal when the tracking system lost contact with the missile. The reentry vehicle landed considerably short of the target area. A restriction in the Stage II gas generator prevented full thrust for Stage II. This flight was rated a partial success as first-stage flight, as well as staging, took place as planned.

The launch schedule picked up pace with N-6 (61-2729) launched on 11 July 1962 from Pad 15. The primary objectives of this flight were to evaluate the performance of missile subsystems, sequencing, and the reentry vehicle. All systems performed properly, although the Pogo oscillation phenomenon occurred again. Impact was 0.01 nautical miles beyond and 0.46 nautical miles left of the intended target at a range of 4,389 nautical miles. While impact accuracy was not a primary objective, this was certainly an excellent start.[6]

Missile N-4 (60-6811) had been scheduled for launch on 27 June 1962. Stage I ignition occurred but at T+2.5 seconds, the Stage I engine shut down, resulting in the first and only aborted launch of a Titan II N-series missile. A combustion instability, also referred to as a severe start transient, in the Stage I Subassembly 2 thrust chamber had resulted in the thrust chamber being literally sheared off just below the fuel manifold torus and blown out the flame deflector several hundred feet. Automatic sequencer instrumentation sensed that the Stage I engines had not come to full thrust and shut down Stage I, saving the missile and the launch pad from a possibly devastating explosion and fire. Subsequent investigation found that the most probable cause of the combustion instability was residual alcohol left from cleaning the engine after an earlier static firing. This residual alcohol had caused a "tangential combustion instability" similar to that which had caused problems in the development of the Stage II engine (discussed earlier in Chapter 3). The resulting high-frequency oscillations had acted as an ultrasonic saw and cut through the thrust chamber wall.[7]

The Stage I engine assembly for N-4 was removed, and a replacement engine was installed on the launch pad. N-4 was successfully launched on 25 July 1962, from Pad 16, toward the same target area as N-6, with the same primary objectives. All missile systems performed satisfactorily through 208 seconds of flight, when a malfunction of the Stage II engine occurred. Telemetry indicated a mechanical pulse in the engine compartment that was followed by an engine hydraulic actuator failure and a 50 percent reduction in thrust. The engine continued to operate at reduced thrust for 111 seconds when fuel depletion occurred. Since the engine shutdown was not generated by the guidance computer, the vernier engines did not operate nor did the reentry vehicle release. Impact occurred 2,888 nautical miles short of the 4,388-nautical-mile range to the target area.[8]

The launch of missile N-5 (60-6812) was a significant step forward in the test program. For the first time, the prototype decoy system was flown. The decoy subsystem was located in the reentry vehicle adapter with doors that were explosively released, followed by decoy deployment. N-5 was successfully launched on 12 September 1962. The flight was routine through Stage I operation with the now "normal" Pogo effect. There was an unexpected

Figure 4.2. Titan II N-4 (60-6811) sits on Pad 16, Cape Canaveral, Florida, 27 June 1962, shortly after the first and only ground abort in the Titan II ICBM research and development program. A combustion instability at the Stage I Subassembly 2 injector cut the thrust chamber off. It came to rest several hundred feet from the flame deflector. N-4 had a new Stage I engine set installed within a month and was successfully launched on 25 July 1962. *Courtesy of Robert Stahl.*

roll oscillation in Stage II flight during vernier operation. The solid-propellant vernier motors were used to make final corrections to Stage II velocity prior to reentry vehicle separation. The decoys failed to deploy, along with improper reentry vehicle separation.[9]

N-9 (61-2732) carried the Mark 6 Mod 2 reentry vehicle and was launched on 12 October 1962 from Pad 16. The primary flight objectives were to test miss distance; to evaluate the guidance contribution to miss distance; and to evaluate propulsion, flight control, guidance, and reentry vehicle separation and the compatibility of the decoy/launcher/missile combination. All flight objectives were met as the reentry vehicle impacted 2.0 nautical miles downrange and 0.1 nautical miles right of the target impact area, 4,389 nautical miles from the launch site.[10]

N-12 (61-2735) was the first of the research and development missiles without the belly bands and was therefore the first flight of a missile closely configured to that of the opera- tional airframe. The primary objectives were to continue testing of the missile subsystems as well as to collect data to determine the weapon system accuracy. Reentry vehicle decoys were again carried and were to be deployed. N-12 was launched on 26 October 1962 from Pad 15 after a 125-minute hold at T-45 minutes left in the countdown. Telemetry indicated that a thrust chamber pressure switch needed replacing as well as a Stage I destruct battery and a Stage II destruct package initiator. A nearly perfect flight was marred by only one of the decoy doors opening, with no evidence of any of the decoys deploying. Impact was 0.16 nautical miles long and 0.77 nautical miles to the left of the intended target.[11]

The first Martin Marietta Company solution to the Pogo problem was the installation of a surge-suppression standpipe in the Stage I oxidizer feed line. Doubling the Stage I fuel tank pressurization reduced the Pogo effect 50 percent, but this was still not sufficient for NASA. The next launch was N-11 (61-2734). In addition to flying the Pogo problem fixes, N-11 carried a Mark 4 Mod 2A reentry vehicle, the reentry vehicle used on Titan I and Atlas F. One of the primary objectives of this flight was to evaluate the interaction of the Mark 4 reentry vehicle with the missile subsystems, such as arming and fuzing, as well as it subsequent flight. The Mark 4 reentry vehicle was less than half the weight of the Mark 6, so this mission was known among the test staff as the "Hot Rod" flight since a greater range could be tested. N-11 was targeted for the "Broad Ocean Area," in the southern Indian Ocean, at a

Figure 4.3. Stage I engine set on N-4 after the ground abort. The combustion instability worked like an ultrasonic cutoff saw, cleanly cutting off the Subassembly 2 thrust chamber bell, just below the fuel torus. The thrust chamber bell was expelled from the flame deflector by the exhaust gases. The airframe suffered no damage. *Courtesy of Robert Stahl.*

range of 6,475 nautical miles. This was the only launch of a Titan II with the Mark 4 reentry vehicle.[12] N-11 was launched on 6 December 1962. Unfortunately, the Pogo oscillation became even greater, reaching ±5g. The oscillations caused the Stage I thrust chamber pressure to fall below the setting for the thrust chamber pressure switch; consequently, the Stage I engines shut down prematurely. Staging occurred successfully but with a reduced velocity due to Stage I malfunction: the guidance system did not function properly, and abnormal maneuvers occurred after staging. Impact was only 703 nautical miles downrange; as a result, most test objectives were not met.[13]

The ninth and final launch of 1962, N-13 (61-2736), took place on 19 December 1962 from Pad 15. N-13 did not have the standpipe installed and had aluminum, instead of steel, oxidizer feed lines, as the second approach to solving the Pogo problem. Stage I fuel tank pressure was also increased. The Pogo effect was significantly reduced from previous flights, and all systems worked well to the point of decoy deployment. Five of six decoys released. The reentry vehicle impacted 1.12 nautical miles beyond and 0.29 nautical miles left of the intended target, 4,389 nautical miles from Cape Canaveral. The Titan II flight program for 1962 ended on this highly successful note.[14]

The first and only Titan II test launch conducted at night took place on 10 January 1963 with the launch of N-15 (61-2738). Stage I performance was normal, and the Pogo effect was reduced to ±0.6g, the lowest yet. Again, while more than acceptable to the Air Force, NASA insisted on reducing the effect to ±0.25g maximum. Stage II, for the second time in 10 launches, failed to achieve full thrust, again due to restrictions in the gas generator, and the reentry vehicle impacted 525 nautical miles down range. Even with the failure of Stage II to achieve full thrust, the Air Force was ready to freeze the missile design for use as an ICBM, since Pogo had been reduced to well below the system specifications by using aluminum oxidizer feed lines and increased fuel tank pressure.[15]

With one partially successful flight from Cape Canaveral, N-16 (61-2739) on 2 February 1963, and a fully successful flight, N-18 (61-2741) on 21 March 1963, the Pogo question, as well as a newly appreciated Stage II ignition instability problem, continued to cause friction between Air Force Ballistic Systems Division (BSD) personnel and the NASA Gemini program staff.[16] BSD did not want to spend any additional funds, nor face development delays to further reduce the Pogo effect. The statistical probability of instability was 2 percent, a number that the Air Force could easily live with for the weapon system. NASA, however, was adamant about Pogo and could not man-rate the missile until both Pogo and the Stage II instability problems were resolved. On 29 March 1963, Gen Bernard Schriever, Commander, Air Force Systems Command, convened a meeting at Andrews Air Force Base that brought together the BSD and the Space Systems Division staff to brief him on Titan II problems that were directly related to the Gemini Program success. NASA representatives were also invited. The intent was to chart a course to resolve the two major problems expeditiously for both parties.[17]

BrigGen John L. McCoy, director of the BSD Titan Program Office, reiterated that the two problems did not threaten further development of the weapon system. Delaying testing any further would delay deployment. Alvin L. Feldman, chief project engineer for Aerojet-General, was confident that the inclusion of mechanical accumulators in the Stage I fuel lines, as well as installation of baffles on the Stage II engine injector plate, would solve both the

Pogo and the combustion instability problems, respectively. By the end of the meeting it was agreed that the Air Force would provide additional funds and continue to work with NASA to find solutions to Pogo and combustion instability. McCoy would head a joint committee that would explore solutions to these problems while, at the same time, ensuring that the weapon system did not "incur undue delays by waiting for Gemini items."[18]

Two months and five flights later (three of them successful), N-20 (61-2743) was ready for launch on 29 May 1963. Modifications to the Stage I engine included surge suppression devices for both fuel and oxidizer lines. Immediately after liftoff, fuel from a leaking thrust chamber fuel valve on the Stage I engine ignited, damaging the flight controls, and N-20 pitched over and broke up 52 seconds into flight. The flight ended too soon to observe the effect of the newly installed Pogo suppression equipment. The fuel valve failure was traced to stress corrosion of the aluminum valve body. (See Chapter 3.)[19]

With the Stage II gas generator partial failure of N-22 (61-2745) at Vandenberg AFB on 20 June 63, only 10 of 20 flights fully successful and with 13 flights remaining in the test program, BrigGen McCoy had no choice but to halt flight testing until there was some assurance of a long string of fully successful flights. Since deployment of the weapon system was to be completed in six months, McCoy halted all further attempts to fix the Pogo problem. The Pogo problem had been responsible for only one of the 10 flight failures. Of the remaining 9, 3 were due to Stage II gas generator restrictions and 6 were due to a failed weld or broken lines. In the eyes of NASA and the Air Force Space Systems Division, these were quality-control issues. A review of Aerojet-General's quality control, as well as redesign of 40 engine parts, was implemented under an augmented engine improvement program. The order to fly no further Pogo fixes was rescinded and, with what was hoped to be the final solution to the gas generator restriction problem in place, Titan II was ready to fly again.

Flight testing resumed on 21 August 1963 with the successful launch of N-24 (61-2747) from Pad 15. While N-24 did not carry the Pogo fixes, the Stage II gas generator performed perfectly. All systems performed to Air Force specifications, and the program was now back on track with the only remaining obstacle the final fix of the Pogo problem.

On 1 November 1963, N-25 (61-2748) was launched from Pad 15. N-25 carried what was hoped to be the final Pogo fixes. This was the twenty-third launch in the Titan II flight test program and was a resounding success by all criteria. Pogo was reduced to +0.11g, the lowest to date and well below the NASA required +0.25g. Stage II thrust was normal, and the missile achieved a full-range flight.

Three additional flights were flown in 1963, and four in 1964, all but one successful. The single failure was due to an erroneous signal transmitted by the tracking system. The flight test program at Cape Canaveral ended in 1964 on a highly successful note.[20]

Scientific Passenger Pod Program

In July 1962 a short-term project named the Scientific Passenger Pod Program began. The program was funded by the Air Force Office of Aerospace Research. While this program "piggy-backed" on the Titan II research and development test launches, its objectives

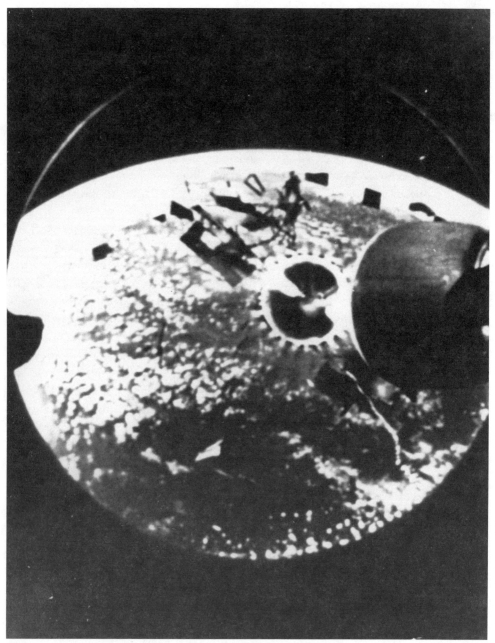

Figure 4.4. Internally mounted cameras were used to film the staging sequence during the research and development flight program. A frame taken from the 16mm film shows the fragmentation pattern of the interstage portion of the airframe between Stage I and II of N-25 (61-2748) on 1 November 1963. The line dangling on the lower right portion of the photograph was a 40-foot staging cable used in five of the flight tests. *Courtesy of Lockheed Martin Astronautics, Denver, Colorado.*

were secondary, and airframe modifications were kept to a minimum. The test objectives were to study engine exhaust plume infrared signatures in tests named Project DEFENDER and to test personnel oxygen breathing equipment for the Apollo Program.

The scientific passenger pod was a structural shell that mounted on the side of Stage II. Experience with the Titan I external decoy pods led to a self-contained electrical and telemetry system as well as data capsule ejection capability. The pod weighed 450 to 525 pounds, depending on the installed experiment. Eight pods were built and five were flown: N-24, N-25, N-29, N-31, and N-33. Four of the five missions were successful.[21]

Western Test Range, Vandenberg AFB

Flight test program objectives at Vandenberg AFB were directed toward evaluating the in-silo launch environment and the compatibility of the operational launch complex equipment. Three launch complexes had been constructed, designated 395-B, 395-C, 395-D. Launch Complex 395-B was used to develop the technical order documentation and to verify maintenance procedures. Launch Complexes 395-C and 395-D were used for the remaining research and development launches. Once this first test program was completed, all three were used for the operational flight test programs.

The first Titan II missile operations began on 21 August 1962 with the arrival of N-7 (61-2730) from Denver. N-7 was an early configuration missile with circumferential aluminum reinforcing bands, known as belly bands, added to Stage I and Stage II. After inspection and several Vandenberg specific modifications for telemetry and range safety, N-7 was installed in Launch Complex 395-C on 5 October 1962. Subsystem checks and instrumentation functional tests took from 10 October to 4 December when the first runs of the Combined Systems Test (CST) program were started. A CST was performed to verify connections of the operating ground equipment and missile ordnance circuitry whenever equipment changes in the silo or launch control center were made. The first CST was completed on 20 December 1962. On 10 January 1963 the only intentional captive firing of a Titan II in the launch duct was successfully conducted. Lasting 7.4 seconds, the resulting damage to the silo and missile components was not extensive, although Harlan Weissenborn, a guidance system engineer for AC Spark Plug, watched the test and had a sinking heart as debris began to fall from the exhaust plume. The above-ground areas looked like a major disaster had struck since pieces of acoustical modules used to line the launch duct and exhaust duct interiors were strewn across the site. The test was considered a success, all the modules were replaced, and the program moved on to the next milestone.[22]

One of the more interesting tests involving a complete Titan II airframe was the "twang" test conducted on 11 February 1963 at Launch Complex 395-D. Airframe N-3 (60-6810) had been installed in the silo on 29 November 1962. After completion of full-scale propellant transfer system design verification tests that lasted from 12 December to 27 December 1962, the missile propellant tanks were purged and filled with water. On 11 February, a series of tests, nicknamed twang tests, were begun to evaluate the missile shock isolation system under dynamic conditions. The missile shock isolation system thrust

Figure 4.5. Titan II N-3A (60-6810) lifts off on 9 April 1964, from Pad 15, Cape Canaveral. This was the last launch of the Titan II research and development flight test program. Visible on the right side of Stage II is an external camera pod, not to be confused with the scientific passenger pod used on five flights. *Courtesy of the History Office, 45th Space Wing, Patrick Air Force Base, Florida.*

Figure 4.6. Titan II N-7 (61-2730) is lowered into Launch Complex 395-C, Vandenberg AFB, on 5 October 1962. The "belly band" reinforcement is clearly visible on the aft and forward dome locations on the Stage I oxidizer and fuel tanks. *Courtesy of Andy Hall.*

Figure 4.7. The captive fire test with Titan N-7 (61-2730) at Launch Complex 395-C, Vandenberg AFB, took place on 10 January 1963. While the exhaust from the Titan II is transparent in flight, the "oxidizer lead," meaning that oxidizer reached the injector plate before the fuel, caused the initial exhaust, already opaque due to the steam generated by vaporizing the sound suppression deluge water, to be tinged with orange nitrogen tetroxide vapor. *Courtesy of Rick Grossman.*

mount, with the water-filled missile in place as if ready to launch, was pulled down or to the side of the silo with chains held by explosive bolts. The bolts were fired, quickly releasing the missile, simulating ground shock conditions from a nearby explosion to be mitigated by the missile shock isolation system. Andy Hall, a Martin Marietta Company engineer, was present during these tests. The sound as the missile skin "oil-canned," the skin puckering and then returning to shape, was a sight to see and a sound to hear.[23]

Elmer Dunn, the Martin Marietta Company engineer in charge of the twang tests, found that while the tests verified the ability of the missile shock isolation system to dampen thrust mount movement and then be able to lock up for launch, a mechanical means of spring centering was needed. In addition, the spreader jack for unlocking the dampers, the mechanisms that locked the thrust mount into a rigid configuration to support actual launch, proved to be structurally insufficient. Dunn reported that adjustments to permit load equalization were difficult and that refurbishment after launch, at the Vandenberg AFB sites, would be time consuming and costly unless the components were better protected from the effects of the engine exhaust.[24]

The twang test resulted in major system changes to all sites, including spring centering devices and new spreader jacks for unlocking the dampers. Ratchet-type positive shuttle lock mechanisms were designed so that the dampers would not unlock from vibration during the time period between engine ignition and liftoff. A special lubricant was found to facilitate damper unlocking and to inhibit corrosion. The original protective devices for the thrust mount shock suppression system springs were inadequate. Dunn's team built fiberglass cocoons that were reusable and proved to require little maintenance. These cocoons were used only at the three Vandenberg AFB sites.[25]

The first launch of a Titan II ICBM from a silo environment was scheduled to take place on 15 February 1963. About midnight, two days before the launch, Don Kundich (a Martin Marietta Company engineer who was the "missile mother," the engineer responsible for expediting and ensuring that all changes to the missile were completed), George Teft (Martin Marietta Company engineer), and John Adamoli (the N-7 flight test conductor), along with several other engineers, were finishing the preflight inspection of N-7. The group walked around the missile at all six levels of work platforms. This inspection was the last chance to look for any unusual connection or situation that might have passed the earlier walkthroughs. The group looked at the way the umbilical lanyards were attached to the launch duct wall. These umbilicals were "flyaway" in that they were pulled free of the missile as it rose off the thrust mount, rather than being pneumatically ejected. The umbilical lanyards were stainless-steel cables attached to the wall at one end and to the umbilical connector at the other. When the missile lifted off the thrust mount, the lanyards were pulled taut, activating the connector release and then pulling the connector and cable free of the missile. The lanyard attachment points on the launch duct wall were just D-rings of metal attached to a galvanized pipe mounted directly on the wall. The entire group commented that they just did not look strong enough. The only other silo launch, that of Titan I VS-1, 3 May 1961, had a completely different umbilical release mechanism. The Titan II launches at Cape Canaveral used booms to support the umbilicals and were not a valid comparison. They decided to reanalyze the rise rates of the vehicle to see if the lanyard would tighten and snap the D-ring. The concern was that the lanyard had to pull tight to activate the plug release mechanism, fingers of metal that pulled out and allowed the airborne half of the connector to be pulled free as the missile lifted off. A phone call to Denver the next day resulted in a recalculation and reassurance that the installation was strong enough.[26]

On the morning of 15 February 1963, SSSgt George Sansone, one of the first instructor missile facility technicians, and SSgt Jim Smith, one of the first ballistic missile analyst technicians (both were missile combat crew positions on a Titan II launch crew), were positioned about a mile from Launch Complex 395-C, waiting for the launch. Sansone and Smith had been in the first training class at Sheppard AFB, Texas, in 1962 and were looking forward to this historic moment. Smith had binoculars and was focused on the silo closure door, giving Sansone a running commentary. As he watched, the silo closure door came to the fully opened position and then closed halfway. It appeared that it had opened correctly but then rebounded into a half-closed position. Both Sansone and Smith knew that this was not the normal sequence. The silo closure door was then closed, and three hours later the launch sequence was again initiated. This time the door did not move at all. With limited airborne battery life remaining, the launch was scrubbed until the next day. This was hardly an auspicious start for the in-silo launch program.

Sansone and Smith returned to their vantage point the next day. The silo closure door was now chained open and a test tool wired into the launch sequence logic to simulate the door opening. On 16 February 1963, the first in-silo launch took place. Sansone was astonished as he saw—and felt—the missile rise out of the silo between the two exhaust plumes. The missile was rotating at approximately eight revolutions per minute as it emerged from the silo. Sansone turned to Smith, and Smith turned to him, both saying simultaneously, "Something isn't right!"[27] Indeed it was not. Three electrical umbilicals on the second stage had failed to disconnect properly. The lanyards had snapped prior to activating the connector release mechanism. The airborne halves of the electrical umbilical connectors were torn from the missile skin, pulling out portions of the missile guidance cabling. N-7 began a continuous, uncontrolled, counterclockwise roll that increased to a rate of 16 revolutions per minute as the missile climbed to approximately 18,000 feet.[28]

Robert Popp, an engineer at AC Spark Plug, the supplier of the inertial guidance system, had driven to the official viewing area to watch this inaugural in-silo launch. He had remained in his car, filming the launch through the long sloping windshield of his Buick. As the missile emerged from the silo, he too noticed the unusual rolling motion. As the missile cleared the silo, the programmed roll and pitch maneuver did not take place. Popp panned up until the roof blocked his view. He started to get out of the car with his camera and then thought better of it when he realized a lot of top Air Force brass were nearby and might not like the idea of his amateur cinematography. Nearly simultaneous with this decision on his part was the breakup of N-7 at 18,000 feet. Popp dove back into the car, realizing that while he was a good two miles from the launch site, debris was starting to spread from the explosion of Stage I.[29]

Kundich and Adamoli were among the Martin Marietta Company employees watching the launch through binoculars from the engineering compound. They, too, noticed that the missile was spinning as it left the silo and immediately knew something was very wrong. Both Kundich and Adamoli saw all too clearly that the Stage II electrical umbilical connectors, normally flush to the surface of the missile, were dangling out at the end of about three feet of wiring.

At 18,800 feet the missile leaned over, and the stages separated due to the weight of Stage II. This activated the inadvertent separation destruct system and destroyed Stage I. The range safety officer had tried to destroy the missile, but the system had not worked because the missile logic still sensed it was on ground power due to the electrical umbilical connector problem. Stage II fell into the water more or less intact, and the expanding cloud of propellant vapor was luckily blown out to sea. The primary objective of the test had been accomplished, that of a Titan II successfully clearing the silo environment intact, but all involved were hardly celebrating.[30]

Figure 4.8a. Titan II N-7 (61-2730) lifts off from Launch Complex 395-C, Vandenberg AFB, 16 February 1963. This was the first in-silo launch in the Titan II program. Observers noted immediately that the missile was spinning at this point. *Courtesy of F. Charlie Radaz.*

Figure 4.8b. N-7 is nearly clear of the silo. Note that the "R" of U.S. Air Force has rotated slightly to the left of the previous photograph. *Courtesy of F. Charlie Radaz.*

Figure 4.8c. Enlarged section of 4.8b showing the torn wiring hanging from the side of the missile. *Courtesy of F. Charlie Radaz.*

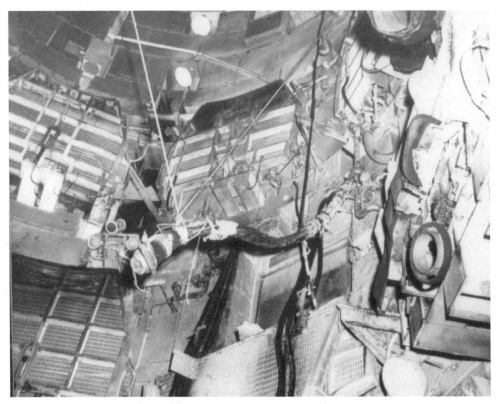

Figure 4.8d. Umbilical 2B1E in the silo after launch. The airborne disconnect is still attached as well as a small piece of the skin of N-7. *Courtesy of Ron Hakanson.*

Later that night, Kundich and Adamoli returned to the silo to find that the Stage I electrical umbilicals had separated properly but that electrical umbilicals 2B1E, 2B2E, and 3B1E of Stage II had remained closed. The airborne half of the connector and a piece of the missile skin were dangling from each umbilical. Due to the problem with umbilical release, the logic circuitry had sensed that the missile had not lifted off, returned the missile to ground power, and left the range safety system disarmed. Basically, the missile had lifted off without any airborne electrical power, and thus no guidance. The force of the umbilicals not releasing properly had started the spinning motion and without electrical power to the missile components, the guidance system could not stop the spin. This spin was fortunate, in a sense, because it imparted some stability to the missile and might have helped it clear the silo intact.

Further investigation showed that the lanyards became taut too quickly and snapped before they could activate the release mechanism in the umbilical connectors. The interim fix was a spring mechanism that cushioned the shock of the umbilical becoming taut. The final fix was to make the D-ring fixture into a J-bar shape that gave enough by bending to absorb the shock and permit the lanyard to pull tight and release the umbilical properly.[31]

Damage to the launch duct equipment and components was extensive, including air conditioning, communication and camera cables, propellant transfer fill and drain lines and

Figure 4.9. Umbilical 2B1E showing the ground and airborne portions still connected. A piece of missile skin is also attached to the airborne portion. The stainless-steel lanyard that would normally have opened the release mechanism was pulled taut and broke a clevis before the release mechanism could be activated. The operational fix was to change from a D-ring to a J-bar attachment point. The J-bar permitted limited spring action as the line was pulled taut, giving the release mechanism time to open. *Courtesy of F. Charlie Radaz.*

valves, vapor detection system components, and umbilicals. While the thrust mount received only superficial damage, the flame deflector was damaged, and 55 acoustic modules in the launch duct and a further 209 in the exhaust ducts needed replacement or repair.[32]

Ed Carson was the Martin Company engineer responsible for refurbishing the silos after each of the contractor launch operations. A protective cocoon was placed on the vertical and horizontal shock isolation damper assemblies. The material was fiberglass with Inconel wire embedded in a butyl rubber coating. This was wrapped, in sheet or tape form, around the various parts that needed protection. The sheets went on first and were sewn together using stainless-steel wires. Then this was wrapped with the tape material and secured. At the bottom of the launch duct was a concrete flame deflector in the shape of a "W." The middle of the "W" was reinforced with steel rebar. After the first launch they drilled out and embedded rebar that had depth markings on it so they could see how much concrete was left and repour when necessary. Initially a rapid cure vacuum process was used to provide for rapid refurbishment turnaround. While the vacuum process worked quite well, it was exorbitantly expensive. Eventually they came up with a material called "Fondue

Fyre" that was embedded in a netting-like arrangement. A layer six inches thick lasted more than one launch. They placed it on the "W" on both sides up to the beginning of the acoustical modules. The acoustic modules were always being blasted loose from both the interior of the launch duct at Level 7, as well as in the exhaust ducts. The entrained air, sucked into the open launch duct as the exhaust ducts directed engine exhaust up and out of the silo, carried the modules into the exhaust stream and flung them well outside of the silo boundaries. Having to replace these frequently, Carson set up a metal shop midway between the three Titan II sites to fabricate the modules and other items necessary for replacement.

During the contractor phase of launch operations, the refurbishment routine began with an immediate photographic assessment of the launch blast damage. Carson and his team would meet with their Air Force counterparts, agree on the necessary repairs, and then get to the task at hand. The governing determinate for the amount of damage was how much the engines had been gimbaled by surface winds as it emerged from the silo. If gimbaling had taken place, the flames literally hosed the sides of the launch duct versus barely touching the sides if the missile rose out of the silo with minimum guidance correction.[33]

Just what happened to the reentry vehicle from N-7 has been the subject of many stories over the past 35 years. The Air Force was extremely interested in finding the reentry vehicle, more specifically the warhead and the guidance system. Though the warhead was not armed with nuclear components, it was still a highly classified piece of equipment as was the guidance system. The Air Force contacted General Motors' Sea Operations Department for assistance. Located one hour south of Vandenberg AFB in Santa Barbara, General Motors' Defense Research Laboratory Sea Operations Department operated the *Swan,* a 137-foot research vessel used under contract to support U.S. Navy submarine operations off the California coast. With her extensive acoustical and oceanographic equipment, *Swan* was uniquely suited to conduct the search for the missing warhead. In addition, she had sufficient divers cleared for SECRET work, which this certainly would be.

Using the radar plot from the failed flight, the Air Force selected an area about four miles from the launch site and in approximately 100 feet of water. On 26 March 1964, divers operating from *Swan* recovered several pieces of the missile but not the warhead. The AC Spark Plug Guidance Lab personnel, Ray Petryk, Howard Wesienborne, and Duane McIntosh, were anxious for the recovery of the guidance system due to the fact that the missile had been rotating as it left the silo. Unfortunately, the guidance system was not among the pieces of debris recovered.

More pressing research for the Navy required *Swan* to abandon the search for 10 days. In early April, *Swan* was back on station ready to resume the search. During the intervening days, further refinement of the radar plot and phototheodolite observations generated a second impact point estimate. Since this point was over the horizon, a helicopter was dispatched to hover over the point as directed by the manned tracking stations. *Swan* was positioned under the hovering helicopter, and a taut-line marker buoy was placed directly under the helicopter.[34]

Unlike in March, sea conditions were severe for search and recovery with swells measuring eight feet. Leo Blickley, Sid Kuphal, and David Potter, divers for the Sea Operations Department, found the 58-degree-Fahrenheit water cold even wearing wetsuits. They used single tanks, meaning they had 20 minutes of air for each dive. The visibility at 100 feet was like pea soup. Due to the extremely limited visibility, two methods of search were used. The first used two sets of anchors and buoys separated by about 150 feet. A line between the two buoys served as a guide, with the divers descending the line from the buoy to the anchor, swimming out along the line and then back up at the second anchor. The anchors would then be moved 10 to 12 feet parallel to the previous search point and the process repeated. This did not prove fruitful and was too time consuming, so a second method was used. A single buoy and anchor were used as a reference point, and the divers would attach a line to the anchor and swim a full 360-degree search before increasing the length of the line 3 feet and repeating the process. At one point during the search Potter smashed his face mask into an outcropping of rock which he had not seen because of the limited visibility. Clearly this was not a trivial undertaking.

About noon a marker buoy popped to the surface indicating that the dive team of Dick Hegeman and Paul Smith had located a large missile part, fortunately reasonably accessible in a shallow canyon, 110 feet deep. Upon surfacing and describing the object, it was concluded that this was indeed the W-53 dummy warhead from N-7. Blickley and Potter were scheduled for the next dive. Blickley was bemused at their high-tech equipment for digging in the bottom mud, two long-handled metal soup ladles from the galley. They were to dig a tunnel underneath the warhead, slip a line around the girth of the warhead, and attach a buoy to bring the line to the surface.

Once they relocated the warhead, Blickley, holding Potter's right wrist in his left hand for stability, began to dig from one side while Potter started on the opposite side of the warhead. Blickley had his right arm more than elbow deep in his tunnel when either the warhead rolled from the wave action or because the tunnel collapsed. An urgent signal on Potter's wrist brought him over to Blickley, and they both started to dig to free his arm. Blickley knew he had perhaps three minutes of air left, so Potter began to prepare to blow and go; that is, leave his bottle with Blickley and head to the surface for help. "Blow and go" is a diving term for an emergency ascent to the surface with only the air in the diver's lungs. As the diver ascends to the surface, the compressed air in his lungs begins to expand, making it necessary to blow out air during the ascent. Blow out too much and you run out of air early, too little and you burst your lungs. Just as the decision had to be made, Blickley managed to free his arm, and they both headed for the surface, somewhat annoyed that they had not secured the hoist line but thankful nonetheless.

The divers eventually were able to rig a harness, but after three attempts to hoist the warhead to the surface, the surface team decided that a fixture was needed to firmly grasp the warhead due to the motion of the water and the mass and shape of the warhead. A team of divers went down and measured it. The *Swan* proceeded to shore at Avila Beach where the dimensions were driven back to the Defense Research Laboratory in Goleta. There Vic Hickey and the lab personnel spent the night to build a device similar to an ice tong except that the arms would enclose the entire cylinder of the warhead. The tongs were lowered to the bottom where the divers manually operated the tongs around the warhead. Once the arms were closed, a line was tied around the upper arms of the tongs to keep them from opening and the entire assembly was hoisted to the surface. After two unsuccessful attempts, finally the warhead made it to the surface and then to the afterdeck of the *Swan.*

Ironically, while the divers had spent nearly 4 days and 25 hours of combined bottom time climbing all over the warhead, once it reached the deck of the *Swan* it was draped in canvas and declared off limits. The Air Force requested that the *Swan* remain at sea until dark, and at 0100 the next day, 12 April 1963, the *Swan* arrived at Stearn's Wharf in Santa Barbara and offloaded the warhead onto an awaiting Navy truck for return to Sandia Laboratory for inspection.[35]

The second launch from Vandenberg AFB was N-8 (61-2731), also from Launch Complex 395-C. Rated as successful, a recurrence of bubbles outgassing in the inertial measurement cooling system caused the reentry vehicle to impact 27 nautical miles past and 0.9 nautical miles to the left of the target point. N-19 (61-2742), launched from Launch Complex 395-D on 13 May 63, had a successful flight except the bubble problem appeared again, accounting for N-19 overshooting the target by 55.6 nautical miles and 1.13 nautical miles to the left. Sealing and pressurizing the inertial measurement unit shell served as a temporary fix, and the problem did not recur. The permanent fix was to seal the fluid system on the platform.[36]

Figure 4.10. This grappling device was used in the recovery of the N-7 W-53 warhead package. The grapple was designed and fabricated overnight. *Courtesy of Vic Hickey.*

Of the six remaining research and development test flights from Vandenberg AFB with the N-series missiles, all but one were successful. N-22 (61-2745), the fourth at the Western Test Range and the twentieth in the program, was launched on 20 June 1963 from Launch Complex 395-C. The flight was partially successful with Stage I flight normal. A Stage II gas generator restriction failure caused thrust to reach only half of the required level and the reentry vehicle–Stage II combination impacted 1,593 nautical miles short of the target.

If this had been a Gemini program launch with astronauts on board, the mission would have been aborted. This was the third failure due to insufficient Stage II thrust that was caused by a problem with the Stage II gas generator system. This launch was the final straw, and flight testing was halted as the system was reviewed. Flight testing at Vandenberg

Figure 4.11. Titan II N-8 (61-2731) lifts off from Launch Complex 395-C, Vandenberg AFB, on 27 April 1963. This was the second silo launch and was completely successful. *Courtesy of Andy Hall.*

Figure 4.12. Titan II B-15 (61-2769) lifts off from Launch Complex 395-B Vandenberg AFB on 17 February 1964. This was a demonstration launch by Martin Marietta Corporation personnel and was the first operationally configured missile to be launched. The flight was successful. *Courtesy of Lockheed Martin Astronautics, Denver, Colorado.*

AFB resumed on 9 September 1963 with the launch of N-23 (61-2746), and the research program using N-series missiles at Vandenberg AFB was successfully completed with four successful flights culminating in the launch of N-30 (61-2753) on 13 March 1964.

The only research and development launch using an operationally configured missile took place on 17 February 1964 with the launch of B-15 (61-2769) from Launch Complex 395-B. Nicknamed "Safe Conduct," this was the first launch from what previously had been the documentation validation silo. The flight was successful and marked the end of contractor-conducted launch operations in the Titan II program.[37]

The research and development program was not yet complete when the Titan II program was declared fully operational on 31 December 1963. The concurrency concepts of General Schriever had worked to perfection as deployment leapfrogged testing, allowing the nation's most powerful strategic weapon to be deployed much earlier then traditional development would have allowed. Missile testing resumed in the summer of 1964 when the first operationally based missiles were removed from their silos and transported to Vandenberg AFB for accuracy and reliability testing. The remaining piece of the larger picture of the Titan II system was the simultaneous construction of the 57 launch complexes beginning in December 1960.

V

TITAN II LAUNCH COMPLEX DESIGN AND CONSTRUCTION

The need for in-silo launch to provide missiles with much greater protection from the effects of nearby nuclear blasts was one of the driving forces for Titan II development. True, tests had shown that a Titan I airframe could be strengthened sufficiently to withstand the acoustic environment of an in-silo launch. However, an in-silo launch implied rapid response; therefore, the inability to store the liquid oxygen oxidizer on board the missle led to the major rationale for development of Titan II.

Many options for highly protected launch facilities were explored. Deep underground facilities, such as abandoned mines, offered superior hardening possibilities with a down-side of slow access to the surface for rapid launch. Natural caves and vaults were another possibility but were not numerous enough and again offered slow access to the surface. Relatively shallow silos with massive concrete structural elements could be designed and would offer far greater site selectivity. Site selection criteria such as availability to highways, railways, and large airfields, a nearby community to support construction, and low-density population areas under the probable launch trajectories, all factored into the design of underground launch facilities.[1]

Design Considerations

Designers of hardened missile silos had to plan for three basic modes of attack: (1) massive bomber attack; (2) small surprise bomber attack; and (3) missile attack followed by bomber attack. Massive bomber attack would probably provide adequate warning to launch the missiles, thereby providing an empty silo as a target. A surprise attack with accurate bombing was best defended by a dispersed missile force that would make such an attack unlikely due to the small number of bombers and the large number of dispersed sites. The third scenario, that of a general missile attack followed by bombers to "mop up" survivors, was the one considered most likely. Hardening and dispersal were the obvious solutions, but hardened to what degree and dispersed how far apart?[2]

War-gaming exercises conducted as part of a missile hardening study in 1958 generated the Soviet threat estimates for the 1961–1965 time frame that are detailed in Table 5.1. These estimates indicated the need for over 10,000 Soviet missiles to destroy 1,359 aim points in the United States. Clearly, the number of hardened and dispersed U.S. missile sites projected by 1965 would make any Soviet attempt to completely eliminate our ICBM forces an apparently insurmountable task. The question by late 1959 was how much hardening was necessary and affordable? Superhardening could make the missile complex nearly impervious to all but a direct hit. Besides being extremely expensive, superhardening meant that the missile was so well protected that a significant amount of time would be required to expose the missile for launch. Wide dispersal would prevent multiple site damage from a single missile hit but would again have increased logistical costs.[3]

A study on missile site separation conducted for the Air Force Ballistic Missile Division completed in October 1959 revisited the problem of optimization of missile site separation. This study concluded that the best definition of "optimum separation distance" was one which considered both system effectiveness and cost in determining separation distances.[4] Clearly the exposure time of the system was critical. Exposure time was defined as the time during a launch sequence and early trajectory when the missile would be vulnerable to the effects of a 2 pounds per square inch (psi) overpressure, enough to cause severe damage to the missile skin, and/or 7 to 100 calories/cm^2 thermal energy, since 100 to 135 calories/cm^2 is sufficient to melt aluminum aircraft skin. A one megaton nuclear weapon detonated as an air burst would generate 100 calories/cm^2 of thermal energy at a distance of slightly over three miles.[5] The attack scenarios for this study were limited to two main categories. In the first scenario, the majority of enemy weapons would reach U.S. forces prior to initiation of a retaliatory launch. In this case, each site would still have its silo door closed and be protected to 100 pounds per square inch overpressure. A minimum 5-nautical-mile separation assured that each site would have to be treated as a separate target. In the second scenario, a majority of enemy weapons would reach the United States after initiation of a retaliatory launch. Silo doors would be open and missiles would be vulnerable in their silos or in the early stages of flight. The recommended basing systems

TABLE 5.1

U.S. PROGRAMMED FORCES AND SOVIET THREAT, 1961 AND 1965

	1961			1965		
	Soviet forces 500 ICBM, 2 MT[a], 4 nm CEP[b]; 170 submarine launched missiles, 3 MT, 3–4 nm CEP			*Soviet forces, 750 ICBMs, 5 MT, 2 nm CEP; 250 ICBMs, 13 MT, 1.4 nm CEP and 280 submarine launched missiles , 3 MT, 2 nm CEP*		
U.S. Forces	Aim points	Hardening (psi)	Soviet Missiles Required	Aim points	Hardening (psi)	Soviet Missiles Required
Bomber bases	62	2.8	434	62	2.8	124
Soft sites	9	2	36	8	2	16
Hard sites	3	25	72	1,289	25-100	10,386
	TOTAL		542	TOTAL		10,526

Source: Adapted from BrigGen W. R. Large Jr., "Ballistic Missile Hardening Study," 10 July 1958, pp. 29 and 35, Air Force Material Center History Office, Wright-Patterson AFB, Ohio.

[a] *MT = megaton.*
[b] *CEP = circular error probable.*

TABLE 5.2

CHARACTERISTICS OF BASING SYSTEMS
WITH RECOMMENDED SEPARATION DISTANCES

SYSTEM	SITE HARDNESS (PSI)	MISSILES/ SITE	SITES/ SQUADRON	EXPOSURE TIME (MIN)	SEPARATION DISTANCE (NM)
Atlas-Titan Silo Lift	100	1	9	5 to 7	14 to 18
Titan In Silo	100[a]	1	9	2 to 3	7 to 10
Minuteman	1000	1	50-100	0.5 to 1.5	5

Source: Adapted from "Missile Site Separation, October, 1959," page vi, Air Force Material Center History Office, Wright-Patterson AFB, Ohio.

[a] *Original estimate was 100 psi, actual construction was to 300 psi hardening.*

resulting from this study for Atlas and Titan silo-lift, Titan II in-silo launch, and Minuteman are listed in Table 5.2. The optimum site separation in the Titan II program was determined to be 7 to 10 nautical miles.[6]

During the cold war, volumes were written addressing the perceived vulnerability of the U.S. and Soviet strategic forces. Reliability, probability of surviving to launch, and missile guidance-system accuracies were all valid points to study. The designers of the hardened silos for Atlas, Titan, Minuteman, and Peacekeeper had to contend not with the probability of a weapon reaching the target but with the effects from its detonation. Any number of criteria can be used to evaluate how close a warhead would have to impact to prevent launch of the targeted missile. The three scenarios that follow illustrate how variable the answer is to the question of "How close was close enough?" In Table 5.3, selected calculated effects from 2-, 5-, or 30-megaton weapons detonated as air or ground bursts are listed, as those yields were used in many of the planning documents for developing the three scenarios.[7]

The first scenario governs the damaging effect of surface overpressure. The Titan II silo closure door was designed to withstand an overpressure of 300 psi.[8] Table 5.3 lists the distance, as a result of an air or ground burst, at which the overpressure would be 300 psi or higher. Depending on warhead yield, the distance varies with a radius of 3,000 to 7,300 feet. The blast lock structures, serving to protect from the effects of a nearby nuclear blast or an explosion in the silo area, were also designed around the 300 psi surface overpressure value.[9]

The second scenario was the physical disruption of the underground portion of the launch complex by the incoming warhead hitting close enough to expose the launch complex to large magnitude ground movement. This did not require a direct hit on top of the silo. Physical damage to the silo would have occurred in the rupture zone, which would contain numerous cracks of various sizes, or in the plastic zone where soil is compressed or deformed permanently.[10] This plastic zone extends out to the edge of the continuous ejecta zone. From Table 5.3 it is readily apparent that the continuous ejecta zone would have a radius from approximately 1,300 to 5,100 feet, depending on soil type and weapon yield.

TABLE 5.3

WEAPON EFFECTS FOR A 2-, 5-, AND 30-MEGATON (MT) AIR OR SURFACE BURST

EFFECT	2 MT	5 MT	30 MT
300 psi overpressure contour (air or surface burst)	3,000 feet	4,100	7,300
Radius plastic zone, surface burst			
dry soil, soft rock	1,670	2,199	3,770
wet soil, wet soft rock	2,245	3,000	5,000
wet hard rock	2,050	2,777	5,100
dry hard rock	1,330	1,800	3,300
Radius of 6-inch layer of debris, surface burst			
dry soil, soft rock	2,873	4,200	8,270
wet soil, wet soft rock	4,104	5,800	7,535
wet hard rock	3,641	5,307	7,056
dry hard rock	2,890	4,212	8,800

Source: Adapted from Air Force Manual for the Design of Hardened Structures, *pp. 35 and 53.*

Note: At overpressures below 30 psi an airburst can be significantly more effective than a ground burst in area coverage. Above 30 psi, both air and ground bursts give nearly the same effect, The Effects of Nuclear Weapons, *pp. 255–56.*

The third scenario involves the depth of ejecta thrown from the crater. The silo closure door was designed to operate with a covering of a maximum of 6 inches of soil.[11] Referring to Table 5.3, depending on soil type and weapon yield, the effective radius—that is, a covering of soil to a depth of 6 inches or more that could be expected—was approximately 2,800 feet to 8,800 feet. In reality, the question of "How close is close enough?" becomes a difficult one to answer completely. Keeping in mind that these calculations are considered to be approximations and, in many cases, scaled from small weapon effect demonstrations, one can see that at some point the designers simply had to accept a cost-effective solution and proceed.[12]

An obvious point remains concerning the design of the Titan II silos. Given that the Soviets knew that these were the "blockbuster" warheads in the U.S. arsenal, carrying the single largest nuclear warhead used in the strategic missile program, they reasonably would have concluded that they would have been launched almost immediately upon confirmation of the beginning of World War III. Except for a sneak attack, that would have meant that the Titan II silos would have been empty by the time the incoming weapons arrived. So why target them? While this appears to be a logical argument now that the cold war is over, 40 years ago protection against a successful first strike was paramount in the silo design considerations.

Components and Layout of an Operational Titan II Launch Complex

The typical operational Titan II launch complex layout is shown in Figure 5.1. The security fence enclosed an area of approximately 3.3 acres, while the actual military reservation area was approximately 10 acres. The major underground structures located within the security fence were the launch control center, blast lock, access portal, cableway, and silo, shown in Figure 5.2. Each of these, except the access portal, were hardened to withstand the effects

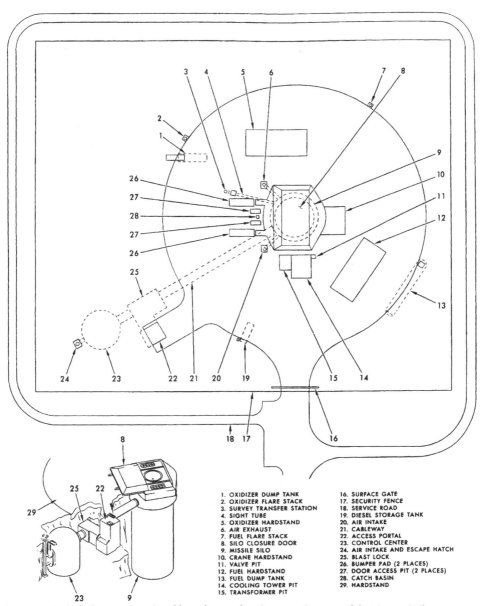

1. OXIDIZER DUMP TANK
2. OXIDIZER FLARE STACK
3. SURVEY TRANSFER STATION
4. SIGHT TUBE
5. OXIDIZER HARDSTAND
6. AIR EXHAUST
7. FUEL FLARE STACK
8. SILO CLOSURE DOOR
9. MISSILE SILO
10. CRANE HARDSTAND
11. VALVE PIT
12. FUEL HARDSTAND
13. FUEL DUMP TANK
14. COOLING TOWER PIT
15. TRANSFORMER PIT
16. SURFACE GATE
17. SECURITY FENCE
18. SERVICE ROAD
19. DIESEL STORAGE TANK
20. AIR INTAKE
21. CABLEWAY
22. ACCESS PORTAL
23. CONTROL CENTER
24. AIR INTAKE AND ESCAPE HATCH
25. BLAST LOCK
26. BUMPER PAD (2 PLACES)
27. DOOR ACCESS PIT (2 PLACES)
28. CATCH BASIN
29. HARDSTAND

Figure 5.1. Typical Titan II operational launch complex site map. *Courtesy of the Titan Missile Museum National Historic Landmark Archives, Sahuarita, Arizona.*

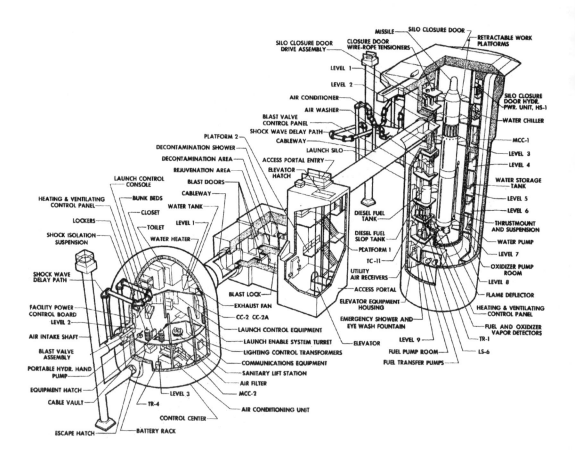

Figure 5.2. Cutaway view of the underground facilities at a Titan II launch complex. *Courtesy of the Titan Missile Museum National Historic Landmark Archives, Sahuarita, Arizona.*

of a nuclear blast. The brief discussion of each of the major components will orient the reader when the design and construction process is described.[13]

The launch control center was a three-story domed cylinder 42 feet in height and 37 feet in diameter (Figure 5.2). The top of the dome was 8 feet below the surface and 18 inches thick. Inside the dome, a steel cage was suspended on eight sets of springs. This served to isolate the launch control center from ground motion. On the top level, Level 1, were the living facilities; four beds, kitchen, shower, and toilet. The launch control and checkout equipment was located on Level 2, and communications, radio filters, backup battery power supply, and utility equipment were located on Level 3. Air intake was through the launch control center air shaft. After passing through a nuclear and biological contamination air filter, the fresh air was dispensed into the launch control center. Air pressure in the below-ground structures was alway highest in the launch control center so that air movement was toward the silo areas in order to prevent contamination from the silo if a propellant spill or fire occurred.

The blast lock was 35 feet underground and served to isolate the underground structures from a nearby nuclear blast at the surface (Figure 5.2). Located between the silo and the launch control center were two sets of 3-ton blast doors. Blast Doors 6 and 7 served to isolate the launch control center from the effects of explosions on the surface. Blast Doors 8 and 9 served to isolate the launch control center from an explosion in the silo. The hydraulic system for locking and unlocking the blast doors was located in Blast Lock Area 202 between Blast Doors 6 and 7. Blast Lock Area 201, between Blast Doors 8 and 9, housed the Vapor Detector Annunciator Panel (VDAP), which was used to scan the various silo levels for hazardous propellant fumes prior to daily inspection of the silo and launch duct.

Two cableways connected the blast lock to the launch control center and the silo. The short cableway, approximately 25 feet long, connected the launch control center to the blast lock area. The long cableway, approximately 180 feet long, connected the silo to the blast lock area. The guidance lighting, communications, and power cables to and from the missile were routed through the cableways, as were the chilled water used to air condition the launch control center and the power for the launch control center equipment.

The access portal was the connection between the underground structures and the surface. A staircase, with an entrapment area to prevent unauthorized entry, and a service elevator were the only ways into and out of the silo on a normal basis. An escape hatch was built into the air intake for the launch control center and served as an emergency exit (Figure 5.2).

The missile silo and launch duct were two concentric cylinders (Figure 5.2) The outer cylinder was the silo, 55 feet inside diameter and 145 feet deep, which housed the silo equipment area and the launch duct, which was 26.5 feet in diameter. There were nine levels in the silo equipment area, including the silo sump. The launch duct interior had six sets of work platforms which folded against the launch duct wall when not in use.

Silo Level 1 housed the silo elevator control equipment and the silo closure door operating equipment (Figure 5.3a). Access to the launch duct for maintenance on the reentry vehicle using work platforms was also provided for on Launch Duct Level 1.

Silo Level 2 was the access point from the launch control center to the silo equipment area (Figure 5.3b). Before its replacement in 1978, the missile guidance Azimuth Alignment Set (AAS) was located on Level 2. The three water chillers which provided cold water for air conditioning throughout the silo and launch control center were located on Level 2. Access to the launch duct for maintenance on the missile guidance system was provided on Launch Duct Level 2.

Silo Level 3 housed the diesel generator, Motor Control Center 1, steam separator, silo fresh air intake and exhaust vents, air filter, supply and exhaust fan, emergency safety shower, and the manhole for access to the hard water tank (Figure 5.3b). Two access points for the launch duct work platforms were located on Level 3. The first was referred to as Launch Duct Level 3 and provided access to the Stage II engine equipment, explosive nuts for staging, and the vernier rocket motors. The second access point was referred to as Launch Duct Level 4. This provided access to the Stage I oxidizer dome and the missile transtage areas.

Silo Level 4 housed the silo air-conditioning equipment as well as outside air intake and exhaust fans (Figure 5.4a). The 100,000-gallon hardened water tank took up nearly

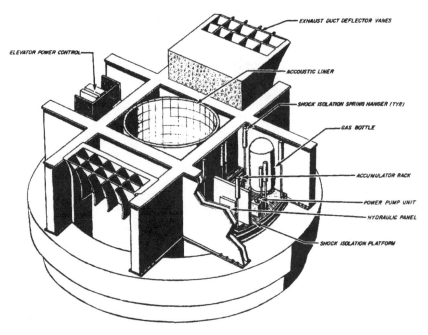

Figure 5.3a. Major equipment at silo and launch duct Level 1. *Courtesy of the Titan Missile Museum National Historic Landmark Archives, Sahuarita, Arizona.*

half of the Level 4 area and continued down to Level 6. No launch duct access was provided at this level.

Silo Level 5 housed the diesel generator service tank and the launch duct air conditioner which kept the launch duct at 60 ±2 degrees Fahrenheit at all times (Figure 5.4b). The oxidizer had a boiling point of 70 degrees Fahrenheit, thus temperature control in the launch duct was critical. The fuel had a boiling point of 148 degrees Fahrenheit and was not as much of a problem. Access to Launch Duct Level 5 work platforms provided maintenance personnel the ability to work in the between-tanks part of the Stage I airframe.

Silo Level 6 housed the vapor detection equipment that continually sampled all levels of the silo and the launch duct exhaust stream, monitoring for hazardous levels of fuel or oxidizer (Figure 5.4c). The silo and launch duct heating and ventilation controls were also located on this level. No access was provided to the launch duct at this level.

Silo Level 7 (Figure 5.5a) housed the water pumps for the launch complex deluge water system and the air compressors for the pneudraulics equipment throughout the launch complex. The hard water tank drain valve was also located on Level 7. Launch Duct Level 7 provided access to the missile thrust mount, shock isolation system, and Stage I engine subassemblies.

Silo Level 8 (Figure 5.5b) housed the fuel and oxidizer pump rooms and the launch duct dehumidifier equipment. Access to the launch duct was provided but without work platforms. A small platform, nicknamed the diving board, was connected to a ladder to give access to the flame deflector area.

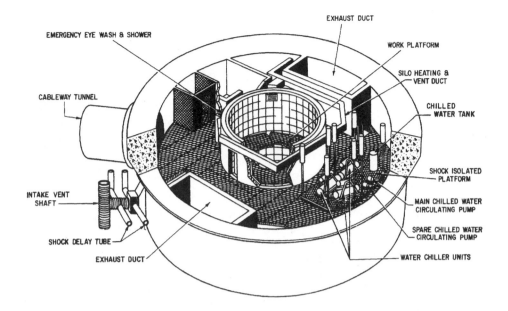

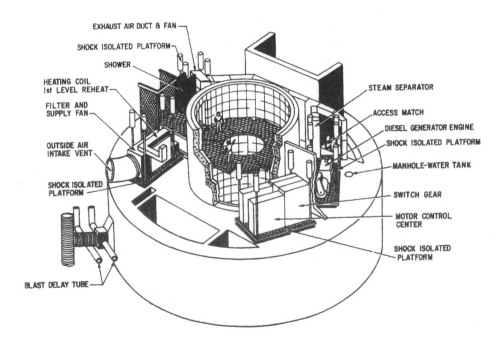

Figure 5.3b. Major equipment at silo Levels 2 (upper) and 3 (lower). *Courtesy of the Titan Missile Museum National Historic Landmark Archives, Sahuarita, Arizona.*

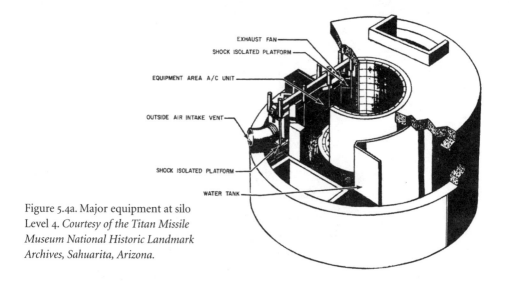

Figure 5.4a. Major equipment at silo Level 4. *Courtesy of the Titan Missile Museum National Historic Landmark Archives, Sahuarita, Arizona.*

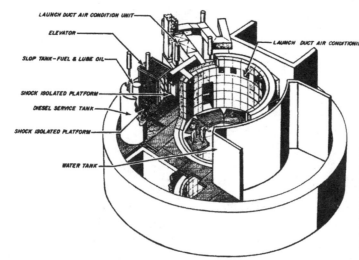

Figure 5.4b. Major equipment at silo Level 5. *Courtesy of the Titan Missile Museum National Historic Landmark Archives, Sahuarita, Arizona.*

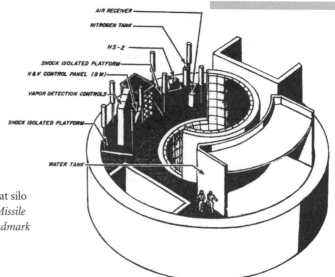

Figure 5.4c. Major equipment at silo Level 6. *Courtesy of the Titan Missile Museum National Historic Landmark Archives, Sahuarita, Arizona.*

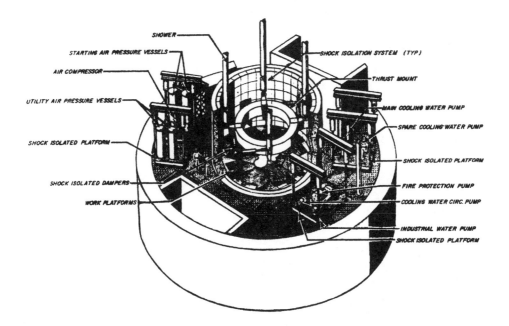

Figure 5.5a. Major equipment at silo Level 7. *Courtesy of the Titan Missile Museum National Historic Landmark Archives, Sahuarita, Arizona.*

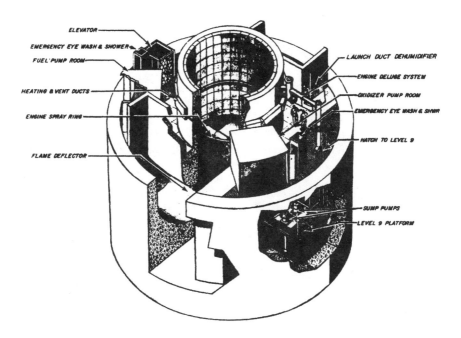

Figure 5.5b. Major equipment at silo Levels 8 and 9. *Courtesy of the Titan Missile Museum National Historic Landmark Archives, Sahuarita, Arizona.*

Silo Level 9 (Figure 5.5b) was the silo and launch duct sump, on the east and west of the flame deflector. These served as the drain points for all of the silo equipment area floor drains and were periodically pumped dry to the surface.

The silo was covered by a massive 740-ton steel and concrete silo closure door. Concurrent with the construction of the Titan II silo facilities was a separate contract to Ralph M. Parsons Company (Parsons) for the design of the silo closure door. The initial contract was for a 1/12th scale model. Stan Goldhaber, at the time Titan II principal project manager and vice president for Ralph M. Parsons Company, recalls that the model was constructed in a large warehouse in downtown Pasadena only a few blocks from the company headquarters and no one knew about it. The test program was successful, and the design was scaled up to the 740-ton door for use on the operational sites.[14]

To test the operation of the door and make modifications before the operational sites reached the door installation phase, a full-scale silo door test model was fabricated by Parsons at Vandenberg AFB, California, Launch Complex 395-B, well away from the construction area. The underground portion of the silo and launch duct was simulated by a combination of wood, steel, and concrete structures which supported the door and tracks and provided easy access to all components. The hydraulic drive system and impulse actuator, which was tested and eventually discarded, were positioned exactly as if this test structure were an operational site. The test program consisted of 169 maintenance runs, 4 operational runs without debris, 4 operational runs with three inches of debris (including 3-inch-diameter rocks), 3 operational runs with six inches of debris (including 6-inch-diameter rocks), and 2 runs with one jack disabled. There were no malfunctions during 179 test runs conducted in April 1962.[15]

Operational Bases

Simultaneous construction of the Titan II ICBM complexes and associated support structures was carried out by the Site Activation Task Force (SATAF). A joint Army Corps of Engineers and Air Force organization, SATAF was a critical factor in the concurrent design and build concept that had been used so successfully in the earlier Titan I program. The concurrency concept meant that Titan II launch complex construction had to begin before the final missile design had been approved; indeed, launch complex Phase I construction was started in late 1960, fully 15 months before the first flight of a Titan II missile.[16]

The Army Corps of Engineers was responsible for coordinating on-site contractors for facility construction with the Air Force and missile design and manufacturer, Martin Marietta Company, during the Corps' work. The major structures on each site were the access portal, blast locks, launch control center, cableway, silo, and launch duct. An all-weather access road from the nearest paved road was ultimately built for each complex. Tables 5.4a–c summarize the quantities of materials excavated and/or used in construction of one of the 18 launch complexes assigned to the 390th Strategic Missile Wing. Construction of a Titan II launch complex took place in three parts, grouped in four phases, designated as Phase I, Phase II, Phase IIA, and Phase III.

TABLE 5.4A

EARTH WORKS QUANTITY SUMMARY, PHASE I CONSTRUCTION,
*390TH SMW, DAVIS-MONTHAN AFB

EARTH WORKS	1 SITE CUBIC YARDS
Open Cut	
Stripping	5,650
Earth Excavation	38,340
Rock Excavation	51,293
Shafting	
Silo	15,313
Launch Control Center	1,400
Total	111,996

Source: U.S. Army Corps of Engineers Ballistic Missile Construction Office, History of Davis-Monthan AFB, October 1960–January 1964.

PHASE I

Phase I took 7 to 14 months, depending on the site soil conditions. Major tasks included clearing of the site; construction of an all-weather access road; excavation of the site to the construction reference elevation; and installation of the major structures such as silo, blast lock, and launch control center.

Construction began with an open cut, which consisted of an excavated area of nearly 106,000 square feet at the top of the cut to an average depth of 31 to 33 feet, depending on the site, sloping down to an area of approximately 26,000 square feet at the reference elevation. Shafting for the silo and launch control center started at the reference elevation level. In the case of the hard rock sites, which occurred at both Davis-Monthan AFB, Arizona, and Little Rock AFB, Arkansas, the open-cut area was greatly reduced.[17]

After the open cut was excavated, a shaft approximately 65 feet in diameter and 115 feet deep was sunk for the launch silo, and another shaft 45 feet in diameter and 22 feet deep was sunk for the launch control center (at Davis-Monthan most of the launch control centers required only 6 to 8 feet of shafting).[18] The shafts were sunk from concrete collar beams set at the open-cut reference elevation. As the excavation proceeded, steel ring beams, supported from the collar beams, were installed every 4 to 5 feet for a total of nearly 127 tons of steel typical per silo. The sides of the shaft were sealed with 2-inch wire mesh and gunite mortar. When the launch silo shaft was completed, gravel drains were installed and piped to the exterior of the silo to remove ground water. A 6-inch gravel bed was laid, and then the 1/4-inch electromagnetic shield was installed and welded to the bottom ring beam. Low pressure grout was pumped in behind the electromagnetic shield. One hundred forty-seven tons of 1/4-inch steel plate were then welded to the ring beams

TABLE 5.4B
STEEL QUANTITY, PHASE I, 390TH SMW, DAVIS-MONTHAN AFB

	TONS
Ring beams	
Silo	116.2
Launch Control Center	11.9
Pit Excavation Support Beams	
Silo	19.1
Launch Control Center	4.2
Structural Steel Girders	575.0
Embedded Steel	68.0
1/4-inch EM Plate	
Silo	147.0
Launch Control Center	45.0
Blast Lock	33.0
Reinforcing Steel	
Silo	845.0
Launch Control Center	188.0
Blast Lock	50.0
Access Portal	44.0
Structural Steel	
Launch Control Center	30.0
Blast Lock	1.1
Access Portal	1.3
Steel Stairs	2.0
Blast Lock Door Jambs	61.2
Blast Doors	12.1
1/4-inch checkered floor plate	1.3
TOTAL	2,255.4

Source: U.S. Army Corps of Engineers Ballistic Missile Construction Office, History of Davis-Monthan AFB, October 1960–January 1964.

TABLE 5.4C
CONCRETE QUANTITY SUMMARY, PHASE I, 390TH SMW, DAVIS-MONTHAN AFB

STRUCTURE	CUBIC YARDS OF CONCRETE
Silo	5,800
Launch Control Center	545
Blast Lock	724
Access Portal	171
TOTAL	7,240

Source: U.S. Army Corps of Engineers Ballistic Missile Construction Office, History of Davis-Monthan AFB, October 1960–January 1964.

Figure 5.6. Excavation of both the launch control center and the silo shafts is well underway at this site of the 390th SMW, Davis-Monthan AFB. The large rectangular vaults on the side of the silo shaft are for the air intake and exhaust shock delay tubes. The electromagnetic shielding was welded to the ring beams visible at the top of the partially excavated silo shaft. *Courtesy of the Titan Missile Museum National Historic Landmark Archives, Sahuarita, Arizona.*

to continue the electromagnetic shield already in place at the base of the silo. The electromagnetic shield was used to protect the silo equipment from the effects produced by a nearby nuclear explosion. Both conventional high explosives and a nuclear weapon produce electromagnetic pulse effects. With nuclear weapons the effect is much greater. A nuclear explosion generates electromagnetic waves over a broad spectrum of frequencies, including the common radio spectrum. When these waves hit metal objects, such as long runs of cable or piping, or structural steel such as girders or rebar, the energy is collected and transferred in the metal as electrical current with disasterous results to radio equipment, electrical control equipment, and computers. The silo electromagnetic shield served as a barrier and effectively grounded out the current before it could damage equipment.

In Phase I construction the silo was completed to the level of the underside of the roof girders, 28 feet below the finished top of the silo. The majority of the silo wall was 4 feet thick, only the upper 28 feet was 8 feet thick. This increased thickness was to permit support of the massive silo closure door and the blast loads on the door from a nearby nuclear blast. The concrete for the silo was placed in four individual pours. First, the silo base, 7

Figure 5.7. The silo is ready for the slipform concrete pour. The ring of rebar at the bottom of the silo is the connection for the launch duct structure. The completed flame deflector is clearly visible. The large vertical steel strips are anchors for the 100,000-gallon hardened water tank structure. *Courtesy of the Titan Missile Museum National Historic Landmark Archives, Sahuarita, Arizona.*

feet thick, used 840 cubic yards. Second was Level 8, which used 1,002 cubic yards, and third was the flame deflector, which used 590 cubic yards for the cross-walls, 6.5 feet thick and 22.3 feet wide with a curved deflector section between the walls. Fourth was the launch silo wall, which involved a slip-form operation using 3,018 cubic yards of concrete to construct the 145 feet deep and 55 feet interior diameter silo with 4-foot-thick walls. The slip-form operation was worked as one continuous pour averaging 75 hours from beginning to end. The internal diameter was maintained at 55 feet, and the wall thickness increased to 8 feet for the top 28 feet of the silo wall. Depending on the contractor, slip forms were suspended by cables and raised by hydraulic motors attached to each cable or by the more conventional method using jack-rods. Platforms were hung beneath the slip-form so that finishers could maintain the interior radius using templates as well as feather the concrete away from the approximately 400 inserts as they became exposed.[19] While use of the slip-form technique varied between operational bases, a placement rate of approximately 40 to 50 cubic yards per hour was usual, resulting in a rise of approximately 1 to 2 feet per hour. A total of nearly 845 tons of rebar was used to reinforce the basic silo structure, varying in diameter from 1.75 inches for the 4-foot-thick wall to 2.25 inches for the 8-foot-thick wall.[20] The sequence of concrete pours for McConnell AFB sites is shown in Figure 5.10.

The final step in Phase I construction for the silo was the installation of four built-up

Figure 5.8. The slipform work is complete. Preparations for beginning the placement of the 0.5-inch steel plate inner and outer launch duct lining is underway. *Courtesy of the Titan Missile Museum National Historic Landmark Archives, Sahuarita, Arizona.*

steel box girders, weighing 144 tons each, on the top of the silo walls. Measuring 19 feet in height, 4 feet in width, and 51 feet in length, the girders were partially prefabricated before delivery to individual sites. Approximately 3,000 man-hours were needed to weld the girder sections together, after which they were filled with 275 cubic yards of concrete. These girders supported the weight of the silo closure door, distributing the weight to the silo wall. The launch duct cylinder did not contact girders, thus preventing nuclear blast overpressure effects on the door from being transmitted directly to the launch duct and missile shock isolation system. Instead, the motion would be transmitted down the silo walls to the foundation. This would sufficiently reduce the motion to a point which the missile shock isolation system could protect the missile from the remaining ground motion.[21]

The 60-foot launch control center shaft excavation was similar to that for the silo, as was the installation of the electromagnetic shielding and rebar. Slip-form construction was not used. The launch control center was a 19-foot-radius concrete dome superimposed on a vertical concrete cylinder structure measuring 37 feet inside diameter and 23 feet, 11 inches high. The foundation was 8 feet thick. Including the foundation and dome, the launch control center was 51 feet, 9 inches tall. The walls were 1.5 feet thick at the crown of the

Figure 5.9. The final work on the box girders is in progress. Note the relative size of the welder in the right center portion of the photograph under the shade cloth. The finished girders were 20 feet high. The cableway entrance to Level 2 of the silo can be seen at the left. The pipe stubs to the left and right of the cableway entrance are the air supply shock delay tubes, temporarily capped off due to construction activity. *Courtesy of the Titan Missile Museum National Historic Landmark Archives, Sahuarita, Arizona.*

dome, 2.25 feet thick at the shock isolation spring line, and 3 feet 1 inch thick at the cableway entrance. A total of 997 cubic yards of concrete were used in the launch control center construction. Before the dome construction could begin, the three-floor structural steel cage, weighing 30 tons, attached to a central column and connected by stairways, was fabricated and installed. Suspension of this cage on eight sets of springs was completed during Phase II construction. The top of the dome was 8 feet below the surface.[22]

The access portal was considered a soft structure, that is, not built to withstand the effects of a nuclear blast. An elevator for transfer of equipment and personnel, as well as a stairway and security entrapment area for personnel, were the major features of the access portal. Because it was not part of the hardened structure, it was not sheathed in 1/4-inch steel electromagnetic shielding.

The blast lock structure design calculations used a surface overpressure of 300 pounds per square inch to calculate the air pressure diffraction patterns at the base of the 35-foot-deep access portal. As the surface shock wave passed over the unhardened access portal entrance, an estimated 130 pounds per square inch shock wave would enter the access portal and undergo complex diffraction as it traveled to the base of the portal. The resulting overpressure that would hit the surface of Blast Door 6, the first blast door, was estimated to be 750 to 800 pounds per square inch. The silo cableway blast lock door load was calculated from the worst case conventional explosion that might occur in the missile silo.

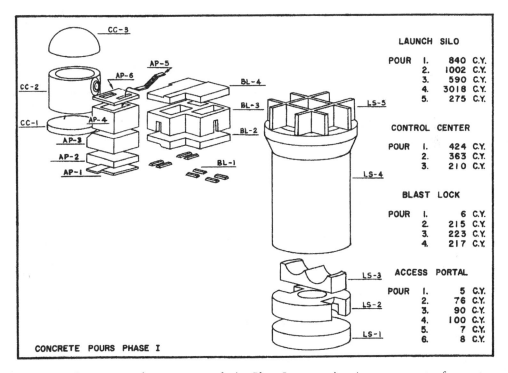

Figure 5.10. The sequence of concrete pours during Phase I construction. Average amounts of concrete for each pour are given in the table on the right. While the launch control center, blast lock, and silo structures each had multiple concrete pours, each was sheathed in 0.25-inch steel electromagnetic pulse shielding. *Courtesy of the Office of History, U.S. Army Corps of Engineers, Alexandria, Virginia.*

This was considered highly unlikely, but provision for it was nonetheless made. The calculations assumed that the explosive energy potential of the Titan II propellants was equivalent to their weight in TNT. If all the propellants mixed completely, the resulting explosion would be equivalent to approximately 400,000 pounds of TNT. The Aerozine 50 fuel was, however, flammable and explosive on its own with an energy equivalent, on a per-pound basis, to 0.38 pounds of TNT. Thus if the total fuel carried on the missile were to leak out and detonate, the energy would be equivalent to approximately 41,000 pounds of TNT. The most likely scenario was considered to be a partial spill of fuel and oxidizer, estimated at 10 percent, that resulted in a complete spill and detonation of the remaining fuel. This scenario gave the equivalent of 77,000 pounds of TNT as the lower limit of an explosion. An upper limit was 100,000 pounds of TNT, equivalent to one-third of the propellant load. The resultant overpressure calculations ranged from 700 to 1,650 psi, depending on the location of the explosion at either the center or the bottom of the launch duct. The final design for the blast lock doors was to withstand a peak overpressure of 1,130 psi.[23]

The design of the blast doors was deceptively simple. The major door components were recurring elements of horizontally welded I-beams and steel plate. The I-beams were 5 feet long and 1 foot tall and had a flange 6.25 inches wide. The steel plate was 12 inches wide and 0.68 inches thick. The U.S. Army Corps of Engineers Ballistic Missile Construction Office documents give blast lock door weights ranging from 6,000 to 18,000 pounds. Calculations from the as-built drawings give an individual blast door weight as approximately 6,000 pounds. The doorjambs weighed an additional 15,300 pounds each, with two doorjambs per door, for a total individual blast door assembly weight of approximately 37,000 pounds.[24] The doors were so well balanced that one person could easily swing the door open and closed.

The blast lock structure construction was routine, consisting of four concrete pours totaling 661 cubic yards of concrete. The walls varied from 3.8 feet to 4.8 feet in thickness; the ceilings and floors were 5 feet thick. The exception to the routine work was the emplacement of the four blast door frame and door assemblies. The blast doors and jambs were initially assembled on site, but difficulty with alignment required the door and jambs to be assembled and emplaced as a unit. Installation of the doors and frames had to be precise to ensure sealing against shock effects as well as perfectly balanced for ease in opening. The blast lock bottom electromagnetic shielding was then welded to each door frame base. In separate pours the blast door floor and walls were completed, followed by roof installation.[25]

PHASE II

Phase II scope of work started where the Phase I contractors had finished. This seems obvious at first, but with the myriad of change orders and modifications due to the concurrency concept, the Phase II contractor often had to backtrack on systems that required further work. Phase II lasted 11 to 13 months and involved fabrication of major silo interior structures such as launch duct, ventilation exhaust ducts, work levels, and cableway connecting the launch control center to the blast locks and silo structures; installation of mechanical and electrical equipment, utility systems, and ventilation shafts; and construc-

Figure 5.11a. Installation of the 3-ton blast doors in their 15.3-ton doorjambs at Launch Complex 374-9, 308th SMW. The doorjambs were first set on concrete foundation pads. The electromagnetic shielding was then welded to the structure to provide a complete seal from the outside environment. Initially the doors and jambs were mated on site, but due to alignment problems, later construction placed the complete door and jamb assembly as one unit. This site is one of the hard rock sites at the 308th SMW. Note how cramped the work area is compared to the open-cut sites at the 390th and 381st SMW. *Courtesy of Don Rawlings.*

tion of the silo closure door and hydraulic system. During this process the site had to be compacted and backfilled to a finished grade.

In Phase II the launch duct was installed in the silo from Level 8 to within 15 inches of the bottom of the box girders on Level 1. The launch duct wall was made up of two 0.5-inch steel plate concentric cylinders spaced 1.5 feet apart with an interior diameter of 26.5 feet. They were prefabricated in heights varying from 5.5 feet to 12.8 feet. The gap between the concentric rings was filled with concrete.[26]

Once the launch duct was installed, level framing was installed from Level 9 to Level 1. Each level had steel flooring supported by steel I-beams, except Level 3, which had a 3-foot-thick reinforced concrete floor placed on top of the sheet steel. This floor was rigidly connected to the silo wall and the exhaust duct walls. A sliding joint was used at the launch duct to permit vertical movement between the two structures while still providing lateral support. On all levels the floor support beams had slip joints at the inner silo wall and outer launch duct wall to permit 1.5 inches to 2.25 inches of movement from ground shock.

Figure 5.11b. Blast lock electromagnetic shielding is being placed prior to the first pour on the blast lock walls at Launch Complex 374-9, 308th SMW. The launch control center three-story cage can be seen behind the short cableway entrance in the center of the picture. *Courtesy of Don Rawlings.*

All major pieces of machinery housed on the various levels of the silo were mounted on shock isolation platforms. For the most part, the Phase II contractors preassembled the large equipment on shock isolation platforms. The resulting assemblies were too large to maneuver through the blast lock area or cableway and were therefore lowered into position as each level was completed and prior to the construction of the floor for the next higher level. The single heaviest piece of equipment, weighing 34,000 pounds, was the 350-kilowatt, 510-horsepower diesel generator installed on Level 3.[27]

The most massive item in the silo equipment area was the 100,000-gallon steel-reinforced water tank, located in the area between the launch duct and the silo wall, between the floor of Levels 3 and 6. The tank was fabricated in four sections, then fitted, welded, and bolted into the inserts in the silo wall. This was the "hard" water tank, hard in the sense that it was hardened against blast effects to ensure the launch complex had sufficient water for both the fire protection and engine deluge. It also served as a heat sink if the diesel engine emergency generator was running and the silo was closed down by a nearby blast.[28]

Fabrication of the silo closure door was one of the unique operations during construction at each complex. The door was massive, 64 feet wide, 42.5 feet long, and 5 feet tall, with 3.5-inch-thick battleship armor plate on the top and bottom surfaces. Reference in the Army Corps of Engineer histories is made to either 3-inch or 3.5-inch-thick battleship

Figure 5.12. The launch duct was composed of two concentric cylinders of 0.5-inch-thick steel plate, separated by 1.5 feet. The space in between was filled with concrete. The launch duct was isolated from direct contact with the box girders that supported the silo closure door. *Courtesy of the Titan Missile Museum National Historic Landmark Archives, Sahuarita, Arizona.*

armor plate. There were initially two designs for the door, with the designs differing mainly in the thickness of the top and bottom steel plate, 3 or 3.5 inches. The operational design was 3.5 inches for both surfaces.[29]

The actual final weight of the silo closure door in the Army Corps of Engineers documents is given as 700 tons, including the wheel trucks, or as 740 tons in the Titan II program technical orders used in maintaining and operating the system. The final design calculation documents generated in 1961 list 749 tons.[30] A design analysis document for sizing the door jacks gives a total weight of the door and wheel trucks as 699 tons. The door was designed to open with the weight of a maximum of 6 inches of debris covering all door surfaces which would have added another 52 tons.[31] Anticipating the possibility of needing to launch after a winter ice storm at McConnell AFB or Little Rock AFB, the door jacks were sized to lift a door that weighed 699 tons, with 52 tons of dirt (6 inches thick) and a 1-inch layer of ice on the top and sides of the door and a 2.75-inch depth at the doorsill. The ice required 597 tons of lifting capacity to break the door free. The four jacks were sized at 325.5-ton capacity each. One modification affecting the weight of the silo closure door was made during Project YARD FENCE in the mid-1960s. Two concrete

PHASE I CONSTRUCTION

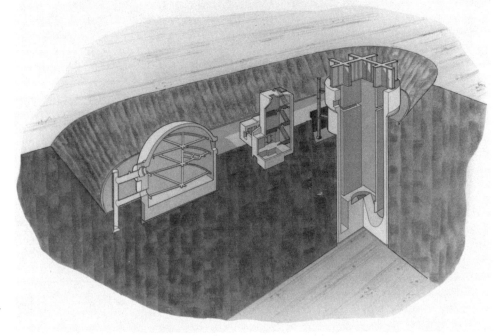

Figure 5.13. Drawing of a typical launch complex at the completion of Phase I construction. The distance between the access portal and silo is fore-shortened for illustrative purposes. *Courtesy of the Parsons Corporation, Pasadena, California.*

neutron shields, referred to as "dog ears," were positioned on the north and south sides of the door to serve as additional shielding over the exhaust ducts when the door was in the closed position. They weighed 9 tons each, for a total additional weight of 18 tons. After modification, the final weight of the silo closure door was 758 tons.[32]

The door was prefabricated in eight sections, which were then assembled on site at the Davis-Monthan AFB complexes; prefabricated in six sections and then shipped to the McConnell AFB sites; and prefabricated in eight sections and shipped to the Little Rock AFB sites.[33] Final assembly took place at each launch complex after the double railroad rails and four double sets of railroad wheel trucks were positioned west of the launch duct where the door would rest in the fully opened position. The rail sections to the west of the launch duct were placed in concrete, and the door sections were placed on cribbing above the rails and welded together. An average of approximately 35 to 40 tons of welding rod was used per door. After the wheel trucks were installed, the entire assembly was lowered onto the rails using jacks. Virtually all the entire door interior welding was manual arc welding since the egg-crate and sandwich-type design precluded use of automated equipment except for the top plates. Welding took approximately 4,000 man-hours. Upon completion of the door assembly, 15 inches of concrete, a total of 148,000 pounds, was poured into the center sections to give final weight and added stability.[34]

Figure 5.14a. The silo closure door was prefabricated in sections and assembled on site due to weight restrictions for transportation on the highway. Note the completed sections at the upper right of the photograph. The egg-crate construction of the door is clearly evident. *Courtesy of the Titan Missile Museum National Historic Landmark Archives, Sahuarita, Arizona.*

Figure 5.14b. The silo closure door sections were mounted on pedestals at the site, then welded together; the door was lowered by jacks onto temporary rail tracks for movement to the permanent tracks near the door. *Courtesy of the Titan Missile Museum National Historic Landmark Archives, Sahuarita, Arizona.*

From the very beginning, construction of the silo closure door broke new ground. Aircraft construction tolerances (e.g., the bottom surface of the door plates had to be flat to within 1/64th of an inch) in an object this massive were difficult to achieve. This initial tolerance was relaxed somewhat as the reality of fabrication problems was seen. Don Boomhower was the Ralph M. Parsons Company program manager at Davis-Monthan AFB when the first silo closure door fabrication began. The first attempt started by welding each of the bottom plates in sequence. This resulted in a warped surface that was not acceptable. The weld between the plates was "V-shaped," approximately 4 inches wide at the top surface and 0.5 inches wide at the lower surface. Welding bead was laid down in lines that filled the weld area. When the bead cooled, the welded metal material raised the outer edges of the plates sufficiently to be out of tolerance. With the silo door fabrication at Davis-Monthan AFB ahead of the other bases, including silo closure door test facilities at Vandenberg AFB, a solution was needed, quickly.

Harry Christman was Phase II and IIA program manager for Fluor Corporation, prime contractor at Davis-Monthan AFB and directly responsible for finding a solution. Fluor Corporation, subcontractors, and several government agencies searched nationwide for a solution to the problem, and many variations were tried. Experts from within the government, the American Welding Society, and the industry as a whole were brought to the site for consultation. Welding at night made little difference. Machine welding with refrigeration was not the solution. The subcontractor to Fluor Corporation for the door, Graver Tank and Manufacturing, a subsidiary of Chicago Bridge and Iron, arrived at a simple and effective solution. By calculating the expected shrinkage of the 3.5-inch V-shaped weld, a jig was used to position the new plate at a precambered angle so that it would come to be nearly flat with the next section at the end of each section weld. A redesign of the doorsill, coupled with a slight lessening of the door flatness tolerance, solved the problem, and door fabrication proceeded on time as the other bases learned and made modifications to their fabrication system.[35]

How did a door this massive open? Four large hydraulic jacks each raised a piece of railroad track into contact with the wheel truck assemblies that hung from the door, lifting the door 2.75 inches off the doorsill, while at the same time aligning the short sections of railroad track with the main door track. The door was then rolled back, using a hydraulic winch and wire rope, a distance of 30 feet in 17 to 21 seconds. The original design used impulse actuators to start the door moving, but these were found to be unnecessary and were not used on the operational sites.

PHASE IIA

Phase IIA lasted 3 to 10 months and integrated into the Phase II schedule. Phase IIA involved installation of work platforms; hydraulic and pneumatic operations systems; blast valves, elevators, and air conditioning; as well as acoustical lining of the launch duct and exhaust ducts. Completion of the propellant transfer system was also part of Phase IIA. The completion of Phase IIA work was supplemental to Phase II and proceeded during Phase III work as well.

Figure 5.15. The silo closure door is complete and awaiting final installation once the silo headworks are completed. The square objects around the launch duct and exhaust ducts are the bearing plates for the silo closure door. *Courtesy of Don Rawlings.*

Two major modifications were made during the Phase IIA installation of equipment. The first involved the launch duct work platforms located at Levels 1, 2, 3, 4, 5, and 7. The platforms consisted of eight segments on all levels except Level 5, which had four segments. The platform segments were heavier than the original design calculations, and the rotary actuators for raising and lowering the platform segments were unable to perform properly. The solution was the addition of a pneumatic booster cylinder mounted on the underside of the platform which assisted the rotary actuator through the first 45 degrees of work platform movement. These were used for all platforms except those on Level 7 and four of those on Level 2. The second modification was to the silo blast valve operating system. The original design of the blast valves provided marginal operation and reliability. Further complicating this issue was the fact that the silo blast valves and work platforms worked off of the same hydraulic system.[36]

PHASE III

Phase III, the installation of ground operating equipment, the missile checkout equipment, and all cabling and missile installation was conducted with Martin Marietta Company as the prime contractor to the Air Force, during and after the end of the Corps of Engineers' efforts. Phase III took 8 to 16 months and overlapped with Phases II and IIA.

Phase III began with joint occupancy of sites with the Phase II and Phase IIA contractors and Martin Marietta Company. Initially joint occupancy was a smooth operation, as most of the Martin Marietta Company work was focused in the launch control center, well away from the work being done on the surface and at the launch duct and silo equipment areas. As the work progressed, scheduling several contractors to work simultaneously in the confined spaces of the launch complex led to temporary delays. Further complicating the issue was the fact that the Fluor Corporation was testing and validating systems prior to site turnover to Martin Marietta Company.

Phase III included installation of the thrust mount shock isolation system; installing all the control cabling to and from the silo and the launch control center; installation of all of the missile monitoring and support equipment; and finally the installation of the missile and check out of the missile and propellant transfer system.

A task as formidable as the construction of the 18 launch complexes surrounding Tucson is hard to comprehend. Final reports make the process appear straightforward and do not convey the day-to-day problems that contractors had to contend with. The experiences of Christman, program manager for the Fluor Corporation of Tucson, sheds some light on a variety of management considerations.

Literally a 12-inch stack of bid documents was necessary. Short turnaround on the bid, coupled with frequent changes in bidding strategy and subcontractor bid values right up to the opening of the bids, made for a hectic time in 1961. Fluor Corporation had worked on large oil and gas industry projects around the world and was involved in construction of Atlas missile bases in Washington State as well as Atomic Energy Commission facilities in Idaho. Fluor was prepared for the lump-sum contract, the government's right to call for two- or three-shift operation with a six- or seven-day work week if the schedule started to slip. What they were not prepared for was the massive amount of design changes as the program evolved. By the end of their work, 750 multiple-part modifications were implemented and well over 25,000 separate action items resulted from those directed modifications. According to Christman, negotiations for the values (time and money) did not keep pace with the progress of the work. The result was that sites got behind the master schedules. On top of these changes was the need to schedule all 18 sites with government-furnished equipment deliveries that, while for the most part were on time, nonetheless caused delays.

Col Clayton A. Rust and LtCol A. P. Richmond were Christman's counterparts at the Army Corps of Engineers and SATAF field office in Tucson in the beginning of the work. With offices housed in adjacent barracks on Davis-Monthan AFB, Fluor Corporation, the Corps of Engineers, and SATAF interaction was professional during the work week and that of friends during the weekends. Of the many goverment projects that Christman worked on with Fluor Corporation, the Titan II Phase II-IIA work was perhaps the most challenging and satisfying.

Communications was obviously a critical item during construction, yet none of the sites had telephone lines, only high-frequency radio installed by the prime contractor. Travel distance to all 18 sites was a total of 420 miles from Davis-Monthan AFB, with Tucson being the hub of the wheel. Keeping in mind that these roads were not the superhighways of today, one did not just hop in a car and visit a site! The Motorola Corporation designed, supplied, and maintained not only the site radio equipment but also the repeater station on Mt. Lemmon that facilitated contact with all remote sites. One late Saturday evening Christman received an urgent call on his radio. Due to weather conditions at the time, a long-haul lettuce trucker was calling for directions on his frequency from Harlingen, Texas! The radio system did work remarkably well and was an expedient tool in emergency situations. Prior to turnover of each complex at the end of the Phase IIA work, all hydraulic systems had to be flushed and acid washed. All the hydraulic lines were jumpered together, and the process had begun at one site when a hose broke and sprayed acid and acid fumes onto a worker in the lower level of the silo equipment area. Instant radio contact to Christman allowed him to facilitate the dispatch of an Air Force helicopter to evacuate the worker who was treated quickly, suffering only minor burns.

Lack of water in the desert conditions of the Tucson construction was a constant problem. Iced drinking water, water for washing, backfilling, concrete—all required a constant supply of a commodity already in short supply in the desert. Few sites if any had operational wells during the Phase II-IIA construction. Iced water was a contract in itself with ice trucked to each site and transferred into a tank that was the potable water source. Winter wasn't all that bad, but in the summer this was a high-priority labor agreement item not to be trifled with. San Xavier Sand and Gravel, a locally owned concrete and earth material supplier, had the master supply contract for ready-mix concrete work. They had portable batch plants located at centrally located sites and access to local ranchers' wells. They would lay 6-inch irrigation pipes along easements and access roads from the wells up to four miles from a site to provide water for the batch plants. Substantial concrete remained to be poured in Phase II-IIA as well as backfill from the Phase I reference elevation that had to be compacted at a specified moisture content.

Labor relations on a project with a complex web of subcontractors and multiple levels of union representation was another critical area that could make or break this large-scale construction effort. Easily overlooked but highly important items such as the basic work week, work jurisdiction, overtime conditions, travel time to and from sites, manpower supply, termination and no-strike clauses, safety, and security were all issues that required constant attention. Christman was fortunate to have Tom Richardson, who was the labor relations manager for the project. Richardson, a former union business agent, had been a union worker on several Atlas missile sites in Washington State. Richardson did have his hands full, since the scope of the project meant that most of the unions had to put out a "Call for Men" nationwide. The warm winters, long hours of work, and length of the project attracted many qualified craftsmen as well as those not so qualified. Richardson had to deal simultaneously with all of the trade union representatives. In especially short supply were boilermakers, electricians, and pipefitter-welders. As an experienced iron-worker, Richardson, with an even temperament and the ability to listen to both sides of arbitration issues objectively, was key to the success of keeping labor-relation problems to a minimum.[37]

395th Strategic Missile Squadron, Vandenberg AFB, California

Vandenberg AFB had three complete Training Facilities for Titan II, known collectively as TF-II, Complex B, C, and D. They were more commonly referred to as Launch Complex 395-B, 395-C, and 395-D, respectively. The original Titan I program Silo Launch Test Facility (SLTF) had been considered for rebuilding as a modified Titan II launch complex but was instead modified slightly, redesignated as the QMT, and used a Titan II propellant-handling trainer, substituting water for actual missile propellants. Ralph M. Parsons Company was the architect/engineer for the design/build work on Titan II complexes, while M. M. Sundt was the prime contractor. Martin Marietta Company was the prime contractor for operating equipment and missile installation.

Just as had been the case with the Titan I facilities at Vandenberg AFB, the high cost of building three Titan II launch complexes was of concern to the Air Force. The issue boiled down to the ability to conduct Category I flight operations which were contractor launches using contractor procedures; Category II flight operations which were contractor launches using procedures to be used in the operational bases; and Category III flight operations which were Air Force operations using procedures validated during Category II testing, in a timely manner. The Strategic Air Command (SAC) and Ballistic Systems Division (BSD) both had strong arguments for their respective needs: SAC needed to be able to conduct operational readiness training launches, and BSD had to verify the capabilities of the missile in its launch and readiness environment.

Originally the modified SLTF was designated as a Category II test facility, while Launch Complexes 395-B, C, and D were to be used by SAC. In February 1961 BSD realized that by the time the SLTF could be modified for use with Titan II, Launch Complex 395-C would be nearly ready to support flight operations. The decision was therefore made to abandon SLTF conversion and utilize Launch Complex 395-C for the Category II testing. BSD suggested that a fourth launcher be built to maintain the three launch complex needs of SAC. A search for possible sites on Vandenberg AFB property came up empty-handed, and the nearby Navy facilities at Point Arguello were not available. The problem was resolved when the Air Force rejected the proposal for Launch Complex 395-E. At the same time, it agreed with the use of Launch Complex 395-C for the Category II training.

The final plan was to use Launch Complex 395-B initially to verify, certify, and demonstrate all aspects of the technical data, manuals, and silo maintenance procedures. This held true for the receipt-to-launch of a missile. This effectively made it unavailable for launch operations until this task had been completed. Launch Complex 395-C would now be the workhorse complex for the research and development Category I flight test program, while Launch Complex 395-D would be available for the Category II flight test program. Category III flights would utilize all three complexes once the research and development work had been completed. Silo construction at Vandenberg AFB began simultaneously with the three operational bases but progressed much faster, reaching the point where missiles could be emplaced months ahead of the other bases (see Table 5.5).[38]

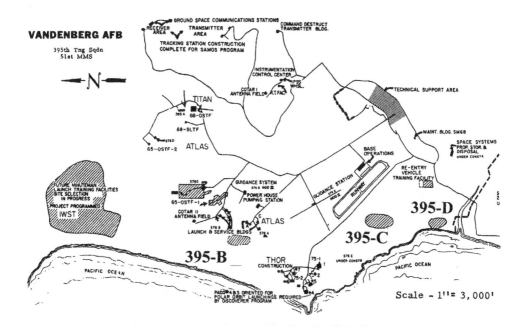

Figure 5.16. A 1963 vintage map of the Vandenberg AFB, California, Titan II ICBM launch facilities. Nine research and development flights of Titan II, as well as 49 operational missile test launches, were conducted at these sites from 1963 to 1976. *Author's Collection.*

TABLE 5.5

TITAN II TRAINING FACILITY CONSTRUCTION
TIME LINE FOR VANDENBERG AFB

EVENT	395-B	395-C	395-D
Start of Phase I Construction	22 Dec 1960	29 Nov 60	5 Dec 1960
End of Phase II, IIA Construction	9 Jul 1962	14 Jul 62	18 Jun 62
First Missile Installation	2 Nov 62	5 Oct 62	29 Nov 62
Complex Turnkey to SAC	29 Feb 64	18 May 64	3 Apr 64

Source: Titan II Master Schedule, 31 March 1964.

390th Strategic Missile Wing, Davis-Monthan AFB, Arizona

The first Titan II program activity related to the 390th Strategic Missile Wing (SMW), Davis-Monthan AFB, Arizona, took place on 20 April 1960 as Headquarters SAC selected Davis-Monthan AFB as the site for the first Titan II ICBM wing assigned to the Fifteenth Air Force. The Davis-Monthan Area, U.S. Army Corps of Engineers, Ballistic Missile Construction Office was established on 3 October 1960, with Col Clayton A. Rust, area engineer. Two months later, 7 December 1960, Col Strother B. Hardwick, commander of the Titan II Site Activation Task Force, dug the ceremonial first shovel of dirt, at Titan II ICBM Launch Complex 570-2.[39] Phase I construction started two days later. Table 5.6 lists the milestone dates for all phases of construction for each launch complex as well as the date the launch complex first went on strategic alert.[40]

On 1 January 1962, the 390th SMW was activated, along with three subordinate units, the 570th Strategic Missile Squadron (SMS), the 390th Missile Maintenance Squadron

TABLE 5.6
390TH SMW LAUNCH COMPLEX
CONSTRUCTION/ACTIVATION/DEACTIVATION DATES

Site #	Complex #	PHASE I			PHASE II		
		Start	Complete	Months	Start	Complete	Months
9	571-8	9 Dec 60	22 Jul 61	7	7 Jul 61	22 Jun 62	12
10	571-9	9 Dec 60	14 Jul 61	7	9 Jul 61	16 Jun 62	11
11	570-2	9 Dec 60	8 Jul 61	7	28 Jun 61	23 May 62	11
15	570-6	21 Dec 60	22 Jul 61	7	1 Aug 61	5 Jul 62	11
12	570-3	22 Dec 60	6 Oct 61	9	10 Oct 61	29 Sep 62	12
17	570-8	22 Dec 60	24 Sep 61	9	25 Sep 61	17 Sep 62	12
14	570-5	27 Dec 60	26 Jul 61	7	3 Aug 61	13 Jul 62	11
16	570-7	28 Dec 60	17 Aug 61	8	18 Aug 61	30 Jul 62	11
13	570-4	5 Jan 61	9 Aug 61	7	14 Aug 61	2 Aug 62	12
18	570-9	9 Jan 61	10 Sep 61	8	10 Sep 61	18 Aug 62	11
1	570-1	10 Jan 61	30 Aug 61	8	31 Aug 61	10 Aug 62	11
5	571-4	24 Jan 61	16 Dec 61	11	21 Dec 61	11 Dec 62	12
6	571-5	2 Feb 61	21 Dec 61	11	27 Nov 61	31 Oct 62	11
8	571-7	2 Feb 61	5 Nov 61	9	9 Nov 61	26 Oct 62	12
4	571-3	3 Feb 61	3 Dec 61	10	4 Dec 61	5 Dec 62	12
3	571-2	8 Feb 61	7 Nov 61	9	13 Nov 61	13 Nov 62	12
2	571-1	10 Feb 61	11 Dec 61	10	21 Dec 61	26 Nov 62	11
7	571-6	10 Feb 61	22 Oct 61	8	30 Nov 61	20 Oct 62	11

Site number refers to the Army Corps of Engineers designator. Complex number refers to the SAC operational launcher designator.

(MIMS), and the 390th Headquarters Squadron, all assigned to the 12th Strategic Aerospace Division, Fifteenth Air Force. Col Raymond D. Sampson assumed command of the 390th SMW. An estimated 1,200 officers and airmen were transferred to Davis-Monthan AFB with the bulk of the personnel not arriving until mid-1963, shortly before the Titan II ICBM complexes were to become available to receive missiles and become operational. The 571st SMS was activated 1 May 1962, completing the operational units assigned to the 390th SMW.[41]

In August 1962 a Senate Preparedness Subcommittee team of investigators conducted an inquiry in Tucson, Arizona, into charges of waste, safety problems, and contractors with unpaid claims in the construction of the ICBM launch complexes across the country. The charges of waste came from an apparent misconception in the media between construction costs and operational system costs. Cost of construction had been publicized as $80 million. Operational costs were reported as $410 million. Operational costs included not only the cost of construction, which was now nearly $113 million, but also the cost of missiles installed, $56.4 million; ground equipment, $55.8 million; installation and checkout of

			PHASE IIA			
Start	Complete	Months	Months To Alert	On Alert	First Off Alert	Years On Alert
1 Feb 62	6 Jul 62	5	28	15 Apr 63	8 Feb 84	21
28 Jan 62	1 Jul 62	5	28	6 Apr 63	4 Jan 84	21
7 Jan 62	23 May 62	4	28	31 Mar 63	4 Apr 83	20
5 Feb 62	8 Aug 62	6	28	22 Apr 63	27 Feb 84	21
16 Jun 62	5 Oct 62	4	29	22 May 63	27 Mar 84	21
8 Jun 62	28 Sep 62	4	30	25 Jun 63	27 Jun 83	20
11 Feb 62	9 Aug 62	6	28	3 May 63	21 May 84	21
5 Mar 62	4 Sep 62	6	29	24 May 63	3 Jan 83	20
16 Feb 62	21 Aug 62	6	29	1 Jun 63	7 May 84	21
4 Jun 62	14 Sep 62	3	29	14 Jun 63	2 Jul 82	19
7 Jun 62	19 Sep 62	3	29	8 Jun 63	12 May 83	20
16 Aug 62	14 Dec 62	4	33	24 Oct 63	10 Aug 83	20
11 Jul 62	31 Oct 62	4	30	24 Jul 63	3 Mar 83	20
29 Jun 62	25 Oct 62	4	29	15 Jul 63	12 Nov 82	19
9 Aug 62	6 Dec 62	4	32	10 Oct 63	4 Oct 83	20
2 Jul 62	13 Nov 62	4	31	17 Sep 63	21 Nov 83	20
29 Jul 62	20 Nov 62	4	29	18 Jul 63	2 Aug 83	20
23 Jun 62	17 Oct 62	4	29	2 Jul 63	29 Sep 82	19

Note: The respective Army Corps of Engineers wing history was used for Phase I, II, and Phase IIA dates. Phase III dates are not available. The 1965 Ballistics System Division Management Plan was use as the source for when SAC accepted each launch complex and placed it on alert.

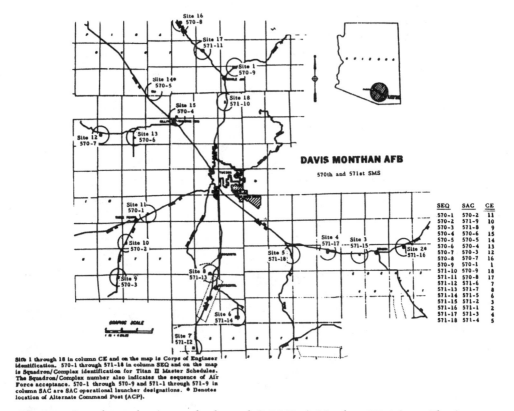

SEQ	SAC	CE
570-1	570-2	11
570-2	571-9	10
570-3	571-8	9
570-4	570-6	15
570-5	570-5	14
570-6	570-4	13
570-7	570-3	12
570-8	570-7	16
570-9	570-1	1
571-10	570-9	18
571-11	570-8	17
571-12	571-6	7
571-13	571-7	8
571-14	571-5	6
571-15	571-2	3
571-16	571-1	2
571-17	571-3	4
571-18	571-4	5

Site 1 through 18 in column CE and on the map is Corps of Engineer
identification. 570-1 through 571-18 in column SEQ and on the map
is Squadron/Complex identification for Titan II Master Schedules.
The Squadron/Complex number also indicates the sequence of Air
Force acceptance. 570-1 through 570-9 and 571-1 through 571-9 in
column SAC are SAC operational launcher designations. ● Denotes
location of Alternate Command Post (ACP).

Figure 5.17. Launch complex site map for the 390th SMW, Davis-Monthan AFB, Arizona. The sites were at least 7 nautical miles apart. This distance was a balance of blast protection and economy of logistics. *Author's Collection.*

missile equipment, $70.6 million; training, $19.6 million; industrial facilities, $9.6 million; updating the missiles, $18.4 million; a five-year operating cost of $64.5 million; and other unit costs of $2.4 million. The subcommittee report indicated that the increase in construction costs paralleled that at the other Titan II wings. The committee concluded that the cost overruns were an inevitable result of the concurrency concept where both the launch facilities and the missile itself were designed and built simultaneously. As for the safety concerns, which were primarily a question of adequate ventilation in the 145-foot-deep silos, the committee found that the problems had been remedied.[42]

The 390th SMW began to function as a Titan II wing on 27 November 1962 with the arrival of its first Titan II missile, B-5 (61-2759). Construction at the wing's launch complexes was ahead of schedule, and B-5 was readied for transportation to Launch Complex 570-2. On 7 December 1962, the convoy left Davis-Monthan AFB but only got as far as Valencia Road when it had to stop for over an hour due to a hydraulic failure in one of the tow trucks. Early in the morning of 8 December 1962, installation of missile B-5 began with 390th MIMS personnel directed by Martin Marietta Company personnel. Installation was hampered only by a malfunctioning switch on the Coles crane, a problem that was to plague missile maintenance personnel for many years. Two years and one day after the first shovel of dirt had been turned at Launch Complex 570-2, the first Titan II operational missile had been installed in its silo.[43] With the heightened cold war tensions due to the Cuban Missile Crisis still fresh, the beginning of the Titan II system deployment illustrated to all concerned that the United States was committed to deterring nuclear war. January 1963 saw the delivery of five more missiles as the 390th SMW was designated by SAC as the primary recipient of completed missiles since it was the base that was the furthest along in construction.

On 11 February 1963, a TOP SECRET message was received at Headquarters SAC from the Joint Chiefs of Staff. SAC was directed to prepare a study that would result in the launching of several operational ICBMs from bases across the United States. A preliminary study was developed and presented to the secretary of defense on 11 March 1963, concluding that peacetime launches from operational sites, properly selected, was indeed feasible. Five days later, Headquarters SAC was directed to develop a plan for a peacetime launch from the 390th SMW which would impose the minimum risk to the public. The plan, SACOP SM-68-2, was completed and published by Headquarters SAC and given the code name "On Target." By 15 April 1963, the plan was complete. The Titan II would be launched from one of the westernmost launch complexes (the exact launch complex remains classified). There was considerable concern about the possibility of demonstrators protesting the launch since the flight path would extend over land for approximately 285 nautical miles, in both the United States and Mexico, before arching out over the Pacific Ocean. Also noted in the plan was the fact that the 390th SMW launch would be the third in a series of four, with the other three being a Titan I launch from Beale AFB, California; an Atlas F launch from Plattsburg AFB, New York; and a Minuteman I launch from Malstrom AFB, Montana. None of the launches took place because of complaints from the state governments concerned with overflight of cities and towns. The Mexican government was likewise reluctant to have a Titan II overflight of even their sparsely populated territory under the flight path.[44]

On 31 March 1963, Launch Complex 570-2 became the first Titan II ICBM on strategic alert. This event was a defining moment for the concurrency concept championed by General Schriever many years before. The first test launch out of a full-scale silo had taken place just 45 days earlier on 16 February 1963 at Vandenberg AFB. When construction started at Launch Complex 570-2, on 9 December 1960, the first test launch of Titan II was still 15 months in the future. Yet here, 27 1/2 months later, the missile was ready and on operational alert. It would be another 9 months until the entire Titan II force was on alert.[45]

For the most part the impact of launch complex construction was minimal in the daily life of the citizens of Tucson. The majority of the launch facilities were well out of town. However, for one 12-hour period in May 1963, the reality of the Titan II program's possible effect on the average citizen became all too clear. In the afternoon of 6 May 1963 an Intermountain Express Corporation tanker truck, en route to Launch Complex 570-7 with 3,500 gallons of nitrogen tetroxide, flipped over on its side at the intersection of Miracle Mile Strip and Oracle Road, a busy intersection in Tucson. No one was injured, but the tanker lay in a ditch by the side of the road for nearly 11 hours. Residents and business owners were advised to be ready to evacuate at a moment's notice should the tanker begin to leak. Early the next morning, two cranes from Davis-Monthan lifted the tank trailer onto a flatbed truck. The tanker was taken to the complex and the propellant downloaded into the waiting missile without further mishap.

As might be expected, many Tucson citizens were upset at the possibility of a release of toxic chemicals within the city limits. In a written response to questions posed by the *Tucson Citizen*, BrigGen John L. McCoy at Ballistic Systems Division, Norton Air Force Base, California, pointed out that while the accident was unfortunate and steps would be implemented to ensure police escort of propellant trucks traveling through the city in the future, such vehicles had been hauling nitrogen tetroxide for nearly 30 years without serious incident. Specially designed tankers with 4-inch-thick walls made the operation reasonably safe.[46] Ironically, sulfuric acid–laden trucks followed the same route to the mines outside of Tucson without undue fanfare from the press.

On 13 June 1963, the first squadron of Titan II missiles, the 570th SMS, became operational. At the time of acceptance by SAC, each site was considered to be on alert. There were two Emergency War Order categories in 1963. The first was a ready posture at all times, referred to as an alert posture, and throughout this book as on alert. The other was the Emergency Combat Capability, which was used for facilities that could be quickly placed on alert but are not at the time technically on alert.[47]

The 570th SMS was a composite of seven launch complexes from the 570th (570-1, 570-2, 570-3, 570-4, 570-5, 570-6, 570-7) and two from the 571st SMS (571-8 and 571-9) that were temporarily assigned to the 570th SMS. By mid-July, an additional four complexes, 570-9, 571-6, 571-7, and 571-5, were placed on alert.

Three weeks later, all but two complexes, 570-1 and 571-5, were declared off alert due to oxidizer leaks. McConnell AFB and Little Rock AFB had experienced this problem several weeks in advance of the 390th SMW due to the higher humidity earlier in the summer season. The problem varied from minute, barely detectable leaks in weld areas on the oxidizer tanks, to three missiles that had leaks of much higher magnitude at the oxidizer valve fittings. SAC advised all three Titan II wings that no additional launch complexes would be accepted until the problem was resolved.[48]

The reason for the leaks was clear. Along with the installation of the first missile in December 1962 came the first prolonged exposure of a fully loaded missile to the relatively high humidity of the launch duct. Minute oxidizer leaks of nitrogen tetroxide led to the formation of highly corrosive nitric acid at some of the missile tank welds and valve connections. Since the research and development program at Cape Canaveral did not use silos, and the missiles launched from Vandenberg AFB had not sat in the silo for prolonged periods, this was simply a problem that finally had the right conditions to reveal itself. The arrival of warmer weather and higher humidity at all three missile wings exacerbated the problem to the point where a solution had to be found.

In August 1963, representatives from the BSD and Martin Marietta Company agreed on a program to fix the leaks. Named "Operation Wrap Up," missiles were to be repaired at the operational base if possible; if not, they would be shipped to Denver (see Chapter 3 for a more detailed description of Operation Wrap Up). On 4 September 1963, SAC announced that the problem was resolved and launch complex acceptance resumed. By the time the Titan II system was declared operational at the end of 1963, 78 missiles had been installed fleetwide only to have 24 removed for repair. Missiles built after October 1963, when manufacturing modifications were made, did not leak due to problems with tank welds.[49]

With the completion of Operation Wrap Up, the 570th SMS and 571st SMS were turned over to SAC on 14 October 1963 and 29 November 1963, respectively. While all of the sites had been turned over to SAC earlier, the actual squadron transfer was delayed by Operation Wrap Up. While all 18 launch complexes were on alert as of 30 November 1963, Headquarters SAC did not declare either squadron combat ready until 19 and 31 December for the 570th SMS and 571st SMS, respectively.[50]

381st Strategic Missile Wing, McConnell AFB, Kansas

The first Titan II ICBM program activity related to the 381st SMW, McConnell AFB, Kansas, took place on 20 April 1960 as Headquarters SAC selected McConnell AFB as the site for the second Titan II ICBM wing. On 7 October 1960, the McConnell Area Office, U.S. Army Corps of Engineers Ballistic Missile Construction Office was established with Col Lawrence M. Hoover, area engineer.

Phase I construction began at Launch Complexes 532-6 and 532-8 on 19 December 1960. All 18 sites were located in sedimentary rock with ground water depths ranging from 7 to 62 feet. Only one site had ground water in sufficient quantity to present more than minor excavation problems. Two types of construction were used at the 381st SMW sites. The western, or "dirt" sites, were constructed using the open-cut excavation technique described in Chapter 5 since the soil and sedimentary rock were relatively easily excavated. Unlike similar sites at the 390th and 308th SMWs, the 381st SMW "rock" sites were excavated using a modified open-cut excavation procedure which provided greater space for maneuvering heavy equipment and permitted greater overall efficiency during Phase I and II construction. The exception to this was Site 13 (532-4): there 12 six-inch diameter wells were bored and water was pumped at 300 gallons per minute initially, subsiding to 120 gallons per minute after several weeks. Excavation was routine until a material similar to quicksand was found at a depth of 6 feet. The contractor switched to a dredging operation, and the open-cut operation was successfully completed.[51] Table 5.7 lists the milestone dates for all phases of construction for each launch complex as well as the date each complex first went on alert.[52]

On 29 November 1961, the 381st SMW was activated and given bestowal of the lineage and honors of the 381st Bombardment Group. On 1 March 1962, the 381st SMW was assigned to the Second Air Force, 42nd Air Division, SAC, to be organized at McConnell AFB, Kansas, with Col George W. von Arb as commanding officer. The 532nd Bombardment Squadron was reactivated and redesignated as the 532nd SMS and assigned to the 381st SMW on the same day. On 1 August 1962 the 533d Bombardment Squadron was reactivated, redesignated as the 533nd SMS, and assigned to the 381st SMW as was the 381st MIMS.[53]

By November 1962, seventeen 381st SMW missile combat crews were in training (a wing would normally have at least 3 crews per complex when fully operational, a minimum of

TABLE 5.7
381ST SMW LAUNCH COMPLEX
CONSTRUCTION/ACTIVATION/DEACTIVATION DATES

		PHASE I			PHASE II		
Site #	Complex #	Start	Complete	Months	Start	Complete	Months
15	532-6	19 Dec 60	3 Aug 61	8	3 Jul 61	19 Jul 62	12
14	532-5	22 Dec 60	6 Aug 61	7	7 Jul 61	28 Jul 62	13
17	532-8	19 Dec 60	6 Aug 61	8	11 Aug 61	8 Aug 62	12
18	532-9	31 Dec 60	22 Aug 61	8	22 Aug 61	16 Aug 62	12
16	532-7	14 Jan 61	5 Sep 61	8	9 Sep 61	31 Aug 62	12
12	532-3	17 Jan 61	12 Sep 61	8	12 Sep 61	7 Sep 62	12
11	532-2	8 Feb 61	25 Sep 61	8	25 Sep 61	4 Oct 62	12
13	532-4	22 Jan 61	5 Oct 61	8	5 Oct 61	1 Oct 62	12
10	532-1	10 Feb 61	10 Oct 61	8	10 Oct 61	11 Oct 62	12
9	533-9	27 Jan 61	23 Oct 61	9	23 Oct 61	22 Oct 62	12
8	533-8	30 Dec 60	30 Oct 61	10	30 Oct 61	8 Nov 62	12
7	533-7	19 Jan 61	15 Nov 61	10	9 Nov 61	4 Nov 62	12
6	533-6	3 Jan 61	27 Nov 61	11	29 Nov 61	12 Nov 62	11
5	533-5	26 Jan 61	23 Dec 61	11	11 Dec 61	1 Dec 62	12
4	533-4	27 Jan 61	29 Dec 61	11	2 Jan 62	28 Dec 62	12
3	533-3	27 Jan 61	16 Jan 62	12	15 Jan 62	4 Jan 63	12
2	533-2	27 Jan 61	1 Feb 62	12	2 Feb 62	11 Jan 63	11
1	533-1	27 Jan 61	9 Feb 62	12	15 Feb 62	24 Jan 63	11

Note: The respective Army Corps of Engineers wing history was use for Phase I, II, and Phase IIA dates. Phase III dates are not available. The 1965 Ballistics System Division Management Plan was use as the source for when SAC accepted each launch complex and placed it on alert..

54 crews). Six crews had completed the Operational Readiness Training (ORT) program at Vandenberg AFB and were awaiting final upgrade training and certification. Eleven crews were still at Vandenberg AFB. Maintenance personnel were involved in training both at Cape Canaveral, where 10 members of the maintenance squadron were working side-by-side with Martin Marietta Company personnel in the test flight program, and on base, attending courses conducted by Martin Marietta Company representatives. On 30 November 1962 the final stages of an agreement between SAC, the Site Activation Task Force, and Martin Company were worked out where Martin Marietta Company would use SAC personnel for operations and maintenance at the nearly completed sites, enhancing training capabilities and speeding up the takeover process.[54]

On 11 December 1962, a proposed change was announced for the activation process at the 381st SMW because of missile engine and airframe production delays. The BSD proposed that the 390th SMW remain on schedule using the available missile engines and airframes; the program at the 308th SMW, Little Rock AFB, Arkansas, remain as scheduled;

			Months To Alert	On Alert	First Off Alert	Years On Alert
Start	**Complete**	**Months**				
5 Dec 61	3 Jul 62	7	34	31 Oct 63	16 Jan 86	22
15 Jan 62	15 Jul 62	6	35	10 Nov 63	10 Aug 84	21
8 Jan 62	26 Jul 62	7	34	23 Oct 63	29 Jan 86	22
4 Dec 61	30 Jul 62	8	33	15 Oct 63	25 Mar 86	22
20 Dec 61	21 Aug 62	8	34	14 Nov 63	29 Oct 84	21
11 Jan 62	22 Aug 62	7	35	4 Dec 63	17 Sep 84	21
7 Feb 62	10 Sep 62	7	34	29 Nov 63	16 Nov 85	22
12 Feb 62	15 Sep 62	7	34	13 Nov 63	20 Feb 86	22
24 Feb 62	4 Oct 62	7	33	21 Nov 63	8 Jan 85	21
26 Feb 62	17 Oct 62	8	33	21 Oct 63	27 May 86	23
6 Mar 62	24 Oct 62	8	30	5 Jul 63	2 Jul 84	21
16 Mar 62	1 Nov 62	8	30	20 Jul 63	24 Aug 78	15
23 May 62	31 Oct 62	5	33	14 Oct 63	29 Apr 85	22
6 Jun 62	21 Nov 62	6	33	30 Oct 63	14 Mar 85	21
15 Jun 62	30 Nov 62	6	33	4 Nov 63	29 Jul 85	22
25 Jun 62	11 Dec 62	6	33	7 Nov 63	3 Oct 85	22
25 Jun 62	27 Dec 62	6	34	16 Nov 63	5 Jun 85	22
27 Jun 62	11 Jan 63	7	34	27 Nov 63	22 Aug 85	22

while the 381st SMW would be slipped behind the 308th SMW's schedule. This changed the scheduled operational date for the 533d SMS from June 1963 to October 1963 and would probably slip complete operational status for the 381st SMW to January 1964. As a result of this schedule change, the 381st SMW requested that one complex, with a missile installed, be made available to permit final upgrade training and missile combat crew certification to take place despite the delayed arrival of operational missiles. A second complex, with a missile simulator and facilities control simulator, would permit continuation of the pre-ORT that would be seriously set back due to the proposed delays. The Installation and Checkout phase of the training program was also hindered by the lack of available missiles installed in complexes.[55]

The first missile arrived at the 381st SMW on 4 January 1963 but was referred to as a "tool" for use in checking out the launch complexes as they were completed. This missile and a second one that arrived on 9 January 1963 were installed in Launch Complexes 532-6 and 532-5, respectively. Missile numbers are not available.[56] The first attempt to run an

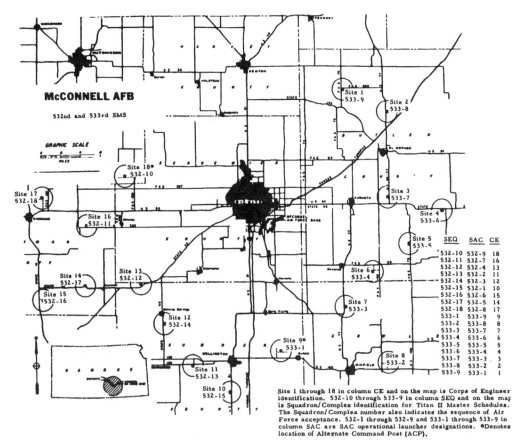

Figure 5.18. Launch complex map for the 381st SMW, McConnell AFB, Kansas. *Author's Collection.*

activation exercise procedure with the new missiles resulted in a temporary delay when the maintenance teams found that the test missiles had 250 differences from the technical orders written for use at the complexes! These missiles had not yet been modified to the operational missile standard. Local procedures were written to permit the exercise to continue.[57]

In March 1963, selected missile combat crews that had passed ORT but were not yet certified as combat ready were assigned to the Technical Acceptance and Demonstration Program. With one crew per complex, they were certified through a three-week school provided by Martin Marietta Company and then became an integral part of the contractor effort in turning over that particular complex to the wing as scheduled. Once the complex was turned over to SAC, these crews would be assigned to that complex as well as serve as the initial cadre of instructor and standboard crews. In early April 1963, an agreement was signed with Martin Marietta Company to modify Launch Complex 532-3 with missile and facilities simulators to expedite missile combat crew proficiency training. The facility was to remain in this converted form until the launch complex was scheduled for activation in September 1963. The first class of six crews completed training at the new facility on 23 April 1963.[58]

The first operational missile, B-28 (62-009), arrived at Launch Complex 533-8 on 13 April 1963.[59] Acceptance and checkout proceeded smoothly and Launch Complex 533-8 was placed on alert 5 July 1963. As was the case at the other operational wings, oxidizer leaks were soon experienced in the first missiles that went on alert. The first such missile from the 381st SMW was removed on 14 June 1963 from Launch Complex 533-7 a week after being placed on alert. It was repaired at the base and quickly returned to alert status. The missile at Launch Complex 533-8 soon followed and again was repaired on base. Two more missiles were removed in July and shipped to Denver as part of Operation Wrap Up. Due to delays in delivery of missiles and the time to sequence missiles through Operation Wrap Up, the actual turnover date for the 533d SMS was 4 November 1963; for the 532d SMS it was 4 December 1963. The 381st SMW was fully activated, and all 18 missiles were on alert by 31 December 1963.[60]

308th Strategic Missile Wing, Little Rock AFB, Arkansas

The Titan II era in Arkansas began on 23 June 1960 when Headquarters, SAC, designated the general area around Jacksonville, Arkansas, as the site for the third Titan II ICBM wing. On 5 October 1960, the Little Rock Area Office of the U.S. Army Corps of Engineers Ballistic Missile Construction Office was established at Little Rock AFB, Jacksonville, Arkansas, with Col R. E. Snetzer, area engineer. Four months earlier the Army Corps of Engineers and an Air Force siting team had surveyed a general area around Jacksonville, selecting 23 possible sites for the 18 silo emplacements. A more detailed investigation, including subsurface and topographic surveys, as well as surface water studies, took place in June 1960. By the end of August 1960 final decisions on the 18 sites had been made. In December 1961 SAC assigned the 308th SMW to the 825th Strategic Aerospace Division, Second Air Force. Construction began on 3 January 1961 at Launch Complex 373-4. Table 5.8 lists the sites with major construction milestone dates as well as the date each site was first placed on alert.[61]

Two sites presented significant problems for the Corps of Engineers. At Launch Complex 374-1, 66 feet of soil over bedrock required excavation to start in earth for both the silo and launch control center shafts, followed by blasting once bedrock was reached. Two minor earth slides in the upper portion of the shaft required special excavation and shoring methods. Launch Complex 374-2 experienced a major rock slide after excavation had penetrated 30 feet below the collar beam level.

Midland Contractor, the Phase I contractor, and the Corps of Engineers developed a new engineering technique for use in the excavation of the silo shafts. To precisely control the blasting for shaft excavation in the hard rock sites and prevent excessive overbreaking, 9-inch-diameter holes on 2-foot centers were drilled at the circumference of the shaft, anywhere from 60 to 100 holes, depending on the site. Two 2.75-inch holes were drilled between the 9-inch holes. The holes were drilled to the full 117.5-foot depth of the silo excavation. The drill was mounted on a D-9 caterpillar chassis, and four drills, working in pairs on opposite sides of the silo, worked simultaneously, averaging slightly over one hour per hole. Holes for the explosive charges were on 4-foot centers in 4-foot concentric rings radiating

TABLE 5.8

308TH SMW LAUNCH COMPLEX
CONSTRUCTION/ACTIVATION/DEACTIVATION DATES

Site #	Complex #	PHASE I			PHASE II		
		Start	Complete	Months	Start	Complete	Months
4	373-4	3 Jan 61	1 Sep 61	8	2 Sep 61	31 Jul 62	11
5	373-5	3 Jan 61	5 Sep 61	8	5 Sep 61	14 Aug 62	11
8	373-8	3 Jan 61	8 Sep 61	8	9 Sep 61	28 Aug 62	12
7	373-7	5 Jan 61	19 Sep 61	8	20 Sep 61	16 Sep 62	12
3	373-3	7 Jan 61	29 Sep 61	9	2 Oct 61	25 Sep 62	12
2	373-2	9 Jan 61	11 Oct 61	9	16 Oct 61	9 Oct 62	12
18	374-9	16 Jan 61	25 Oct 61	9	27 Oct 61	23 Oct 62	12
1	373-1	23 Jan 61	26 Oct 61	9	13 Nov 61	6 Nov 62	12
16	373-9	23 Jan 61	1 Nov 61	9	2 Dec 61	20 Nov 62	12
6	373-6	19 Jan 61	9 Nov 61	10	4 Dec 61	29 Nov 62	12
14	374-6	30 Jan 61	17 Nov 61	10	21 Dec 61	13 Dec 62	12
15	374-7	30 Jan 61	28 Nov 61	10	1 Jan 62	27 Dec 62	12
17	374-8	31 Jan 61	7 Dec 61	10	8 Jan 62	15 Jan 63	12
12	374-4	6 Feb 61	14 Dec 61	10	20 Jan 62	24 Jan 63	12
11	374-3	6 Feb 61	26 Dec 61	11	20 Jan 62	7 Feb 63	13
13	374-5	15 Feb 61	19 Jan 62	11	8 Feb 62	26 Feb 63	13
9	374-1	9 Feb 61	3 Mar 62	13	1 Mar 62	12 Mar 63	12
10	374-2	14 Feb 61	23 Apr 62	14	23 Apr 62	26 Mar 63	11

outward from the center of the excavation to the silo perimeter. Approximately 182 holes containing 1.33 to 2.66 pounds of dynamite were used.

Unlike the McConnell AFB and Davis-Monthan AFB geology, the sites around Little Rock were characterized, with one exception, with shallow overburden (soil layer above bedrock). This required extensive blasting in order to excavate a "working bench" area at the collar beam level. Spaces were cramped at both the launch control center and silo shaft area, and much of the work ordinarily done at this work bench level was done elsewhere on the site. Due to this narrow work bench area, and the varying height from the working bench level to the collar beam, a special structural steel bridge pier was fabricated at each of the sites to enable operation of the 50-ton crane. This bridge spanned the silo and also served as support for the personnel elevator and fresh air supply tube. It also served to carry the main load of the slip-forms.[62]

On 1 April 1962, the 308th SMW was activated with Col Charles P. Sullivan, acting

Start	Complete	Months	Months To Alert	On Alert	First Off Alert	Years On Alert
16 Feb 62	31 Jul 62	5	28	16 May 63	18 Feb 87	24
5 Feb 62	14 Aug 62	6	29	15 Jun 63	20 Oct 86	23
2 Feb 62	28 Aug 62	7	35	8 Dec 63	6 May 87	23
14 Mar 62	16 Sep 62	6	30	26 Jun 63	3 Apr 86	23
9 Feb 62	25 Sep 62	8	33	19 Oct 63	18 Mar 87	23
21 Mar 62	9 Oct 62	7	35	29 Nov 63	4 May 87	23
1 Apr 62	23 Oct 62	7	33	28 Oct 63	3 Oct 85	22
29 Jan 62	6 Nov 62	9	34	15 Nov 63	5 Jan 87	23
22 May 62	20 Nov 62	6	34	15 Nov 63	6 Feb 86	22
15 May 62	29 Nov 62	7	34	23 Nov 63	20 Jun 85	22
15 May 62	13 Dec 62	7	35	18 Dec 63	25 Jun 86	23
9 Jul 62	27 Dec 62	6	35	18 Dec 63	21 Sep 80	17
12 Jul 62	10 Jan 63	6	35	20 Dec 63	17 Mar 85	21
31 Jul 62	24 Jan 63	6	35	28 Dec 63	27 Aug 86	23
25 Sep 62	7 Feb 63	4	34	19 Dec 63	5 Aug 86	23
26 Sep 62	26 Feb 63	5	34	26 Dec 63	19 May 86	22
4 Oct 62	12 Mar 63	5	34	23 Dec 63	15 Aug 85	22
8 Jun 62	26 Mar 63	10	34	19 Dec 63	15 Sep 86	23

Note: The respective Army Corps of Engineers wing history was use for Phase I, II, and Phase IIA dates. Phase III dates are not available. The 1965 Ballistics System Division Management Plan was use as the source for when SAC accepted each launch complex and placed it on alert.

commander. Two squadrons were assigned, the 373d SMS and the 308th MIMS. On 1 June 1962, the 308th SMW received its first commander, Col Collier H. Davison. On 1 September 1962, the 374th SMS was activated, along with the 308th Headquarters Squadron Section.

TSgt Don Rawlings was assigned to the 308th SMW as a ballistic missile analyst technician (BMAT) in the fall of 1962 after completing electronics school at Sheppard AFB, Texas, the training base for all Titan II crew positions. The instructors and missile combat crew members had to be resourceful during the early days at the 308th SMW while they were waiting for the training equipment to arrive and the launch complexes to be completed. Many an hour was spent honing newly learned BMAT skills in front of cardboard mockups of the control panels with photographs representing the real equipment as they trained on the routine of launch complex operations.

As construction progressed and the launch control centers became fully equipped, missile combat crew members would report for training at approximately 1700 as the

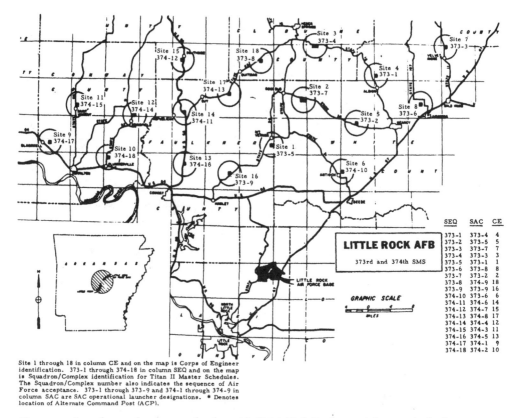

SEQ	SAC	CE
373-1	373-4	4
373-2	373-5	5
373-3	373-7	7
373-4	373-3	3
373-5	373-1	1
373-6	373-8	8
373-7	373-2	2
373-8	374-9	18
373-9	373-9	16
374-10	373-6	6
374-11	374-6	14
374-12	374-7	15
374-13	374-8	17
374-14	374-4	12
374-15	374-3	11
374-16	374-5	13
374-17	374-1	9
374-18	374-2	10

Site 1 through 18 in column CE and on the map is Corps of Engineer identification. 373-1 through 374-18 in column SEQ and on the map is Squadron/Complex identification for Titan II Master Schedules. The Squadron/Complex number also indicates the sequence of Air Force acceptance. 373-1 through 373-9 and 374-1 through 374-9 in column SAC are SAC operational launcher designations. ✦ Denotes location of Alternate Command Post (ACP).

Figure 5.19. Launch complex site map for the 308th SMW, Little Rock AFB, Arkansas. *Author's Collection.*

contractor's workers left for the evening. All-night training sessions would then follow, and finally each crew member began the qualification process. The exams were difficult. For example, the written portion was 150 to 200 questions. If you missed no more than one, you were highly qualified, two wrong and you were qualified; if you missed three, you failed the exam. In order for a crew to become highly qualified or qualified, all crew members had to have the same rating.[63]

On 6 February 1963, the 308th SMW received its first Titan II missile, B-8 (61-2762). The missile was unloaded and taken to the Missile Assembly and Maintenance Shop, where major discrepancies were discovered, including a thrust chamber leak in the Stage I engine and seal failures in the oxidizer pump and gearbox of Stage II. These discrepancies threatened to delay the programmed installation of the missile into Launch Complex 373-4, near Searcy, Arkansas. Fortunately, a second missile, B-20 (61-2774), arrived on 22 February and was installed at Launch Complex 373-4 on 28 February 1963 as scheduled.[64] Three months later, on 16 May 1963, Launch Complex 373-4 became the first 308th SMW Titan II launch complex to be placed on strategic alert with the missile combat crew composed of Maj John R. Rhoads, missile combat crew commander (MCCC); 1Lt James E. Vannoppen, deputy missile combat crew commander (DMCCC); SMSgt Walter Kundis, ballistic missile analyst technician (BMAT); and A2C U. Frank Ainsworth, missile facilities technician (MFT).

Kundis and Ainsworth found out that while it certainly was an honor to be the first crew assigned to alert duty at the 308th SMW, it was also just a little bit nerve racking. Both were in the launch duct with the launch platforms down on Level 2 inspecting the umbilicals attached to Stage II. Suddenly the voice signaling system crackled to life with a tense message from Vannoppen for Kundis to contact him immediately. Kundis did so, and Vannoppen asked if they had moved the umbilicals because the UMBILICAL MONITOR indicator on the PDC-2 Power Sequence chassis had gone out. They checked all of the Level 2 umbilicals again and reported to Vannoppen that everything looked perfectly fine in the launch duct. Kundis asked if Vannoppen had performed a lamp check. Somewhat sheepishly, Vannoppen replied that he had not but would do so right away. Sure enough, the bulb was bad and the first trouble-shooting experience for a missile combat crew on alert at the 308th SMW was to replace a burnt-out light bulb.[65]

On 6 June 1963, missile B-36 (62-017) arrived at the 308th SMW, the ninth Titan II delivered to the wing. By the end of June all nine were installed, but on 26 June 1963, missile B-20 was removed from alert status because of an oxidizer leak. Stage I had to be removed and returned to Denver for repairs, the first of the Operation Wrap Up missiles to depart the 308th.[66] By 30 June 1963 only two complexes, 373-5 and 373-7, were in an emergency combat capability status due to oxidizer leaks in the remaining missiles.[67] By 30 September 1963, the 308th SMW had only four complexes, 373-4, 374-4, 373-5, and 373-7, in emergency combat capability status due to the continuing oxidizer leakage problems. Complexes continued to be turned over to SAC through the fall of 1963. Actual turnover of all sites to the 373d SMS took place on 29 November and for the 374th SMS, 28 December 1963.[68] Three days later, the 308th SMW had 18 missile complexes in an emergency combat capability status. Even in late December, oxidizer leaks were still causing some missiles to be removed from emergency combat capability status, but the Air Force had directed all Titan II complexes to be turned over to SAC by the end of 1963. Only extensive efforts by the Site Activation Task Force personnel and those of the 308th SMW permitted this directive to be accomplished.[69]

Modification Programs

Throughout the life of the Titan II program, both the missile and the launch facilities were monitored for signs of aging and deterioration. A large number of modification programs were conducted to improve the conditions for storing the missile and for the working environment of the missile combat crew members. Examples of several of the programs are given to illustrate the types of problems that needed to be addressed.

PROJECT GREEN JUG

Project GREEN JUG was the first modification effort in the Titan II system. Primarily a direct response to the oxidizer leaks that had appeared throughout the Titan II missile force once missiles were emplaced in the launch duct environment, GREEN JUG also included modifications to the silo ventilation system, cooling towers, and diesel generators at the complexes. Dehumidification equipment for the launch duct atmosphere was installed from March 1964 to September 1964 throughout the Titan II fleet, substantially reducing the possibility of oxidizer-induced corrosion. The modification work was carried out six silos at a time, and each silo remained on alert as the work was accomplished.[70]

PROJECT TOP BANANA

The second modification project was nicknamed TOP BANANA. TOP BANANA was separated into three phases due to shortage of parts. Phase I consisted of updating 32 flight-critical items as a result of the completion of the research and development program test flights. Phase II dealt with components that affected operational and maintenance reliability such as the launch enabler system, modification of the thrust mount shock isolation system, modification of the fixed propellant transfer system equipment, and improvements to the silo closure door systems. Phase III pertained to the guidance system with depot-level modifications to the inertial measurement unit, making it less complex from both a maintenance and operational standpoint, resulting in higher reliability. Project TOP BANANA started on 6 July 1964 at Davis-Monthan AFB and was completed in 1965 throughout the system.[71]

LONG-TERM READINESS EVALUATION, 1965–67

On 22 December 1964, Headquarters, U.S. Air Force, approved a three-year program, designated Project TAKE HOLD, also known as the Long-Term Readiness Evaluation (LTRE) program, for evaluation of the Titan II weapon system. Two objectives were listed. First, prolonged exposure of the fully loaded missile to the high humidity of the launch duct in the operating environment had revealed corrosion and leak problems due to the oxidizer. Project GREEN JUG had installed dehumidification equipment, and the LTRE would now determine its capabilities. Second, the missile design specification was for one-year storage life in the operating environment on alert without major maintenance and three years in a nonoperating environment.[72] The LTRE determined the effects of age and operational environment on total system performance—that is, both missile and ground equipment; identified components and subsytems which were major contributors to system degradation; and determined the recycle requirements for all major system components. At selected sites, missiles were removed for shipment to Denver for a thorough inspection of all systems. A total of 12 missiles had been evaluated when the program ended in 1967.[73]

The results from the LTRE program were encouraging. Three of the last four missiles removed and inspected at Denver were judged fully capable of completing a flight mission; the fourth was not, due to a faulty electrical repair. Some degradation was found in all four of the missiles' flight controls, but nothing was found that would have prevented completion of the mission. No structural defects were found in the missile welds due to prolonged storage. Electrical system degradation, when found, was a result of faulty application of the Hysol-Butyl propellant proofing material used to protect electrical connections from the corrosive effects of propellant fumes. The operational ground equipment was found to be in generally good condition, with mechanical system problems due primarily to adjustment and maintenance. The silo closure door hydraulic system was found to be a high-maintenance item that needed further review. All in all, the Titan II program equipment and facilities were in good condition.[74]

PROJECT YARD FENCE

On 14 February 1965 the largest modification project in the history of the Titan II program, Project YARD FENCE, began at the 390th SMW. After two years of operation, design changes and upgrades to a number of facility systems were necessary. These included an additional 18 tons of concrete shielding, known as "dog ears," to the exterior of the silo closure door; modifying parts of the silo equipment area to improve hardness capability; improving hydraulic system reliability; increasing blast door reliability; modifying the acoustical liner modules used in the launch and exhaust duct; and correcting deficiencies in the utility systems. Projected to cost $13 million, YARD FENCE proceeded simultaneously at the 390th and 381st SMW, but did not start at the 308th SMW until early summer. The missiles stayed in the silos, but with the reentry vehicle removed, fully loaded with propellants, since no modifications to the launch duct were in the program schedule.[75] Project YARD FENCE was completed for the entire program on 2 February 1967 with the turnover of Launch Complex 374-6 at the 308th SMW.

PROJECT EXTENDED LIFE

In 1975 the anticipated need to prolong the service life of the Titan II launch complexes into the late 1980s resulted in Project EXTENDED LIFE. The first phase involved installation of new pumps, ducting, exhaust fans, and compressor after coolers. The second phase addressed the need for new exterior buried piping and phase three replaced environmental controls, installed new interior water and sewer piping, installed new hydraulic lines, and made structural repairs as necessary on Levels 8 and 9 of the silo. Project EXTENDED LIFE was accomplished on an as-needed priority basis but was not completed before the program deactivation began, at which point the program was discontinued.[76]

SOUND SUPPRESSION PROGRAM

The concrete and steel construction of the launch control center was not conducive to a quiet environment. Level 3 held a large ventilation fan as well as the motor generator which provided clean DC power to the guidance system. Numerous small fans in the equipment racks on the launch control panels on Level 2 also contributed to the din. The noise was bothersome in a variety of ways with the major effect being a constant irritant that was found to be a morale factor for some combat crew members. Many crew members reported temporary hearing loss after a 24-hour alert tour.

In 1977 the 390th SMW Civil Engineering Staff developed a sound suppression study with the approval of Headquarters SAC. Over a three-week period, Launch Complex 571-4

TABLE 5.9

AVERAGE LAUNCH COMPLEX CONSTRUCTION COSTS,
INCLUDING MISSILE AND MISSILE EQUIPMENT

ITEM	LITTLE ROCK AFB	MCCONNELL AFB	DAVIS-MONTHAN AFB
Construction (Phase 1, II, and IIA)	4,829,000	5,994,000	6,140,000
Missile	2,550,000	2,550,000	2,550,000
Missile Support Equipment	3,922,000	3,922,000	3,922,000
Total	$11,301,00	$12,466,000	$12,612,000

Source: Adapted from respective Army Corps of Engineers reports.

modifications included ventilation ducts rerouting, carpet installed on Level 2 of the launch control center, more efficient shock isolators on the motor generator, and sound-absorbing blankets draped along the staircase connecting the three launch control center levels. The results were impressive as the modifications reduced ambient noise in the launch control center to that of a normal household environment. The Sound Suppression Program was adopted fleetwide, much to the relief of all the crew members.[77]

The construction of the 57 Titan II launch complexes, 18 at each of three operational bases and 3 training facilities at Vandenberg AFB, was a monumental task. Construction took 37 months from the start of Launch Complex 395-C at Vandenberg AFB on 29 November 1960 to the acceptance of Launch Complex 374-4 at Little Rock AFB on 28 December 1963 at a cost of nearly $700 million.[78] Table 5.9 summarizes the costs of construction for a single launch complex at each of the operational wings. Who could have foreseen that the Titan II program would perform well past its expected 5-year lifetime, standing over 24 years as the blockbuster component of our strategic nuclear arsenal? Within 2 years Titan II would be the only liquid-propellant ICBM missile left in the nation's arsenal. The Soviet Union had not only the multitude of the Minuteman program missiles to contend with, albeit with relatively small warheads, but also the knowledge that 54 Titan II missiles capable of destroying all but the most hardened structures were poised and on alert, to be launched just as readily as Minuteman.

VI

TITAN II IN CONTEXT

With the full deployment of Titan II in January 1964, the nation had a new and powerful strategic weapon system on alert. The details of the role of Titan II in the strategic war plans of the United States remain classified, but a picture of its role can be drawn by considering possible target areas, the general war plan concepts and how they evolved during the life span of Titan II, and the advent of strategic arms negotiations and their effect on the status and usefulness of Titan II.

Titan II was a weapon of considerable reach. Carrying the Mark 6 reentry vehicle, Titan II had a maximum range of 5,500 nautical miles. Targets in most of the Soviet Union and the northeastern part of China were within reach when all three operational wings were on alert. While the specific targets for Titan II remain classified, in all likelihood they would have been directed toward either specific missile complexes, Figure 6.1, or "core area" targets, such as those shown in Figure 6.2 for the Soviet Union. Core areas are "geographically distinct aggregations that contain targets of great military, political, economic, and/or cultural significance, the seizure, retention, destruction, or control of which would afford marked advantage to any opponent."[1] Core area targets can also be described as wide area targets and/or targets with closely associated aim points that a single large yield weapon can effectively destroy.

Prior to the advent of strategic missile systems, long-range bombers were the sole strategic nuclear weapon delivery system. In the later 1940s and early 1950s, nuclear war was considered feasible by some military planners because the relatively few bombs available limited targeting to major cities. In the mid-1950s the number of nuclear weapons began to climb into the hundreds and then thousands. Long-range bombers were still the prevalent strategic delivery system and response time was measured in hours. A Soviet Union or United States bomber attack over the North Pole would be detectable at relatively long range and air defenses could be raised, perhaps mitigating the effects of such an attack on both adversaries.

The quantum leap in nuclear warfare came with the deployment of ICBMs. Initially, the most important aspect was the time to target with ICBMs, coupled with the total lack of defense against this type of weapon. Flight times of 30 to 35 minutes gave much less warning than bombers flying for hours to reach their targets. There was no known defense against such missile attacks, and none was projected to be available in the foreseeable future. While accuracy was not the same as that of bombers, ICBMs were suitable for use against bomber-filled air bases and large area military facilities such as army depots and naval fleet anchorages. ICBMs became a true first-strike, counter-force weapon, able to so damage the war fighting capability of the enemy that they would capitulate. If not, after absorbing the retaliatory strike of the enemy, a second-strike, counter-value attack,

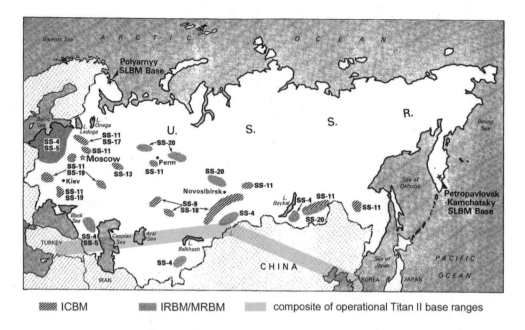

| ICBM | IRBM/MRBM | composite of operational Titan II base ranges |

Figure 6.1. Soviet missile base locations versus Titan II range. While the Titan II ICBM was most likely not used to target specific missile complexes, clearly most of the Soviet sites were within range, something that the Soviets had to consider. *Adapted with permission from John M. Collins,* U.S.-Soviet Military Balance: 1960–1980.

targeting industrial capabilities, government centers, and economic infrastructure could take place using bombers and the remaining missile forces. This is where the concept of mutually assured destruction received its name. No imagination was required to see that such a war could easily end up with the total devastation of both sides.

The 54 Titan II missiles achieved full alert status on 31 December 1963. They joined 123 Atlas, 54 Titan Is, and 310 Minuteman missiles in a land-base force totaling 487 ICBMs. By 3 May 1967 the land-based ICBM force reached its maximum, with 54 Titan II and 1,000 Minuteman missiles; the Atlas and Titan I missiles had been deactivated between 1964 and 1965.[2]

The role of ICBMs in the nation's war plans were not clear at the onset of their deployment. Prior to December 1960, the commanders in chief of the nuclear capable U.S. military commands prepared their own plans for use of nuclear weapons within their areas of operation. Considerable overlap in targets resulted. In August 1960, Secretary of Defense Thomas Gates initiated a new concept in nuclear weapons targeting, the Joint Strategic Target Planning Staff (JSTPS). The JSTPS was instructed to generate a coordinated plan for nuclear war, the Single Integrated Operations Plan (SIOP), using all four armed services nuclear strike capabilities. Two conditions for targeting were to be evaluated. First, eliminate the Chinese-Soviet strategic nuclear delivery capability, including military and governmental command and control centers. Second, attack major industrialized, hence urban, areas of both China and the Soviet bloc countries.[3] A national strategic target list

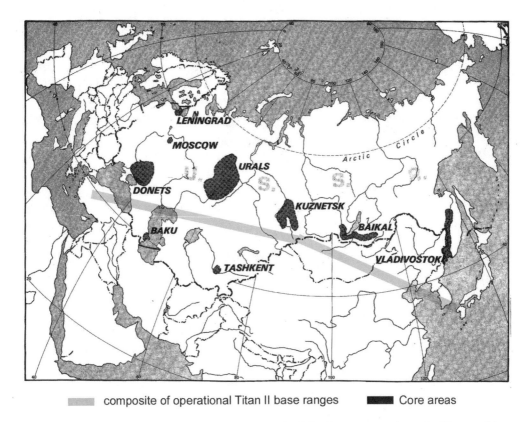

composite of operational Titan II base ranges   Core areas

Figure 6.2. Soviet "core area" as of 1980. Core areas can be defined as concentrations of military, political, or economic activity vital to the welfare of the Soviet Union. *Adapted with permission from John M. Collins*, U.S.-Soviet Military Balance: 1960–1980.

was prepared with over 80,000 potential targets. These were further screened and filtered down to 3,729 critical facilities. A total of 1,060 designated ground zeros or aiming points were selected, which meant that in many cases the targets were co-located and one nuclear weapon of sufficient yield could successfully destroy more than one target.[4]

The resulting plan was entitled SIOP for Fiscal Year 1962 (SIOP-62). The final preparation of SIOP-62 was completed before the end of the Eisenhower administration. The chief of naval operations, Adm Arleigh Burke, was a staunch critic of the result. The Strategic Air Command had damage criteria that required the equivalent of 500 kilotons of nuclear weapons to achieve the results of the 13-kiloton bomb damage at Hiroshima. The Navy strategic planners pointed out that fallout alone from these massive over-targeting policies would endanger areas far removed from the Soviet Union or China. Professor George B. Kistiakowsky, President Eisenhower's science advisor, reviewed the preliminary SIOP-62 and was in agreement with the initial attack premise but felt that the follow-on portions of the plan resulted in the ability "to kill 4 to 5 times over somebody who is already dead."[5] The Eisenhower administration did not implement SIOP-62, preferring to leave that task to the incoming Kennedy administration.

SIOP-62 became effective on 15 April 1961 and was briefed to President John F. Kennedy on 13 September 1961. Of the 1,060 aim points covered by SIOP-62, 800 were considered primarily military targets with the remainder urban-industrial. Fourteen "options" were listed, starting with the alert forces, those defined capable as being launched toward their targets within 15 minutes. Option 1 was called the Alert-Option and could generate 1,004 delivery systems, bombers and missiles combined, with 1,685 nuclear weapons. The final option, Option 14, utilized the entire 2,244 delivery systems with a total of 3,267 nuclear weapons. Although 14 options were presented, giving a sense of flexible response, in reality this was still a massive attack since the remaining 13 options built upon Option 1. More correctly, each option was really the result of time of preparation for launching the attack. The longer the time to prepare, the more forces that would be available and launched. The major concern was to get the nuclear forces launched before they could be destroyed by a Soviet attack. While changes to withhold attack on one or more of the Soviet-bloc satellite countries could be generated from the plan given enough time, these were not among the 14 options.[6]

Kennedy was not satisfied with the lack of flexibility in SIOP-62. In the somewhat surreal world of nuclear war strategy, these essentially all-or-none options would have made it difficult for the Soviet leadership to distinguish between a military facilities or urban/industrial centers oriented attack, making their likely response one of total retaliation. The chairman of the Joint Chiefs of Staff, Gen Lyman L. Lemnitzer, countered that even if the distinction could be made, the JSTPS felt that the Soviet response would be equally devastating. Lemnitzer again emphasized the benefit of using our weapons rather than lose them in the retaliatory attack. A simplified plan, such as SIOP-62, made successful retaliation against a surprise attack much more probable if communications were already disrupted.[7]

Secretary of Defense Robert S. McNamara also urged greater flexibility, and the result was SIOP-63, briefed to President Kennedy on 14 September 1962. Now there were three tasks: strategic nuclear forces, conventional military forces, and urban-industrial targets. Five primary attack options were available, with the ability to withhold attacks against each of the three tasks as well as any of the Soviet-bloc countries. The five major options were broken down into (1) Soviet strategic nuclear forces; (2) military facilities located away from cities, such as air defenses that covered bomber routes; (3) Soviet conventional forces in close proximity to urban areas; (4) Soviet command and control centers; and (5) total attack, including all of the above. Within each of the five major options were sub-options. Ironically, while these options and sub-options gave the appearance of greater flexibility, they still used thousands of nuclear warheads.[8]

In October 1964, when President Lyndon B. Johnson requested an update briefing on the status of the nation's nuclear forces, the alert force was listed as consisting of 1,798 megatons worth of nuclear warheads. Using a value of 9 megatons for the Titan II's W-53 thermonuclear warhead, a value cited in a number of authoritative references but never declassified by the Air Force, Titan II accounted for approximately 27 percent of the nuclear capability of the alert forces in 1964.[9]

From 1963 to 1976 Titan II missile served within SIOP-63 as the premier strategic weapon "for use against a collection of targets that can all be damaged by a single, high yield weapon."[10] During this period, 48 operational test launches within the Western Test Range had confirmed that Titan II met the original design criteria accuracy with a circular error probable of less than 0.8 nautical miles. The 1963–76 time period encompasses the era dubbed mutually assured destruction by the press. The intent of the American response to a nuclear attack was, according to McNamara, "to inflict an unacceptable degree of damage upon any single aggressor, or combination of aggressors, even after absorbing a surprise first strike."[11] SIOP-63, with some modifications, remained in effect through to the Nixon administration.

In late January 1969, Henry Kissinger, President Richard M. Nixon's national security advisor, began a review of SIOP-63. Comprehensive in nature, the review resulted in a multi-year process that culminated in a new plan, SIOP-5, which took effect on 1 January 1976.[12] SIOP-5 relied on a series of limited attack scenarios to prevent escalation to all-out nuclear war. Target classes continued to be all encompassing but were segregated into four main areas: (1) Soviet strategic nuclear forces; (2) conventional forces and bases; (3) mili - tary and civilian command and control centers; and (4) economic/industrial base facilities. Each target class was then broken down further into a type which governed which of four attack options would be used against it. This refinement into an increasing number of options, ranging from selective, limited attacks to all-out coverage, was due in part to improvements in the U.S. strategic forces. Significant increases in warhead delivery capability with Minuteman III and Poseidon and Trident submarine launched ballistic missiles, capable of a circular error probable of 0.1 to 0.5 nautical mile, made such a shift in policy possible. Titan II presented planners with a large area weapon that was useful in only a few of the attack scenarios. True, the high yield of the warhead made pinpoint accuracy unnecessary, but the scope of the updated SIOP was rapidly moving away from area-type weapons. In addition, the capability to use multiple independently targetable reentry vehicles (MIRVs) on Minuteman III, Poseidon, and Trident permitted coverage of the previously co-located targets of Titan II more efficiently with far less collateral damage.[13]

A number of weapon system developments caused both the United States and the Soviet Union to pause during the strategic arms race of the late 1960s. The development of MIRVs provided the ability to overwhelm ballistic missile defenses that were being considered by both sides. Efforts began to develop an agreement to halt yet another round of nuclear one-upmanship, culminating on 26 May 1972 when President Richard Nixon and General Secretary Leonid Brezhnev signed two treaties which comprised what came to be known as Strategic Arms Limitation Talks I (SALT I). The first document was the Treaty between the United States of America and the Union of Soviet Socialist Republics on the Limitation of Anti-Ballistic Missile Systems, also known as the Anti-Ballistic Missile Treaty. The second was the Treaty between the United States of America and the Union of Soviet Socialist Republics on Certain Measures with Respect to the Limitation of Strategic Offensive Arms, also known as the Interim Agreement.

While any antiballistic missile defense system was important to the Titan II system's effectiveness, of greater concern was the Interim Agreement, which generated a five-year freeze on construction of new ICBM launchers as well as placed numerical limits on ballistic missile submarines and the number of missiles they could carry.[14]

Article I of the Interim Agreement stated that both countries agreed that no further construction of additional fixed land-based ICBMs silos would be permitted after 1 July 1972. Article II stated that both countries agreed not to convert current land-based launchers for small ICBMs into large ICBM launchers, nor would launchers for ICBMs deployed prior to 1964 be converted into launchers for heavy ICBMs. Modernization of existing silos was permitted within the above restrictions, which were further detailed in a series of initialed agreements and interpretative statements. Article III stated that deactivation of older, pre-1964 heavy ICBM launchers would permit construction of an equivalent number of submarine launchers. Descriptions of the difference between heavy and light ICBMs were intentionally vague.[15]

Titan II was the only pre-1964 heavy ICBM in the United States' arsenal at the time, with 54 launchers, thus three new ballistic missile submarines, each with 16 launchers for a total of 48, could be built within the treaty guidelines. The immediate effect of the Interim Agreement was the possibility of deactivation of the Titan II program in order to enhance the submarine ballistic missile capability. However, as late as September 1979, the Congressional Budget Office plans included funding for 54 Titan II silos into the 1980–84 fiscal years.[16]

The intent of the Interim Agreement had been that it would serve until a more detailed agreement, SALT II, could be reached in the next year or two. World events, technological advances, and the simple inability to agree kept the United States and the Soviet Union circling around the issue for seven more years. Finally on 18 June 1979 President Jimmy Carter and General Secretary Leonid Brezhnev signed the Treaty between the United States of America and the Union of Soviet Socialist Republics on the Limitation of Strategic Offensive Arms, known as SALT II. SALT II built upon the agreements of SALT I, thus the effect on the Titan II program remained the same and it remained a possible bargaining chip.[17]

Over the 24-year life span of the Titan II ICBM weapon system, the context for its use, as detailed and implemented in the SIOP, evolved in parallel with improved missile technology. Solid-propellant missiles such as Minuteman, Poseidon, and Trident were easier to maintain and had greater accuracy, yet the large warhead carried by Titan II was something that the war planners were reluctant to discard.

Minuteman and Titan II were to be the two lasting components of the land-based strategic offensive nuclear capability. Minuteman and its solid-propellant technology led to a single two-man launch crew monitoring 10 missiles which were several miles away from the launch control center. Minuteman crews had little if any contact with the weapon system they were monitoring and ready to launch. Titan II was the complete opposite, with a larger, four-man crew that had daily contact with the missile, its silo, and facility support equipment. To fully understand the Titan II program, one needs to place the daily routine and experiences of the missile combat crews in context.

VII

MANNING THE TITAN II

A Titan II missile combat crew was composed of two commissioned officers (second lieutenant up to colonel) and two enlisted personnel (airman second class to senior master sergeant). The officer in charge was designated as the missile combat crew commander (MCCC). He or she managed the crew and launch complex in all operations, including site security, missile launch, launch complex facility operation, system testing, malfunction isolation, maintenance efforts, and emergency situations. All visitors to the site, be they maintenance personnel or dignitaries, were the responsibility of the MCCC. The second officer was designated as the deputy missile combat crew commander (DMCCC). He or she assisted the MCCC in their duties, being able to take over directing operations at the launch complex as needed. In addition, the DMCCC monitored the communications system and the air balance of the complex and coordinated all checklist-related activities such as daily shift verification and maintenance both above and below ground. This included updating the locator board which tabulated the location of all personnel on the complex. The MCCC and DMCCC were responsible for receiving and decoding the Emergency War Order (EWO) fast reaction messages and the positive control procedures involved with nuclear armed missile operations.

The two enlisted personnel were the ballistic missile analyst technician (BMAT), also referred to for a time as the missile systems analyst technician (MSAT) and the missile facilities technician (MFT). The BMAT maintained a constant check on the missile status and support equipment such as the guidance system and missile fault locator. If a malfunction occurred, the BMAT would isolate the cause as far as possible and either fix it or provide expert advice to the MCCC and DMCCC as to the required course of action. The MFT was responsible for the facilities that supported the crew and missile functions. The MFT provided the MCCC and DMCCC with information on the status of the launch complex support equipment such as power, ventilation, and water sytems and could troubleshoot and fix malfunctions of the support equipment or recommend the next course of action.[1]

A typical day of operations for a Titan II crew in any squadron was similar to that for crews operating from Davis-Monthan AFB and evolved over the 24-year life span of the program. The alert day started with pre-departure briefing. Prior to pre-departure briefing the BMAT and MFT would go to the motor pool and pick up the vehicle to drive to the site. Once the vehicle was checked and inspected they would stop by linen supply for sheets and pillow cases, then go to the mess hall and pick up the crew's food (in later years the crew members were responsible for their own rations). During this time, prior to the briefing, the officers would check the crew mailbox and see if there were any special events that were going to be accomplished at the complex during the alert period.

There were 18 crews at pre-departure briefing, 9 crews from each operational squadron. The operation squadrons were divided into three sectors, and each sector was headed by a sector commander who was responsible for the crews and sites in their sector. The sector commander was responsible for the overall cleanliness and well-being of each of his sites. The sector commanders would attend the pre-departure briefings and would assign tasks that they wanted the crews to accomplish during the alert period. These sector commanders also spent a lot of time at their sites helping in any way they could.

Once the pre-departure briefing started, the doors were locked and no one else was admitted. During the briefing, the crew would receive the latest weather conditions and the latest road conditions and be notified if there were any exercises going on or going to start and any Emergency War Order instructions or changes that needed to be passed on. There was usually a classified briefing relating to military activities going on in the world that might affect alert status—in the sixties and early seventies most briefings related to the Viet Nam conflict. A normal briefing would last about 10 to 15 minutes.

Once the briefing was over and all of the crew's personal belongings were loaded in the vehicle, the crew was ready to depart for the site. Each crew member carried a briefcase that contained a copy of Technical Order 21M-LGM25C-1, known as the "-1" (dash one), the technical manual for overall operation of a Titan II launch complex, overnight toiletries, and any other personal belongings that were needed to sustain an alert tour.

Not every pre-departure was this simple. If classified documents had to be couriered to the site they had to be signed for after checkout. This was done one launch complex and one crew at a time. If an exercise, inspector general, or 3901st Strategic Missile Evaluation Squadron visit was taking place, the status was briefed. Time added to the briefing just delayed departure time from the base. Driving time to each site depended on the distance the site was from the support base. As the crew approached the surface gate to the complex, the real work was about to begin. The only thing that might put a damper on a smooth alert would be a standardization crew (also known as the standboard) waiting at the entrance, meaning your crew was going to be evaluated.

Every crew member required a standardization check once each year to keep current. These checks consisted of a site check with a minimum of observing a daily shift verification starting at the launch control center and at least three levels in the silo equipment area; an initial check for either the BMAT or the MFT consisted of a complete daily shift verification walk through the site on all levels in the silo equipment area. There was also a check in the Missile Procedures Trainer where the crew's knowledge of the system was tested.

The crew on alert knew within 5 to 10 minutes when the relief crew was going to show up at the gate. In the sixties and early seventies, crew members knew all of the other crews that pulled alert at their complex by sight, and entrance was by personal recognition. The MCCC of the oncoming crew would call from the gate phone, and then the crew would proceed to the access portal. Once the oncoming MCCC was identified, the TPS-39, a radar-based intruder detection system placed around the silo closure door and at the main air intake, would be tested then placed in the ACCESS mode so the oncoming crew could perform daily entry verification (DEV) topside. In the late seventies and eighties, each MCCC had to pick up a code word before departing for the site. The MCCC used the code word

Figure CI.1. Titan II ICBM B-69 (63-7724) finally lifts off on 5 October 1973 from Launch Complex 395-C, Vandenberg AFB, California, nearly six years after the first launch attempt. *Author's Collection.*

Figure CI.2. Titan II N-11 (61-2734) lifts off from Cape Canaveral, Pad 16, on 6 December 1962. The original Titan II design called for either the Mark 4 or Mark 6 reentry vehicle. However, only the Mark 6 reentry vehicle was actually deployed. This was the only flight with a Mark 4 and was unsuccessful due to a failure in Stage II performance. *Courtesy of the San Diego Aerospace Museum, San Diego, California.*

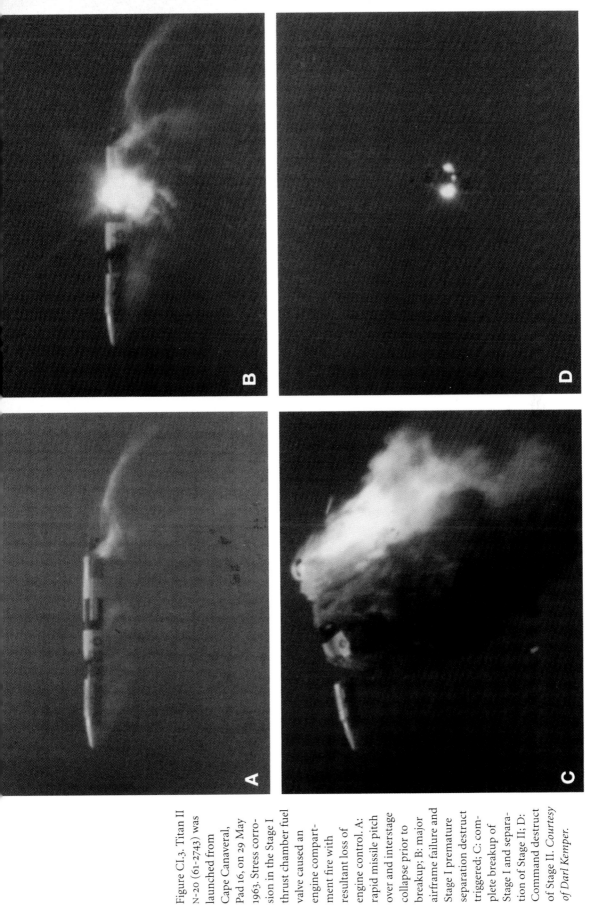

Figure CI.3. Titan II N-20 (61-2743) was launched from Cape Canaveral, Pad 16, on 29 May 1963. Stress corrosion in the Stage I thrust chamber fuel valve caused an engine compartment fire with resultant loss of engine control. A: rapid missile pitch over and interstage collapse prior to breakup; B: major airframe failure and Stage I premature separation destruct triggered; C: complete breakup of Stage I and separation of Stage II; D: Command destruct of Stage II. *Courtesy of Darl Kemper.*

Figure CI.4. This picture was taken two minutes after the explosion of a TX46/W53 device code-named OAK, 28 June 1958, Enewetak, during the Operation HARDTACK I test series. The reported yield for this test was 8.9 megatons. Congressional sources list the W-53 as 9 megatons. DNA-6038F. *Image JE-00-161 courtesy of the Los Alamos National Laboratory, Albuquerque, New Mexico.*

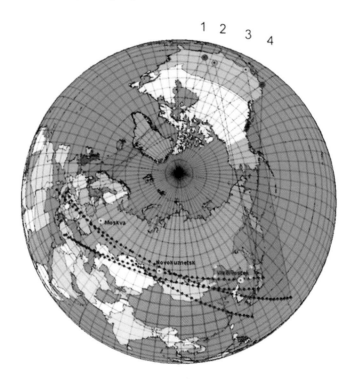

Figure CI.5. The Titan II ICBM had a range of 5,500 nautical miles. This polar projection map shows the reach of Titan IIs launched from the three operational bases and Vandenberg AFB, where it was on operational alert in the 1960s. Launch sites read, 1) 308th SMW, Little Rock, Arkansas; 2) 381st SMW, McConnell AFB, Kansas; 3) 390th SMW, Davis-Monthan AFB, Arizona; and 4) 395th SMS, Vandenberg AFB, California. *Map courtesy of Ernest Sinohui.*

Figure CI.6. Titan II ICBM B-52 (62-12295) climbs skyward from Launch Complex 395-C, Vandenberg AFB, California, on 7 August 1975. This was the last missile launched by the 381st SMW. The flight was completely successful. *Courtesy of Gene Scoular.*

Figure CI.7. Launch Complex 395-C, Vandenberg AFB, California, during propellant transfer operations in preparation for the launch of Titan II B-17 (62-2771). The bicentennial emblem is clearly evident on the silo closure door. Initial plans were to launch B-17 on 4 July 1976, but range safety considerations required moving the launch date to 27 June 1976. *Courtesy of Randy Welch.*

Figure CI.8. View looking north at the entrance to Launch Complex 571-7, the Titan Missile Museum National Historic Landmark. The structure in the center of the photograph is the silo closure door with the environmental cover over the launch duct. *Courtesy of the Titan Missile Museum National Historic Landmark Archives, Sahuarita, Arizona.*

Figure CI.9. Addition of the first portion of 80,000 gallons of water to the nitrogen tetroxide accumulated in the flame deflector at Launch Complex 533-7 caused a violent reaction as the solution boiled and sent clouds of steam and nitrogen dioxide skyward. *Courtesy of Mark Clark.*

Figure CI.10. Aerial view of Launch Complex 374-7 looking southeast several days after the explosion. The 740-ton silo closure door is in the woods at the upper right of the photograph. The W-53 warhead was found in the ditch beside the road to the helicopter pad at the lower center of the photograph. *Courtesy of Bob Eagle.*

Figure CI.11. Titan II ICBM B-46 (62-027) lifts off between two exhaust plumes from Launch Complex 395-C, Vandenberg AFB on 24 May 1972. This launch was part of the Safeguard Test Target Program which was evaluating the Safeguard Antiballistic Missile System's capability to acquire and lock onto incoming ballistic missile reentry vehicles. *Courtesy of Robert Dreyling.*

in an entrapment area in the access portal. Once verified, the rest of the crew could proceed into the underground portion of the launch complex.

The off-going MCCC would cycle the topside warning device so the oncoming crew could check to see if the lights cycled properly and the siren sounded. Usually the MFT and BMAT performed the DEV checklist, while the two off-going enlisted members transferred the ongoing crews' food and belongings to the access portal elevator and put the off-going crews' belongings and trash in the crew vehicle. While the off-going crew members transferred everything into the launch control center, the BMAT and MFT finished up the topside portion of DEV. Sometimes the MCCC or the DMCCC along with the MFT performed the DEV. If any of the areas checked during the DEV required cleaning or repair, this was noted and would be taken care of later during the alert. On the way down to the launch control center that elevator shaft was checked, and if the access portal exhaust fan was running it was checked for proper operation. The elevator shaft sump and sump pump were inspected. Between Blast Doors 6 and 7, comprising the first blast lock, the oncoming MFT would check the HS-3 system for leaks and proper operation and ensure that the emergency phone worked. The oncoming crew would then proceed to the control center where they were greeted by the off-going crew.

Before the carpets were put on the control center floor and all of the sound suppression blankets were hung, visitors who entered the control center for the first time would comment on the cleanliness of the floors and the control center in general. Every morning the off-going crew would clean, mop, and buff the control center floors in preparation of leaving a clean site for the oncoming crew. The early program crew members took pride in their job and the facility that housed a highly destructive weapon system.

Once in the control center and all greetings were over, the MFTs were given a silo entry briefing and dispatched to the silo equipment area to check the launch duct door on Level 1 and the junction boxes on Levels 3 and 7 of the silo equipment area. When the junction boxes were checked and the MFTs were ready to return to the launch control center from Level 7, they would call the launch control center and have the MCCC turn on the silo klaxon; that klaxon was checked from Level 7 to Level 2. When on Level 2, the klaxon was turned off, and the MFTs returned to the control center. The oncoming MFT would report the status on the launch duct door on level 1 and the junction boxes on Levels 3 and 7. While the MFTs were in the silo, the BMAT would check the presence of the EWO toolbox and ensure that the Control Monitory Group One (CMG-1) Launch Sequence Drawer was locked and that Circuit Breaker 103 was off. The MCCC and the DMCCC would discuss what maintenance teams were coded out and which ones were expected, check weapons, check that the site seal crimping tool and die were in place, ensure that all of the controlled keys were in the key box, and start the inventory of classified documents.

The officers would close launch control center Level 2 while the classified EWO docu-ments were inventoried, the launch keys transferred, and the locks changed on the safe. The enlisted members would go upstairs and any site peculiarities would be discussed, coffee was made, and the oncoming crew's food would be put away. When the officers were finished, they would transfer sidearms, annotate the crew log, and then the MCCCs would contact the command post controller and let them know that crew changeover was completed. The off-going crew would then depart the complex and begin their trip back to the base.

At an alert facility on strategic alert there had to be two EWO-qualified crew members on Level 2 of the launch control center at all times and one had to be an officer. All four crew members were EWO-qualified, able to copy messages (but only officers could decode the launch messages) and to understand the workings of the launch control center equipment, including all of the communications.

On a weekend or when the maintenance crews hadn't arrived on the complex, the crew would sit around the table on Level 2 of the launch control center, have a cup of coffee, and discuss what activities would go on during the day. This was the informal start of the crew operations briefing given by the MCCC. If maintenance teams had already arrived at the complex, then the MCCC and the DMCCC gave them their maintenance and silo entry briefings, while the BMAT and the MFT checked out the protective and emergency equipment —that is, canister masks, CHEMOX units, portable vapor detectors (PVDs), and the radio-type maintenance network communications. All this was done in an effort to get the maintenance teams dispatched without delay. If a change to the "-1" needed to be posted it was done during this briefing and was one of the most boring chores a crew member had to contend with.

The normal activities of daily shift verification (DSV), crew operations briefing, technical data posting, and site cleanup were all planned around the maintenance activities that would be on the complex that day. Crews usually pulled alert at the same complex and knew its idiosyncrasies. They had certain areas assigned to them, and it was their responsibility to ensure these areas were kept clean and well maintained. A training package was sent to every crew each month, and this package had to be completed by a certain time frame, so sometime during the alert all crew members would work on their portion of the package. Once the alert itinerary was laid out, the crew would then finish the crew operations briefing by obtaining the all important time check of the EWO clock.

With the briefing completed, the crew would usually start the DSV checklist. The launch control center portion of DSV was partially done by the MCCC and the DMCCC performing the checkout on the two launch consoles and the communications equipment. The BMAT and MFT would check out the other equipment in the launch control center; on an average this would take a conscientious crew an hour or better to complete.

While the MCCC and the DMCCC finished up DSV and checkout of the communications equipment, the BMAT and the MFT would go to Level 3 of the launch control center and finish up their portion of the control center DSV. Here they would check the sewage pump station, the power supplies, the battery power supplies, the motor generator, the communications equipment, Motor Control Center 2, the location of the power input for the launch control center, and the overall housekeeping and general repair of all equipment. If you were pulling alert at the alternate command post or the squadron command post, there was extra communications equipment that had to be checked. This equipment was connected directly to SAC or a numbered Air Force.

Once DSV in the launch control center was accomplished, the crew would again assess the maintenance on the complex before proceeding with DSV in the cableways and silo equipment area. Maintenance that would usually delay DSV would be Propellant Transfer System (PTS) operations, Combined System Tests, missile stage removals or replacements, diesel engine repair, heavy hydraulic system repairs, or launch duct maintenance. Only a certain number of personnel could be in the silo equipment area at any given time; if that number was reached then DSV would have to wait. The DMCCC kept track of all personnel in the silo equipment area or launch duct and their locations. If team members left one level and went to another level, they would call the DMCCC when they left, tell where they were at, and again call when they arrived at their new level; this included the team's return to the launch control center.

If there was a PTS operation going on at the complex, all other maintenance and DSV were put on hold until the operation was completed. The missile combat crew monitored the operation over the radio-type maintenance network communication system anytime propellants were being transferred or were about to be transferred. Maintenance teams were usually on the complex during daylight hours, unless a team was dispatched for an off-alert malfunction the crew couldn't repair. The crew was on the complex at least 24 hours so that DSV could be performed after the maintenance teams left the complex. DSV had to be accomplished during the crews alert tour, and sometimes it was accomplished in the evening. A normal DSV in the silo equipment area would take a minimum of 4 hours if it was performed correctly and if all equipment was checked and cleaned properly.

During DSV all equipment was checked, all spills were wiped up or cleaned, and general housekeeping chores were attended to. Minor maintenance was also accomplished, and those things the crew could not fix or repair were annotated in the -1 and later transferred to the site maintenance files. DSV was normally performed by the MFT and the BMAT, but a lot of times the MCCC or the DMCCC would go with the MFT. A crew commander would go with the MFT to see how the MFT accomplished the task, and a DMCCC would go to learn more about the equipment that one day would be his or her responsibility when he or she was a crew commander.

When there were no maintenance teams on the site, the crew would usually start DSV around 10 A.M. and would be back in the control center between 2 and 3 P.M. After receiving a silo entry briefing the BMAT and MFT would proceed to the short cableway and continue their DSV checklist. The silo entry briefing was accomplished to ensure that every person who entered the silo area had the proper safety equipment: hardhats, flashlights, earplugs, and PVd and canister masks, if needed. All jewelry and watches had to be removed, flame-producing devices had to be left in the control center, and most important SYSTEM LOCK-OUT had to be entered by the MCCC on the Launch Control Complex Facilities Console (LCCFC).

In the short cableway between the launch control center and Blast Door 8, the crew would check the emergency phone and the portable hydraulic pump and ensure that the pump connections were clean. They would make sure the blast damper was open so there would be normal airflow through the complex. Between Blast Doors 8 and 9 the crew checked another portable hydraulic pump, another blast damper, another emergency phone, and most important the vapor detection annunciator panel. This panel monitored the presence of any hazardous fuel or oxidizer vapors on any level of the silo and in the propellant pump rooms. Once every level was monitored, a minimum of two minutes total, the crew would proceed through Blast Door 9 and begin checking items in the long cableway. Just past Blast Door 9 the crew would grab a handful of rags and check the first of six emergency shower and eye wash stations for proper operation. At the end of the long cableway just prior to entering Level 2 of the silo equipment area, the crew would turn on the launch duct lights and begin an inspection of each silo equipment area level, beginning with the missile air conditioner on Level 2. After the new guidance modification was completed the missile air conditioner was removed. Next, up a ladder to Level 1A, the west portion of Level 1, the crew had to check the silo closure door operating equipment for hydraulic leaks, and the silo elevator panel was checked.

Then it was back down to Level 2 to check the water chillers and the hydraulic platform panels for Levels 1 and 2 and to ensure that the escape route was clear. Back up to Level 1B, the east portion, the crew checked the HS-1 silo closure door hydraulic system equipment for leaks, including all of the accumulators, the reservoir, and the pump. There were always small hydraulic fluid leaks around the reservoir that had to be wiped up. The crew then went back down to Level 2 and proceeded to Level 3 to perform lamp checks, fan and motor inspections, diesel engine checks and inspection, the Motor Control Center 1 check, and another hydraulic work platform panel. On Level 3 was an exhaust and supply fan with corresponding dampers that needed to be checked to ensure proper airflow throughout the complex. The fans had to be checked for proper operation and to ensure there was proper tension on the belts. Once the housekeeping chores were taken care of, the BMAT and the MFT would proceed to Level 4. On Levels 4 through 6 you could not walk completely around the launch duct because the hard water tank occupied the space between the launch duct and the silo wall. On Level 4 there was an air washer that was abandoned in place, but crews still had to check the intake and exhaust fans along with the corresponding dampers. The hard water tank level controller also had to be checked on Level 4. On Level 5 crews checked the silo equipment area air conditioner and another exhaust fan. Crews also had the diesel fuel service tank, diesel fuel pumps, and diesel slop tank. A dam was built around the whole area in the event there was a diesel fuel leak. This would contain diesel fuel to prevent its spilling all over the silo equipment area. After the platform valve panel and hydraulic lines were checked and all the housekeeping was in order, the crew moved on to Level 6.

On Level 6 a hydraulic system operated all of the work platforms and the silo blast valves. Here was an accumulation of valves and hydraulic lines that needed to be inspected and all leaks had to be wiped up. The air sampling equipment that monitored the fuel and oxidizer vapors throughout the launch duct, and pump rooms were also located on Level 6 and had to be checked thoroughly. If this equipment was malfunctioning then every team that entered the silo equipment area or launch duct needed to carry a portable vapor detector to monitor fuel and oxidizer vapors. The hard water tank level indicator was visually checked to ensure that the hard water tank was full. Air system flow manometers were also located on this level. These manometers checked the pressure between the silo equipment area and the launch duct to ensure that the airflow was in the proper direction—that is, high in the silo equipment area so that air would flow into the launch duct from the launch control center to the silo equipment area and finally into the launch duct where it would be emitted to the atmosphere.

Level 7 was underneath the hard water tank and again reached around the entire launch duct. Here were the water pumps that supplied industrial and firefighting water throughout the complex; these pumps had to be inspected to ensure that the packings were not leaking profusely. The air compressors and air receivers that supplied all of the compressed air throughout the complex were located on this level. They were inspected thoroughly, and moisture in the air receivers was drained. Once the pressure indicators were checked and the housekeeping accomplished, the crew moved on to Level 8.

Level 8 was the lowest level in the silo equipment area that the elevator reached. Access to Level 9, the silo sump, was by ladder. Two large sump pumps, located on Level 9B, would pump this water topside when the sump was partially full. The oxidizer and fuel pump rooms were located on Level 8 and were inspected for cleanliness and general conditions. Access to the lowest level of the launch duct was from Level 8. This was known as the flame deflector or "W." The "W" was inspected to ensure that there was no accumulated debris or water. From the "W" one of the most impressive sites on the complex could be observed. Crews looking up into the thrust chambers of the Stage I engine saw a missile that stood 132 feet straight above them. When all the equipment on Level 8 was inspected, the launch duct lights were checked to make sure that none of them were out, the access door was closed, and the team headed back to the launch control center for lunch, turning off the launch duct lights on the way back down the long cableway.

After DSV the crew would go over the condition of the site and the write-ups that were noted during DSV. The MCCC would usually lay out what he or she wanted accomplished on the training package or if there was any special site research they needed to accomplish. Research into the technical orders and SAC civil engineering manuals kept the crew members current, and this research would enhance the knowledge of the crew itself.

The crew sleeping quarters were on Level 1 of the launch control center. Most crews slept a split/straight shift. Two crew members would lie down after the maintenance teams left and would sleep until about 8 P.M.; the other two crew members would then sleep until about 3:30 A.M. Then the first two crew members would go back and lie down until about 7 A.M. At this time breakfast was fixed and site cleanup accomplished.

During the evening hours the crew members would work on the training package, research the technical orders, or study school lessons available to all crew members through the Base Education Program. There was a television downstairs, so the crew was not cut completely off from the outside world. A lot of crew members had hobbies that they also worked on during the late evening hours. Training crews usually had students and would go over their training plans/folders with them. Standardization crews always had paperwork that needed to be done or records that needed to be checked. Very few crew members had a lack of things to do in the evening. If one had nothing to do but read the books and magazines that were sent to the site, then it turned into a long night.

In the morning after breakfast everyone pitched in to ready the site for the oncoming crew. Cupboards and racks were dusted; floors were swept, moped, and buffed; and the crew's belongings were packed and ready for the trip back to base.

Once back at the support base the MCCC and the DMCCC were dropped of at the debriefing area, and the BMAT and the MFT would return the food box, the crew linen, and the vehicle. The entire crew alert would last approximately 28 to 30 hours by the time members got home or back to the barracks. Members then had 48 hours before the next alert. The 24 hours immediately after alert were not available for training, but on the second day after an alert, crew training on the Missile Procedures Trainer would take place. A normal line crew would pull eight or nine alerts a month, along with a Missile Procedures Trainer exercise and an EWO class. An instructor crew would pull approximately five alerts a month, and the standardization crews would pull two alerts a month. The instructor crews, as well as the standardization crews, put a lot of time on the Missile Procedures Trainer in lieu of pulling alerts.[2]

A crew would pull approximately 90 alerts a year, and each one is a story in itself. The experience varied from mind-numbingly dull to some; routine to others, providing a chance to learn more about the system each time they went on duty; or, in the case of when real problems developed, adrenalin-pumping activity that tested the system knowledge of the crews and maintenance personnel alike under pressure.

Life in the Titan II Program

Looking at a Titan II launch complex, one could easily imagine that it was all but impervious to the outside world. In actual fact, a natural disaster thousands of miles away was felt by missile combat crews at both the 390th Strategic Missile Wing (SMW), Davis-Monthan AFB, Arizona, and the 381st SMW, McConnell AFB, Kansas. On the afternoon of 27 March 1964, 571st Strategic Missile Squadron (SMS) Instructor Crew 105 was on duty at Launch Complex 571-5, 390th SMW, Davis-Monthan AFB, Arizona. Maj Nathan Hewitt, MCCC, and MSgt Miguel DeZarraga, BMAT, were on duty on Level 2 of the launch control center when suddenly the LAUNCH NO GO indicator illuminated on the Launch Control Complex Facilities Console. Both Hewitt and DeZarraga looked over at the Missile Guidance Alignment Checkout Group (MGACG) panel where the AAS TEST indicator (Azimuth Alignment Set) was illuminated white. Hewitt asked DeZarraga for suggestions since at the pre-departure briefing that morning they had been instructed not to use the READY pushbutton to advance the guidance system into a READY mode if a problem occurred. This instruction was a result of an incident a month earlier at the 381st SMW, McConnell AFB, Kansas, where a translation rocket had inadvertently fired in the launch due to a power surge associated with a guidance system mode change. No one had been hurt, and a systemwide check of the associated equipment was underway.

DeZarraga replied that a malfunction had been detected by the MGACG and that the AAS test was in progress, a normal chain of events. DeZarraga recommended allowing the test to run to completion. If the test failed, they would have to press the HEAT switch and call the command post to inform them that Launch Complex 571-5 was in a NOT READY guidance condition. The inertial guidance platform gyroscopes had to be kept at a temperature between 138 and 142 degrees Fahrenheit; pressing the HEAT switch placed the system in a maintenance mode, hence in a NOT READY condition. If, however, the test was successfully completed, the MGACG would automatically begin a sequence to get back to READY. Several minutes later, the AAS test was successfully completed; the MGACG went to ALIGN 7 and ALIGN 8 (these are two steps in the automatic realignment of the missile's inertial guidance system) and then to READY green. Both Hewitt and DeZarraga agreed that Launch Complex 571-5 was back on alert, and the incident was noted in the log. The crew resumed their normal daily routine.

Shortly thereafter the 390th SMW Command Post called Hewitt and asked why of all the 18 Titan II launch complexes, 571-5 was in the READY mode, while the rest were still in the process of returning to the READY mode. DeZarraga described his procedures and satisfied the command post that the instructions received at the pre-departure briefings had not been violated and that the system had properly returned itself to alert status, so no policy violation had occurred. The command post agreed and then informed the crew of the devastating earthquake at Anchorage, Alaska. The effect at Tucson had been to cause the highly sensitive missile guidance systems to move out of alignment. The remaining 17 missiles had experienced the same ground motion conditions and were back on alert in a matter of minutes to hours, depending on the site and the response of each crew to the guidance system indicators. The two sites near Launch Complex 571-5, namely 571-6 and 571-7, were located near operating copper mines. Blasting from the mining operations often had the same effect, so the crews manning these sites were perhaps more prepared than others for this type of event.[3]

In December 1970, 1Lt Earl Blackaby and his crew at the 381st SMW had pulled their normal alert duty and were readying the complex for crew changeover. However, this was not to be a normal day, as Blackaby received a message that the snow and ice on the roads would prevent the relief crew from coming out until the next day at the earliest. Not overly thrilled at having to remain on alert an additional 24 hours, Blackaby recalls that the upside of the situation was that there would likewise be no maintenance activity, so the next 24 hours would be relatively quiet. Nothing much really changed, the same crew shifts were used and the same daily work was completed. Since the crew had brought only one day's worth of food, they had to break into the K-rations stored on Level 3 of the launch control center for breakfast, lunch, and dinner and then for breakfast on the third day prior to relief. The K-rations were not all that delectable, but at least there was something to eat. The best part of the meal were the cookies for dessert.

Actually the worst part of the extended stay occurred on base upon their return. They received a bill for the K-rations they had consumed. With different prices for each wonderful gourmet meal listed on the bill, one would have thought they had dined at a fine restaurant. The crew decided that since it was not their fault they had not anticipated the additional 24 hours of duty, they would fill in only the breakfast portion of the list and submit it, hoping the responsible party would find justice in some manner. The bill was returned with the full complement of meals listed, and the crew did pay the complete tab.[4]

In February 1971 another missile combat crew from the 381st SMW had a snow story that ended with them in jail, sort of. Crew E-151 was ready to return to McConnell AFB from Launch Complex 533-8. Snow had just begun to fall as 1Lt Steven A. Alonzo, MCCC;

2Lt Rudolph R. Federman, DMCCC; A1C Paul G. Herder, BMAT; and A1C Eugene P. Thibodeau, MFT, started the drive back to base. The crew made it as far as the Kansas Turnpike and then spent the next nine hours in a toll booth as they watched the snow pile up to a final accumulation of 24 inches. Troops from the Kansas National Guard finally rescued them, and they proceeded to Wellington, Kansas. There the sheriff offered to put them up for the night—in jail. After spending the night in jail cells, the crew returned to McConnell AFB, thanks to the courtesy of Mr. and Mrs. John Littleton, who just happened to be going their way.[5]

Actual modifications to the launch facilities were not routinely a task for the missile combat crew members. Maintenance personnel from the operational wing maintenance squadron or from other resources in the Air Force would work the modification from one silo to the next in a predetermined sequence. Sometimes the seemingly routine work would have humorous, or so it seemed afterward, consequences. One afternoon in October 1964, 2Lt Bill Kelley, an engineer for the San Bernardino Air Materiel Area, was assigned the task of installing centering devices on the thrust-mount shock isolation system springs in the launch duct at one of the launch complexes of the 381st SMW, Kansas, as part of Project TOP BANANA. The vertical spring component of the thrust-mount shock isolation system was 30 feet high. Kelley and a technician were perched in a "tree climber," a platform suspended by a single cable wound on an electric winch that was mounted on the underside of the Level 2 work platform above one of the thrust-mount shock spring units. The platform allowed the workers to traverse up and down the length of the spring as needed to install the centering devices. It was the time of day when the electrical team was preparing for the end-of-day power up of the emergency monitoring system. The electrical team was doing modifications to the launch and facility racks in the launch control center. On any given day the power-up procedure was different and hair raising. Kelley recalls being at the top of the spring unit when the lights went out in the launch duct. With the silo closure door closed and all of the launch duct access doors shut, everything was truly pitch black. Previous experience told Kelley that with the power off and the warning klaxon blaring, missile deluge water was not far behind. He yelled to the technician to hang on for dear life as he grabbed the spring with both hands. They waited for what seemed like 15 minutes but what was probably only one or two, figuring any second the cold water would be hitting them like Niagara Falls. The lights came back on, the water never appeared, and they quickly lowered themselves to Level 7 and left the silo.[6]

Troubling-shooting a problem in the complex facilities was the task of the MFT. A1C Tom Herring remembers his second alert as an MFT in March 1970 all too well. His home site was Launch Complex 571-5, located near the entrance to Madera Canyon, south of Tucson. During the DSV, he exited the elevator on Level 8 and found an amazing sight, a fountain of water spewing from a floor drain in the emergency shower outside the fuel pump room. Herring remembers that his first thought was "Why me, this is only my sec - ond alert!" Returning to the launch control center with the BMAT, Herring called A1C Larry Brooks, the MFT on duty at Launch Complex 571-7. This was not uncommon; classmates from training were often a first resource for this type of problem. Brooks suggested investigating the check valve located in the drain pipe in the Level 9B sump. Since this

Figure 7.1. View of the launch duct from Level 3 of the silo showing the top of one of the four vertical shock isolation springs. The top of the spring is 30 feet above the Level 7 work platforms. *Author's Collection.*

shower was in the lowest part of the drain plumbing, if the check valve was stuck closed, water draining from the levels above could flow down through the piping and back up to the lowest point, the fuel pump room shower. Checking this out would at first appearance seem easy enough, but once one saw the location of the check valve in a pipe near the ceiling of the silo equipment area sump on Level 9B, one would realize that Herring had an interesting problem to solve. Area 9B was 15 feet high and the only "bottom" was the 10-foot by 15-foot platform where the sump pumps were mounted.

First Herring had to verify that the check valve was actually the problem. Armed with a heavy-duty three-cell flashlight, Herring crawled out on the 8-inch-diameter drain line located 15 feet above the bottom of the sump. Once he reached the check valve, he could not hear the water running, exactly what he expected. Since he was already out on the pipe, 15 feet above the floor, he decided to try to free the valve by tapping on the pipe with the flashlight. The flashlight worked in this manner for everything else that was stuck, so why not now? Tapping did not work so he decided to take a couple of good swings and, if that didn't work, return to the launch control center for tools. Fortunately, two swings later, he could hear the sound of water running, and he was able to retreat along the drain line to safety.[7]

On 5 September 1978, a team of facility electricians were working at Launch Complex 570-6 on a modification to Motor Control Center 1 (MCC-1), located on Level 3 of the silo. SSgt John Runholt was drilling new mounting bracket holes, in accordance with the applicable Time Compliance Technical Order, when his drill bit, with proper depth stop attached, hit a bus bar. A shower of sparks erupted, and as he dropped the drill, power went out and smoke started coming from MCC-1. As he retreated to a safer area, Runholt turned off the diesel key switch so the diesel wouldn't start. Being an experienced technician, Runholt knew that if the diesel started it would apply power to the smoking MCC-1, making matters even worse. He called the launch control center and reported the situation.

The missile combat crew on duty was Capt Charles Hanks, MCCC; 1Lt Frank Zappata, DMCCC; A1C William Leslie Jr., BMAT; and SrA John Bibler, MFT. Just before noon, the lights in the launch control center began to flash on and off and then went completely dead for several seconds. The emergency lighting system came on immediately, and now the site was operating on battery power everywhere. They got a call from the silo that the team working on the MCC-1 had shorted the electrical circuits. Hanks directed all maintenance personnel back to the launch control center. Runholt reported what had happened and what he had done to prevent further damage. Hanks directed Leslie and Bibler to don the emergency self-contained breathing apparatus and go out into the silo to put out the fire if there was one.

They found light smoke and no fire on Level 2 and proceeded down the emergency ladder to Level 3. Again no fire was present, but the smoke was much heavier. Leslie and Bibler carefully checked each circuit breaker panel for hot spots, turning off each breaker in the process. They could not see the large copper bus which had melted. They returned to the launch control center for new air purification canisters for the CHEMOX breathing apparatus and went back to the silo to safe the missile systems and preserve air pressure for starting the diesel generator when the time came.

TSgt William Shaff was on duty in the Titan II Technical Engineering Division offices at Davis-Monthan AFB. Shaff was directed to get out to 570-6 immediately; a helicopter was waiting for him on the flight line. When Shaff arrived, the launch complex was incredibly quiet. All power was off except that supplied by the emergency lighting units and the 28-volt DC batteries which had sufficient power to keep the launch critical monitoring systems energized and maintain the missile on alert status for a limited time. After checking in with Hanks, Shaff went down to MCC-1 using the emergency escape ladders since the elevator was not operational. This was a new experience in a completely silent silo equipment area and in pitch darkness save for the flashlight he was carrying. Arriving at MCC-1, Shaff saw that the fire had been small, but there was considerable damage. A large bus bar that carried the 480-volt AC power for distribution amongst the other circuits in MCC-1 had been partially melted by the fire, showering wire bundles in an open tray with molten copper. These wire bundles were the control circuit cabling for the silo and launch duct but were now just a mass of copper and burnt insulation. Shaff clearly remembers how fortunate everyone had been that Runholdt had turned the diesel off as he evacuated the silo. If the diesel had been supplying high voltage power to the motor control center as the fire was progressing, damage would have been much worse.

Shaff returned to the launch control center and called the base, briefing the deputy commander for maintenance on the damage and the tenuous power situation at the site. The 28-volt DC batteries would not last much longer. He requested a portable generator to supply emergency power to the launch control center. He also requested additional help as well as wire and wire connectors so that they could start splicing out the damaged circuits and get the missile back on commercial power as soon as possible. The Davis-Monthan AFB Martin Marietta Company field engineering representative, Wayne Kalymago, and TSgt Robert Bash were sent out, along with the requested supplies.

Now the interesting part began. Shaff, Kalymago, and Bash began what was to be nearly 21 hours of work to splice around the damaged wiring bundle. A 2-foot section had melted together, so the team had to identify specific wires in the bundle of 50 and splice in jumpers. Space was limited and since the generator had not yet arrived, there was little air circulation. The only illumination was from the emergency lighting units and the small hand-held flashlights they carried with them.

When the portable generator arrived at the complex it was connected via long cables down the launch control center air intake shaft, through the open escape hatch, and over to the Motor Control Center 2 (MCC-2) on Level 3 of the launch control center. Leslie had the unenviable task of crawling up the 55-foot air shaft, holding a broom to knock down the spiderwebs. Once he got to the top, he tried to open the grid over the air shaft intake only to find that the wire seal that was supposed to break was too strong. Turning himself upside down on the ladder, he tried to kick the grid open but was again unsuccessful. He resigned himself to the 55-foot climb down to get a pair of wire cutters and upon return was finally able to get the grid open.

Returning power to the site was somewhat complicated. Shaff removed the MCC-2 cir-cuit breaker on MCC-1 on Level 3 of the silo equipment area, isolating both units from each other, and attached the power cable directly to MCC-2 on Level 3 of the launch control center. MCC-2 could now supply 480 volts AC power to the launch control center to run the launch control center air-conditioning unit and to recharge the 28-volt DC power supplies. The batteries could be recharged and, for all intents and purposes, the launch control center was operating as if it were on diesel power from Level 3 of the silo. With the power restored to the launch control center, extension cords were run to the silo equipment area on Level 3 to supply better lighting and air circulation fans to remove some of the stale air in the silo as work continued to repair the damaged cabling.

Once Shaff, Kalymago, and Bash had all the wiring splices in place, they cautiously reconnected the diesel generator, configured to power only the silo and not the launch control center. When they were satisfied that all the connections were good and the equipment was working properly, they shut down the diesel, disconnected the topside generator, and then reconnected the normal complex power. This temporary fix worked well while the permanent fix cabling and bus bars were being fabricated.[8]

On 5 October 1978, Launch Complex 570-6, coincidentally manned with the same missile combat crew that had been on alert on 5 September, experienced its second fire in MCC-1. Maintenance technicians were working on the air compressor circuit breaker in MCC-1. Unknown to them, a nut had worked its way loose from one of the support bolts. When they finished their work and pushed the breaker mechanism back into its mounting on MCC-1, the bolt fell out and shorted another bus bar. The loss of power triggered the diesel generator to come on line, but by this time the technicians were already out of the silo equipment area and did not turn off the generator. With the generator now supplying its full power to MCC-1, the shorted bus bar again melted, fusing large diameter main power cables and causing them to overheat and melt back toward the launch control center.

Leslie and Bibler were on the Level 3 launch duct work platforms conducting a missile leak check. When the fire started, Leslie and Bibler assisted the maintenance personnel up emergency escape ladders, and Leslie tried to turn off the diesel but the smoke was too dense. As they returned to the launch control center, they found that Blast Door 9 would not open due to the rapidly failing power situation. They used the nearby manually operated hydraulic pump to get Blast Door 9 and then Blast Door 8 open. After briefing Hanks, they donned the CHEMOX units in the launch control center, and turning off the diesel generator, they returned to the silo equipment area and descended through the dense smoke to Level 3 where they found their way to the MCC-1. With Bibler holding the electric safety stick and flashlight, Leslie opened each panel and extinguished any flames using the fire extinguisher. Several trips were made to and from the launch control center to resupply the CHEMOX units and finally the missile systems were safed. Repair of this damage took better than four days, but the launch complex remained on alert through the use of the portable generator arrangement from one month earlier.

A1C William Leslie Jr. and SrA John Bibler received the Airman's Medal for Heroism in putting out the 5 October 1978 fire. Their crew was SAC Crew of the Quarter and appeared on the cover of several SAC publications. The entire crew was upgraded to "S" status and kept that designation, the highest a missile combat crew could achieve, until a new MCCC and DMCCC were placed on the crew months later.[9]

Sgt Tom Veltman, MFT, was assigned to Launch Complex 533-8 at the beginning of his career with Titan II in 1979 and spent many weekends there on alert. Weekends were actually the best days since there was usually no maintenance scheduled, allowing the crews to get the DSV done early and have time for other pursuits. The crew, composed of Capt Wally Smart, MCCC; 1Lt Tommy Kramer, DMCCC; A1C John Earnest, BMAT; and Sergeant Veltman, was a creative group. One weekend they found a quantity of paint that had been left behind from a painting project. Since they were frequently flown to their site by the helicopter rescue team based at McConnell AFB, this paint and their creativity resulted in a bright, shiny, and highly visible paint job on the helipad next to the launch complex. The newly emblazoned "Welcome to Complex 533-8, Elevation 1129, Population 4" made for a readily visible landing area for future helicopter trips. This made them quite popular with the helicopter pilots.[10]

Sometimes special guests such as local and national political figures, as well as high-ranking Strategic Air Command or Air Force officers, would visit a Titan II complex. On such occassions, elite crews were scheduled to man the site chosen for the visit. Known as "select" crews, these men and women had shown exceptional skill and teamwork as a regular crew by successfully passing three consecutive standardization board checks (which were given approximately every six months or as new crew members rotated in). The year 1965 started out for the 390th SMW with the visit of Vice President Hubert H. Humphrey to Launch Complex 571-7 on 25 January. This complex was conveniently close to Davis-Monthan AFB, readily accessible by road, and was the "showcase" site for the 390th SMW. Crew S-117, Capt John W. Haley III, MCCC; 1Lt Dennis Polumbo, DMCCC; MSgt Gordon L. Anderson, BMAT; and SSgt Richard Sloan, MFT, was selected to host the visit. Also on duty were MSgt Michael Santistevan, MFT, newly assigned to the crew and 2Lt Jay Kelley, a student DMCCC. Crew S-117 usually stood alert at Launch Complex 571-5 but was informed that with Vice President Humphrey visiting, they would pull their next alert at Launch Complex 571-7.

As might be expected, stories abound concerning this event. Several individuals who were at the wing at the time clearly remember the visitor as President Lyndon B. Johnson, which just goes to show just how much of a profile Vice President Humphrey had at the time. Haley was known as a no-nonsense crew commander who was always working to make his crew stand out in a professional manner.

At this time in the program, crew uniforms were white coveralls. Sidearms were carried by the officers in shoulder holsters. These were uncomfortable, so Haley had purchased web belts with highly polished titanium buckles for the officers on his crew and tooled leather belts for the enlisted members who did not carry sidearms. The belts were worn only at the site at the discretion of Haley. Haley felt they added a touch of military custom and dressed up the crew somewhat.

The visit was a classic protocol nightmare from the beginning. The day started off with a visit by three Secret Service agents. The immediate question concerned the Secret Service agents' sidearms. Ordinarily no one but the Secret Service carried loaded firearms in the presence of the vice president. However, since they were visiting a missile complex, all visitors were under the direction of the site commander, Haley, and had to follow his orders. He told the Secret Service agents only the crew officers could have loaded firearms. After a few minutes of heated debate, it was agreed the crew would retain their weapons, the Secret Service

Figure 7.2. On 25 January 1965 Vice President Hubert H. Humphrey visited Launch Complex 571-7, 390th SMW. *Left to right:* Administrative aide Jack Sherman (partially obscured by the vice president's shoulder), Vice President Hubert H. Humphrey, Capt John.W. Haley III (missile combat crew commander), and Col Ralph D. Sampson (390th SMW commander). *Courtesy of John W. Haley III.*

would remove the bullets from their sidearms but retain the sidearms, and most important, they would be positioned on the gun-side of the officers at all times during the visit.

Kelley recalled the defining moment of the tour nearly 33 years later. As a student crew member, he was there to observe and learn but not, at this point, participate in the demonstrations. Haley was explaining the functions of the Launch Control Complex Facilities Console, leaning over and pointing out what the various indicators meant. Vice President Humphrey stopped him to say how impressed he was with Haley and his crew. He then asked when his enlistment was up. Haley, a captain at the time, was surprised and somewhat miffed. He politely explained to him that he was an officer and did not have a term of enlistment. He grabbed his uniform collar near the captain's insignia and showed the vice president. Vice President Humphrey pursued the point further, asking if he was a career man. Haley replied that while he loved the Air Force and particularly missile duty, he was a reservist and not a career man. "Why not become Regular status?" the vice president asked. Haley replied that he had taken and passed the test but for whatever reason had been passed over and was now too old at the age of 33. The vice president turned to his aide and said, "No one is too old at thirty-three." The aide wrote some notes down about Haley, and the rest of the tour was completed. Six weeks later, Haley and many other reserve officers were upgraded to Regular status.[11]

Visitations to Titan II launch complexes were not limited to politicians. In the early years of the program family members were given orientation visits to more fully understand what their spouses were doing. Civic dignitaries were often shown through the sites as well. Visitors included young people too. Capt Kenneth F. Grunewald and his crew were often involved with distinguished visitors taking tours of the launch complexes at the 308th SMW. One of the favorite stories regarding these tours involved the Boy Scout troops that visited yearly. Grunewald would size up the probable gullibility of the troop members and then set up a demonstration of the 467L teletype printer communications equipment which had a small shelf that held the printer output as it cleared the print head. He began by describing the rest of the communications equipment and would then turn in his chair and say to the DMCCC, "I'm a little hungry, would you have the base send out a peanut butter sandwich?" The DMCCC would pick up the phone and call the base, quietly asking for a test message on the 467L. The message would be sent and, just before the paper cutter was activated, Grunewald would ask to make sure it was sliced. The paper cutter would make a chopping sound, the cover to the 467L would be lifted, and there sat a sliced peanut butter sandwich. The kids' eyes would be the size of half dollars until they realized they had been set up by the crew.[12]

In theory, the missile combat crews never knew when a Wing Standboard Evaluation Crew would show up at their launch complex for an on-site review of each missile combat crew member's performance of his duties during alert. With crew members coming and going, and each crew scheduled for evaluation based on the individual crew member's records, it was difficult to predict. On 9 November 1968, the missile combat crew composed of Maj John K. Powers, MCCC; 1Lt Thomas Lafferty, DMCCC; SSgt Thomas Maerker, BMAT; and A1C Steven Ciani, MFT, had just settled into their daily routine when the gate phone rang. Since no maintenance had been scheduled for that day at Launch Complex 571-7 and the 3901st Strategic Missile Evaluation Squadron (SMES) had arrived the day before, Powers was not optimistic. Sure enough, upon answering the phone his worst fears were answered twofold. Not only had a Wing Standboard Crew arrived, the 3901st SMES team was with them to evaluate the Standboard Crew's evaluation of his crew. Powers remembers that while it was pretty crowded on Level 2 of the launch control center, and the sweat was beginning to show after several hours of questions and such, he thought things were going pretty well. As luck would have it, just as he completed the thought, the FUEL VAPOR LAUNCH DUCT indicator illuminated on the Launch Complex Control Facilities Console. Often this was a false alarm, but it required that the crew initiate the Fuel Vapor Launch Duct Emergency Checklist. Following the checklist, Maerker and Ciani donned their CHEMOX safety masks and carrying a portable vapor detector, entered Blast Lock Area 201 (between Blast Doors 8 and 9) to check the Vapor Detector Annunciator Panel for indication of fuel vapors as well as their concentration. After the team cycled through the detection locations, they reported no indication of fuel vapors in the launch duct or silo equipment area. Powers continued the checklist and directed Maerker and Ciani to go to the surface and check the silo exhaust vent for fuel vapors. After several minutes Maerker reported that the tests were

Figure 7.3. 390th SMW command post showing the controller's console on 17 August 1972 prior to Coded Switch System Installation. *Courtesy of Lockheed Martin Astronautics, Denver, Colorado.*

negative. Both men then came back down and continued the checklist actions that involved going to Level 6 of the silo and checking the detector equipment. It was indeed a false alarm. After more than six hours of evaluation, Powers and his crew passed their combined evaluation with flying colors.[13]

Standardization checks were all the more intense if a real emergency took place during the inspection visit to the launch complex. On 24 July 1973, Missile Combat Crew R-005, composed of Capt Bruce J. Stensvad, MCCC; 2Lt James M. Spain, DMCCC; Sgt Daryl L. Ray, BMAT; and Sgt Glenn P. Allen, MFT, was on duty at Launch Complex 570-4, 390th SMW. A two-man Standization Board Crew team, Capt Norval D. Martin, MCCC, and TSgt James E. Werton, MFT, were evaluating Stensvad's crew as they began a routine Olympic Play exercise. Olympic Play was a standard test exercise which required that a missile verification and launch verification be conducted to test the readiness of missile B-31 (62-012). Olympic Play exercises at this time in the program were conducted approximately once per month at each launch complex.

The missile verification, an automated check to ensure all missile components were functioning properly, proceeded without incident. The checklist then called for an ordnance circuit test, which also was passed without abnormal indications. The launch verification, a semi-automatic process that exercised the flight hardware, was started and after the second key-turn, a red light for STAGE I HYDRAULIC PRESSURE illuminated. The launch verification checklist contained steps that configured the missile so that it could not actually be launched. The first key-turn essentially set the system into the launch verification condition, while the second key-turn started the countdown. LAUNCH DISABLE, rather than LAUNCH ENABLE, was the first countdown indicator illuminated on the Launch Control Complex Facilities Console.

After reviewing the appropriate technical orders and consultation with the Standboard Crew members, Stensvad sent Ray and Allen to Level 2 of the launch duct. This was a normal part of the post-launch verification procedure. When Ray and Allen opened the launch duct access door, they immediately noticed the smell of hot electrical insulation and saw smoke accumulating at Levels 1, 2, and 3 of the launch duct. They looked for but did not observe any indication of fire, which would have been readily evident in the dark silo. There was no indication of fire on the Launch Control Complex Facilities Console and the LAUNCH DUCT FIRE sprays had not been activated. The crew members returned to the launch control center, and Stensvad notified the wing command post. Ironically, almost two years earlier, Stensvad had been on a missile combat crew that attempted a launch at Vandenberg AFB that had resulted in an abort due to a somewhat similar situation. Upon designation by the command post as a Fire Hazard Team, Ray and Allen descended to Level 7 of the launch duct, opened the launch duct access door, and again inspected for fire. With the launch duct lights off, even a small fire would have been easily seen. Much to their relief they saw no flames. Once the launch duct lights were turned on, Allen and Ray could hardly believe their eyes. The turbopump assembly on engine Subassembly 1 had disintegrated. Fortunately, the shrapnel from the disintegration had embedded in the acoustical liner of the launch duct, and the missile airframe had not been damaged.

Investigation of the accident, led by Col William H. Bush, Vice Commander, 90th Strategic Missile Wing (Minuteman), F. E. Warren AFB, Wyoming, was concluded one month later. His conclusion was that a short circuit had developed due to wiring harness and engine start cartridge initiator connector deficiencies. An electrical transient, or spike, from the activation of either the Stage I or Stage II hydraulic pump, or the inertial guidance system being advanced to ready, traveled through the short circuit, causing the Stage I Subassembly 1 engine start cartridge squib to fire, igniting the start cartridge. Under normal launch conditions, fuel and oxidizer would have filled the turbopump impeller spaces by this point in the countdown, but since this was a launch verification, no propellant was present in the turbopump assembly. The start cartridge high-pressure gases spun up the dry turbopump, which caused turbine overspeed and resulted in turbine disintegration. Inspection of the start cartridge connector revealed flakes of gold metal lodged between two pins; inspection of the engine electrical junction box also revealed frayed insulation. An intermittent short circuit, detectable by the ordnance circuit test only when the wires were pushed or pulled, was the result of these two deficiencies and had caused the squib to ignite.

The investigation report listed personnel error as a contributing cause to the accident.

Col Eugene D. Scott, wing commander, 390th SMW, did not concur with this finding, noting that all technical orders had been followed correctly. The recommendations from the report urged all ordnance to be disconnected prior to launch verification checks until the electrical connectors and engine junction box wiring was checked fleetwide. Costs of repair to the missile and engine were estimated at $316,835; launch duct equipment, $67.[14]

While many officers and enlisted men are legends in the Titan II program, one story stands out. Col William E. Bifford, the 308th SMW wing commander from 5 July 1971 to 1 September 1972, was a former C-130 pilot from Vietnam. Prior to assuming command of the 308th SMW, Bifford had been deputy commander for operations at the 390th SMW. Bifford had a tremendous sense of humor, evidenced by the canned laughter box that was prominently displayed on his desk. Whenever someone came in with a "pressing" problem, he would say, "Wait," push the laugh box button, enjoy the 15 seconds of raucous laughter, and then ask just what the problem was. It was not long before the 308th SMW learned to appreciate his sense of humor. One morning at the stand-up briefing, maintenance began a slide show of the work in progress at the two squadrons. The second slide was meant to say "Cathodic Banding," but came up on screen as "Catholic Banding." Bifford turned to the deputy commander for maintenance and, barely able to control his laughter, managed with a straight face to inquire why all the Catholics were being gathered and banded. The rest of the officers present broke into laughter simultaneously with Bifford!

Bifford had a serious side that was pure Air Force. LtCol Robert Buzan, chief of training, was present at more than one of the serious moments. When a missile crew failed a standardization check they had to report to the wing commanders office to explain why they had failed and what they were going to do about it to prevent a reoccurence. On one such occasion the crew reported to Bifford's office only to find that the squadron commander, sector commander, chief of standardization, chief of training, and chief of the command post were present. Into this somewhat crowded office came the four crew members. Bifford was an imposing individual and even more so that day as he stood behind his desk with his thumbs under his belt. After the MCCC, a captain, reported to Bifford, the colonel admonished the crew for failing the check. He then dismissed the two enlisted men and continued, soundly lecturing the DMCCC and then dismissing him too. This left the captain alone, at attention, undoubtedly feeling surrounded and certainly steeling himself for what was about to come. Buzan watched as Bifford ate the captain up one side and down the other. From the look on the captain's face, this was certainly not one of his better days. Suddenly, Bifford stopped and the captain started to smile a bit as Bifford said, "I wish I had a dozen more men like you." The smile disappeared as Bifford quickly followed with, "But unfortunately for me I have two dozen!" The captain was clearly in a state of shock as he was dismissed. Bifford turned to the assembled officers and said, "That ought to hold him for a while and I bet you he will never be back in this office again!" He wasn't.[15]

Not every day started out easily for missile combat crews approaching their site, ready to report in. On the morning of 27 January 1978, the oncoming missile combat crew approaching Launch Complex 374-7, 308th SMW, noticed rust-red oxidizer vapors rising from the missile complex. They drove to Damascus and contacted the command post, which in turn notified the Missile Potential Hazard Team (MPHT) members. Within 15 minutes, the MPHT directed the missile combat crew commander at the complex to turn off the circuit breakers to the heaters on the oxidizer transport trailers. The heaters were used to keep the oxidizer between 42 degrees Fahrenheit and 60 degrees Fahrenheit in preparation for transfer into the holding trailer. In the cold January weather of Arkansas something had apparently gone wrong with the heater control circuit. Meanwhile, a helicopter from the 37th Air Rescue and Recovery Squadron was sent to provide aerial surveillance of the situation. The helicopter crew confirmed the presence of oxidizer vapors rising from the trailer and crossing State Highway 65 in a cloud approximately 3,000 feet long, 300 feet wide, and 100 feet in height. The MPHT immediately directed the Van Buren County Sheriff's Department to block Highway 65 and requested evacuation of civilians in the path of the oxidizer cloud, including an elementary school 1.5 miles north of the complex. A second helicopter with propellant transfer personnel in rocket fuel handlers' clothing outfits was immediately dispatched. Upon arrival at the complex, the team reported that the oxidizer trailer tank was at 101 degrees Fahrenheit and leaking around the manhole cover, but that the safety rupture discs had not yet burst. They sprayed water on the tank to cool it off and tightened the manhole cover bolts, decreasing the amount of vapor considerably. After several hours Highway 65 was reopened to traffic and the Missile Potential Hazard situation was terminated. Four civilians displayed some symptoms of contact with the vapors and were transported to the Little Rock AFB hospital for evaluation. Two were released the same day, and two were held overnight for observation, subsequently released, readmitted, and released again on 4 February 1978. A faulty thermostat switch on the transport trailer had caused the heaters to operate continuously. The 308th SMW immediately instigated new procedures for the surveillance of propellant transport trailers stationed on the complex, and this type of accident did not occur again.[16]

One of the more bizarre stories from the Titan II program involved a case of espionage. On 17 December 1980, 2Lt Chris M. Cooke, a DMCCC in the 381st SMW, was observed by the FBI entering the Soviet Embassy in Washington, D.C. At the time, neither the Air Force, SAC, nor the 381st SMW had any idea that an officer on a missile combat crew was perhaps passing extremely sensitive information to the Soviets. During the next crew departure briefing that Cooke was to attend, a trap was set. An officer from the Office of Special Investigations (OSI) gave a generic briefing reviewing the fact that any contact with members of a communist or hostile country must be reported as a matter of routine. While such briefings are standard procedure for personnel in sensitive areas or those stationed in countries where such contacts might occur, this was somewhat unusual for missile combat crews in southern Kansas.

The OSI and the FBI waited for Cooke's next leave. He was again observed entering the Soviet Embassy in Washington, D.C. He was arrested shortly thereafter and taken to Langley AFB for questioning. Assuming that Cooke was the tip of an espionage iceberg and that a major spy ring was involved in penetrating the operational secrets of the Titan II system, Headquarters SAC offered Cooke full immunity from prosecution and a general discharge from the Air Force in exchange for his full and complete cooperation. In addition, SAC wanted to know what classified information Cooke had given to the Soviets. Although Cooke would go free, his naming names and identifying all other personnel involved, and knowing what he may have given to his Soviet contacts, was worth this price, or so the Air Force and SAC believed. Cooke agreed to SAC's proposal and began to name names, but he provided only one—his own. All of the interrogation was done with Cooke hooked up to a lie detector and with one of the FBI's best experts conducting the analysis. The lie detector information indicated that his answers were not deceptive.

It appears that Cooke had been operating completely on his own initiative. There was no spy ring. While transcripts of the interrogation are still classified, the material that was released indicated that Cooke had been living his own fantasy as an international spy in his own amateurish way. Having cooperated as agreed to by all involved, Cooke fully expected to be released. For a variety of reasons, the Air Force and HQ SAC quickly retracted the immunity agreement and transferred Cooke back to McConnell AFB in late May where he was brought up on one charge with 3 specifications on 29 May by his squadron commander, LtCol Ken L. Hollinga. On 18 June 1981, these 3 specifications were repeated along with a separate charge with 10 specifications involving unlawful visitations to the Soviet Embassy in Washington, D.C., and the passing of sensitive information to Soviet Embassy personnel. The first charge was a violation of the Uniform Code of Military Justice, Article 92, and the second was a violation of the Uniform Code of Military Justice, Article 134.

The rest of the summer was consumed by the resulting media circus as attorney F. Lee Bailey was hired by the Cooke family to defend their son. Bailey arrived in Wichita to media fanfare, and the start of the preliminary hearings for the court-martial began. News of the immunity deal and its withdrawal had not yet been released to the public relations staff at McConnell AFB, and they were kept busy educating the news media about the military justice system. Once the immunity deal information was released, the long, hot summer got longer and hotter as the media hounded base personnel for details on Cooke's motive and the rescinded immunity agreement.

On 1 September Cooke was moved to Fort Meade, Maryland, for additional hearings that were to take place at Andrews AFB. In the end, F. Lee Bailey made a motion to the U.S. Court of Military Appeals. The court upheld the original offer of immunity and ordered Cooke discharged immediately. Cooke's constitutional right to due process had been violated when Air Force officials promised him immunity from prosecution if he told the truth and then failed to live up to the bargain. Cooke was discharged from the Air Force on 22 February 1982 and left a free man.[17]

Barely a year went by before another nationwide press relations problem cropped up for the 381st SMW. In early August 1983, Sgt Mark Hess, a media specialist for the Public Affairs Division of the 381st SMW, had a problem on his hands. Mike Ginsberg, a local reporter for the *Wichita Eagle-Beacon,* called about a story he needed to write as a sidebar to the Jack Anderson column that would be appearing in an upcoming *Parade Magazine* story about accidental near launches of ICBMs, specifically a Titan II from the 381st SMW. While Hess had heard about the article and that it concerned McConnell AFB and the 381st SMW, the details of the final story had not been made available. Ginsberg didn't reveal the entire contents of the story to Hess but did say that a crewman was quoted as saying that a Titan II ICBM had almost been launched, fully armed, from a silo near Wichita, due to faulty maintenance. Ginsberg wanted to know if this was true. This was a surprise to Hess, and he told Ginsberg he would pursue the matter through channels and get back to him.

Hess forwarded the inquiry through the 381st SMW command structure, the Eighth Air Force and Headquarters SAC. Hess remembers that there wasn't much time since the story was going to appear in a couple of weeks. With the Rock, Kansas, oxidizer spill of August 1978 and the explosion at Damascus, Arkansas, in September 1980, Hess knew that his office and the Air Force did not really have the option of an unqualified "No, it is not true."

Unable to get a prepublication copy of the article, further conversations with Ginsberg revealed what had apparently happened and how it had been misconstrued to fit the theme of the "exposé." The source of the story was a DMCCC who was being forcibly discharged from the Air Force. He had decided to retaliate by contacting Anderson and relaying selected parts of a Combined Systems Test that had been run by his crew during a Reliability and Aging Surveillance Program (RASP) operation at Launch Complex 532-4. (See Chapter 8 for a description of RASP.) This was a major inspection and test procedure which was routine for the missile selected for RASP inspection and had been conducted without difficulty since the early 1970s.

To conduct the test, first all power to the missile was turned off, including power to the guidance system. Next, all the explosive ordnance was removed from the missile, including disconnecting the reentry vehicle/warhead as specified in the test protocol. Included in the ordnance removed were the three start cartridges used to initiate Stage I and Stage II engine operation. The removal of the ordnance was done by a maintenance crew and verified by the maintenance chief and maintenance officer as part of the operation checklist. A team from the Wing's Plans and Intelligence Division replaced the operational code in the butterfly valve lock with a maintenance code so that the CST could be performed. At no time would the missile combat crew have been directly involved in any of these procedures other than to monitor their progress from the launch control center. Once all of the ordnance was removed, power was again applied to the missile and guidance system.

At the designated point in the RASP procedure, a CST was performed. The CST performed a simulated launch and flight sequence as part of the test procedure but since all of the missile ordnance had already been removed, the prevalves could not be opened and no propellant could possibly flow to the engines. In this case the CST was monitoring electronic signal response with no possibility of launch. Special fuses on the CST test equipment located on Level 3 and Level 7 of the silo equipment area replaced the ordnance items so that the specific signals to activate batteries, open prevalves, begin engine ignition, et cetera, could be sensed by whether or not these fuses "blew." During the CST, the missile combat crew performed a test launch procedure as per the appropriate technical order. They turned the keys and monitored each light to verify that a proper launch sequence was indicated. When the LIFTOFF indicator illuminated, the test continued through the flight test to checkout the Stage I and Stage II hydraulic and flight control systems.

Anderson's staff had not completely researched the validity of the story by the disgruntled DMCCC, choosing instead to publish a contrived and excerpted interpretation of the actual events. The 381st SMW had not been contacted in any manner prior to the inquiry by Ginsberg. Since the DMCCC had been discharged, the Air Force could release little information concerning his service records due to the privacy act. Hess and the Public Affairs staff at the 381st SMW were faced with a dilemma on how to specifically counter a story that they had not seen and still maintain the privacy of the individual involved. Ginsberg was given a copy of the operations log for the day in question. He read through the log and felt that the missile had almost been launched. Clearly Ginsberg had to be fully briefed about this maintenance operation in a manner that did not compromise the operational security of the system.

With the wing commander's approval, Ginsberg met with the deputy commander for operations, the deputy commander for maintenance, a standardization missile combat crew, and Hess. The ground rules were that Ginsberg could ask any question about how to launch a missile, with obvious restrictions about codes and the like, as well as questions concerning the RASP operation. Hess remembers the meeting lasting most of the afternoon, with Ginsberg asking probing questions and being more than satisfied with the answers.

On Sunday, 14 August 1983, *Parade Magazine* was published nationwide with the Anderson story. The Air Force could not negate the Anderson story on a national basis. Hess was pleased to see on page 2 of the Sunday *Wichita Eagle-Beacon* a short article detailing the investigation of the story and how the portions relating to the near launch of a Titan II from the 381st SMW were completely false.[18]

Titan II missile combat crew members on alert, as well as the maintenance personnel that kept the launch complexes running, had a primary focus. The message to launch might come at any time, and the system had to be ready to respond. Unlike aircraft crews that could fly simulated combat missions against adversaries on a frequent basis to hone their skills, missile combat crews had to rely on the Missile Procedures Trainer to simulate problems and drill on responses. The true test of the system, that of a launch from an operational silo in Arizona, Kansas, or Arkansas, did not take place. The next best situation was to launch from the Titan II training facilities at Vandenberg AFB.

VIII

TITAN II OPERATIONAL FLIGHT TEST AND EVALUATION PROGRAMS

With the conclusion of the research and development launches at Vandenberg AFB on 13 March 1964 came the beginning of the Titan II operational flight test program. The flights were grouped into the formal operational test programs and special evaluation programs for the Army antiballistic missile system research program. The Strategic Air Command (SAC) training and evaluation flights consisted of three programs: the Demonstration and Shakedown Operations (DASO), the Operational Test (OT), and Follow-on Operational Test (FOT) programs, conducted from 1964 to 1969 with 39 launches attempted. Thirty-eight of the launches were successful, and one of the attempts was a ground abort. Of the 38 successful launches, 7 ended with in-flight failures. After the FOT program, Titan II evaluation was switched from flight tests to "bench" tests where missiles were evaluated both in the silo and at the manufacturer for ability to carry out a successful mission.

There were 10 additional successful Titan II launches and 2 aborted launches outside the formal SAC training program. Nine were part of the Army Safeguard System Test Target Program (SSTTP) and Ballistic Missile Defense Test Target Program (BMDTTP). The tenth launch was the Special Operational Test—Integration Test Flight-1, code-named RIVET HAWK, which was a SAC program to flight test the upgraded Titan II guidance system. A list of launch crew members is given in Appendix 1.[1]

A typical Titan II flight profile is given in Table 8.1. During an operational launch, the time from receipt of a valid launch message to actual key turn could vary considerably depending on the war plan being executed. From launch sequence key turn to engine ignition took 58 seconds. After the Stage I engine reached 77 percent of thrust, three timers counted down 1.8 seconds as Stage I thrust continued to build toward 100 percent. At the end of 1.8 seconds, the explosive nuts on the hold-down bolts fired and liftoff took place. The missile rose out of the silo, climbing vertically for approximately 15 seconds. Five seconds after liftoff the missile guidance system commanded a roll signal to align the missile with the target bearing, also called the azimuth bearing.

Stage I powered the missile until approximately 148 seconds into flight when either fuel level signaled engine shutdown or oxidizer depletion led to the decay of Stage I thrust below 77 percent. The Stage I thrust chamber pressure switch monitored the decay in thrust, simultaneously signaling the guidance computer to ignite the Stage II engine and fire the explosive nuts holding the two stages together. This fire-in-the-hole technique utilized the decaying thrust of Stage II to maintain positive acceleration forces on the Stage II propellants. This ensured a smooth flow of propellants to the Stage II turbopump at Stage II engine start up.

TABLE 8.1

TITAN II NOMINAL TRAJECTORY (5,500 NAUTICAL MILES)

EVENT	TIME (SEC)	ALTITUDE (NM)	RANGE (NM)	VELOCITY (FT/SEC)
1. Liftoff	0	0	0	0
2. End of Vertical Rise	15	0.22	0	172
3. Stage I Burnout	148	41	41	8,258
4. Stage II Burnout	328	189	384	22,159
5. Vernier Cutoff	343	207	434	22,165
6. Reentry Vehicle Separation	346	210	446	22,164
7. Apogee	1,165	693	2,701	18,477
8. Reentry	2,114	49	5,381	23,420
9. Impact	2,191	0	5,502	670

Source: Adapted from "Final Titan II Operational Data Summary, Rev 3, September 1964, TRW Space Technology Laboratories," p. 3-1, History Office, Air Force Space Command. This document is classified as SECRET. The information used is unclassified.

Stage II continued to fire for approximately 180 seconds; the actual time depended on the target selected. Simultaneous with Stage II cutoff was the ignition of the solid fuel vernier rocket motors which provided final velocity adjustments for Stage II as signaled by the guidance computer. Approximately 16 seconds later, now 343 seconds into the flight, the vernier engines were shut down by the guidance, the reentry vehicle prearmed and reentry vehicle separation signaled at 346 seconds into flight.

In the early part of the program Stage II had a pair of translation rockets that served to push Stage II back from the reentry vehicle at separation. Coupled with a pair of pitch/depitch rockets also on Stage II, the Stage II airframe was backed away to one side of the reentry vehicle to serve as a decoy during reentry. The translation rockets were removed in April 1978. Details of the reentry vehicle functions during reentry remain classified.[2]

Demonstration and Shakedown Operation, 1964

The DASO launches were the first Titan II operations that were completely conducted by SAC crews (Table 8.2). Commencing with the launch of missile B-28 (62-009) on 30 July 1964 from Launch Complex 395-D, Vandenberg AFB, California, the DASO program goal was to verify the operational capability of SAC crews; determine the practical reliability of the system; measure reentry vehicle and penetration aid performance; achieve special operation requirement accuracy in the reentry vehicle impact area; and provide the Advanced Research Projects Agency with "special" data. The program ended with a record of five-for-five with no holds or aborts due to missile subsystems. The first three launches were conducted by crews from Vandenberg's 395th Strategic Missile Squadron (SMS), the last two by operational wing crews from the 308th Strategic Missile Wing (SMW), Little Rock AFB, Arkansas, and the 381st SMW, McConnell AFB, Kansas.[3]

TABLE 8.2 DEMONSTRATION AND SHAKEDOWN LAUNCH HISTORY

Date	Location	Missile #	Result	Air Force Serial #	Code Name	Op. #	Program	Unit
30 Jul 64	395-D	B-28	S	62-009	Cobra Skin	W-5342	DASO	395th SMS
11 Aug 64	395-C	B-9	S	61-2763	Double Talley	W-5427	DASO	395th SMS
13 Aug 64	395-B	B-7	S	61-2761	Gentle Annie	W-5440	DASO	395th SMS
2 Oct 64	395-C	B-1	S	61-2755	Black Widow	W-4907	DASO	308th SMW
4 Nov 64	395-D	B-32	S	62-013	High Rider	W-5897	DASO	381st SMW

S=completely successful; F=failure, A=launch abort.
Titan II Missile Synopsis, Headquarters SAC, 13 August 1985

Preflight operations initially required 16 days from the combined systems test, where all connections were tested after installation of the missile, to missile launch. By the fifth flight, B-32, this time had been reduced to 10 days, illustrating the readiness of SAC missile combat crews and maintenance personnel to field the missile system. The missiles differed from the operational configuration only in that they carried an instrumentation and range safety system destruction package as well as the DASO program package that included instrumentation for continued development of operational reentry vehicle and penetration aids subsystems. Three of the missiles were among the oldest in the inventory: two of the missiles had been on alert for almost a year, and one, B-1, was the first Titan II operational missile built. Two launches required further action. Missile B-28 demonstrated a slow thrust buildup for Stage I operation which did not affect accomplishment of the mission, and a weapon system ordnance problem was identified on B-32 during preparations prior to launch.[4]

Operational Test Program, 1965–66

The OT program began at Vandenberg AFB in March 1965, 15 months after the entire Titan II fleet had reached alert status (Table 8.3). One of the primary objectives of the OT program was to accrue sufficient reentry vehicle impact data to determine the system accuracy. In order to estimate the circular error probable from missile flight data to within a confidence of 15 percent, 45 successful flights had to be conducted.[5] The OT program started this process with 19 launches scheduled. Another objective was to continue study of the reliability of a missile and reentry vehicle that had been stored in an operational environment for an extensive period. Fifteen scoreable launches were necessary for accuracy determination with a confidence level of 25 percent, and each had to be flown with daylight reentry at the Enewetok impact area—one of the Marshall Islands that was at the terminus of the Western Test Range.[6]

Missile selection was conducted on a random basis. The selected missile was evaluated prior to removal from the operational silo, transported by ground or air to

TABLE 8.3 OPERATIONAL TEST LAUNCH SUMMARY

Date	Launch Complex	Missile #	Result	Air Force Serial #	Code Name	Op. #	Program	Unit
24 Mar 65	395-B	B-60	S	63-7715	Arctic Sun	W-7355	OT-1	390th SMW
16 Apr 65	395-C	B-45	S	62-026	Bear Hug	W-7564	OT-2	390th SMW
30 Apr 65	395-D	B-54	F	62-12297	Card Deck	W-7485	OT-3	390th SMW
21 May 65	395-B	B-51	S	62-12294	Front Sight	W-7722	OT-4	381st SMW
14 Jun 65	395-C	B-22	F	62-003	Gold Fish	W-7799	OT-5	381st SMW
30 Jun 65	395-D	B-30(25)	S	62-011	Busy Bee	W-7814	OT-6	381st SMW
21 Jul 65	395-B	B-62	S	63-07717	Long Ball	W-7820	OT-7	390th SMW
16 Aug 65	395-C	B-6	S	61-2760	Magic Lamp	W-7806	OT-8	308th SMW
25 Aug 65	395-D	B-19	S	61-2773	New Role	W-7835	OT-9	308th SMW
21 Sep 65	395-B	B-58	F	62-12301	Bold Guy	W-7859	OT-10	381st SMW
20 Oct 65	395-C	B-33(23)	S	62-014	Power Box	W-7849	OT-11	308th SMW
27 Nov 65	395-D	B-20(14)	S	61-2774	Red Wagon	W-1404	OT-12	308th SMW
30 Nov 65	395-B	B-4	F	61-2758	Cross Fire	W-7844	OT-13	381st SMW
22 Dec 65	395-C	B-73	F	63-7728	Sea Rover	W-7914	OT-14	308th SMW
3 Feb 66	395-D	B-87	S	64-451	Winter Ice	W-3662	OT-15	390th SMW
17 Feb 66	395-B	B-61	S	63-7716	Black Hawk	W-1555	OT-16	381st SMW
25 Mar 66	395-C	B-16(11)	S	61-2770	Close Touch	W-1851	OT-17	390th SMW
5 Apr 66	395-D	B-50	S	62-12294	Gold Ring	W-8020	OT-18	308th SMW
20 Apr 66	395-B	B-55	S	62-12298	Long Light	W-7984	OT-19	381st SMW

Note: Numbers in parenthesis, i.e., B-30(25), indicates a composite missile with Stage I from B-25 and Stage II from B-30. The serial number is for the Stage II component since it carried the guidance system.

S=successful; F=failure; A=abort

Titan II Missile Synopsis, Headquarters SAC 13 August 1985.

Vandenberg AFB, evaluated for effects of transport, and emplaced in one of the three Titan II launch facilities by the accompanying task force from the operational base. In this manner, all aspects of the Titan II wing operation were tested, not just the ability of the launch crews to successfully fire the missile. The first launch, code-named Exercise ARCTIC SUN, was on 24 March 1965, using B-60 (63-7715) from the 390th SMW, Davis-Monthan AFB, Arizona. The program ended with the successful launch of B-55 (62-12298), code-named Exercise LONG LIGHT, on 20 April 1966 by the 308th SMW, Little Rock AFB, Arkansas. Nineteen missiles were selected and launched during the program; 14 successfully reached the target area, and 5 experienced in-flight failures.

Two reentry vehicle configurations were used in this program. The first reentry vehicle type was the Mark 6 with a denuclearized W-53 warhead, containing Grade II high explosive but no nuclear components, and a passive instrumentation package code-named MIXED MARBLE II. This warhead configuration was known as the W-53 Type 2A and carried the Mk 6 Mod 1033 decoy package consisting of eight terminal and six mid-course penetration aids. The W-53 Type 2A package was an airburst option. Planned airburst setting was for 13,800 feet above sea level. Five of the 19 flights, code-named CARD DECK, LONG BALL, MAGIC LAMP, NEW ROLE, and POWER BOX, utilized the W-53 Type 2A configuration.[7] MIXED MARBLE II warheads were complete operational weapons except for the nuclear components. Originally the Air Force had wanted to test an actual W-53 warhead, much as the Navy had tested a live warhead on a Polaris A-2 missile fired from the USS *Ethan Allen* (SSBN-608) on 6 May 1962, but the nuclear test ban treaty prohibiting nuclear weapons tests in the atmosphere, space, or water became effective on 10 October 1963, preventing such a test.[8] The next best option was MIXED MARBLE II, where high explosives replaced the nuclear materials, providing a visual observation of successful detonation. Four of the five flights were successful, with the one failure taking place in Exercise CARD DECK, which had suffered an in-flight propulsion failure. The actual airburst heights varied from 15,796 feet to 16,017 feet.[9]

The second configuration was for surface impact. In this configuration the W-53 was again denuclearized, but instead of carrying explosives, a scoring kit was flown to assist in accurate tracking to point of impact. Impact was monitored by a variety of tracking systems, including both ground-based optical and radar systems as well as aircraft equipped with analytical equipment to study the chemistry of the heat shield during reentry and a hydrophone array on the lagoon bottom in the target area. In the impact tests every effort was made to recover the reentry vehicle. The decoy type for this configuration was either the Mark 6 1033 with a single terminal and single mid-course penetration aid or the Mk 6 1037J with eight terminal and six mid-course penetration aids.[10]

At the end of 1965, 5 of the 14 missiles launched failed to reach the target area and, to make matters worse, the year ended with back-to-back failures. While all missiles had been successfully launched, the resulting flight-to-target success ratio was 64 percent. SAC and the secretary of the Air Force were concerned. There were no discernible patterns in the 5 failures. The first, Exercise CARD DECK, the third flight in the program, suffered a premature Stage I Subassembly 1 shutdown 100 seconds into flight. Telemetry indicated a bearing seizure failure in a turbopump, most likely due to lubrication problems. The corrective action was to generate new servicing and inspection procedures fleetwide, and this failure mode did not recur.

The second failure was the fifth flight in the program, Exercise GOLD FISH. The number-one vernier rocket shut down prematurely. Indications were a failure in the phenolic thrust chamber expansion nozzle, and the fleet was retrofitted with a redesigned nozzle. Again, this failure mode did not recur.

The next four flights were successful, but the tenth flight, Exercise BOLD GUY, failed in normal flight when uncommanded signals from the computer resulted in premature shutdown of the Stage II engine at the point of staging. Review of the telemetry indicated a failure in the teflon insulation in the Stage I engine compartment had resulted in errant signals being sent to the guidance computer. Only one of the several signals was acted upon, the shutdown of the Stage II engine. By monitoring the computer telemetry through the remainder of the flight, it was determined that the computer was operating normally. A heat source in the Stage I engine compartment had melted the insulation, grounding a number of computer input lines. An improved inertial guidance grounding system was issued fleetwide, and the failure mode was not repeated.

The back-to-back failures of Exercise CROSS FIRE and SEA ROVER, the thirteenth and fourteenth flights, between November and December 1965, were due to engine structural and gimbal actuator adjustment problems. CROSS FIRE had a structural failure either in a propellant line or in the thrust chamber of Stage II resulting in an uncontrollable pitch attitude. SEA ROVER developed an uncorrectable yaw motion due to an incorrectly adjusted Stage I engine actuator. In both cases, changes to field technical orders remedied the situation, and these failure modes were not repeated.

While all failure modes were understood, a joint Titan II Air Force/contractor task force was formed to thoroughly review the problem areas and make recommendations. On 3 December 1965 the task force was assembled, and on 5 January 1966 the first meeting took place. One month later the task force found that no systematic causes were found among the failures. They were random events that required attention and resolution on a case-by-case basis.[11]

With the last five flights completed successfully, the Titan II Operational Test program achieved a 100 percent successful launch rate and a 74 percent in-flight success rate. This was considered by SAC to be a successful test program in light of the fact that the failures had not been due to a consistent structural or guidance problem and did not reoccur. The CEP was determined to be 0.785 nautical miles, well within the system's original specifications of within one nautical mile.[12]

Follow-on Operational Test Program, 1966–69

The FOT program took over where the OT program left off with 14 launches scheduled (Table 8.4). The program goals were the same as for the OT program. Unlike the OT program, missile combat crews from the 395th SMS, on strategic alert from 1967–69 at one or more of the Titan II launch complexes at Vandenberg AFB, conducted 2 of the launch operations. Thirteen missiles were successfully launched. Exercise GLOWING BRIGHT 49 was aborted due to a failure to fire the explosive hold-down bolts to release the missile. The Stage I engine fired for 24 seconds before being shut down by the launch crew.

Of the 13 missiles launched, 2 had in-flight failures. The first launch in the program, Exercise SILVER BULLET, experienced a failure in the reentry vehicle release mechanism. The signal for release was sent by the guidance system and the pitch/depitch rockets fired, but their effectiveness, as relayed by telemetry data, indicated that the warhead did not release.[13] The fifth launch in the program, Exercise GLAMOR GIRL, experienced a failure in the Stage II yaw rate gyro.[14] Neither failure mode reoccurred in the program. With 13 out of 14 launch attempts successful, a 92 percent successful rate, and 11 of 13 successful flights to impact, an 84 percent success rate, the Titan II program was clearly a reliable component of the strategic missile force. The official CEP for the FOT program is still considered classified. Calculation based on declassified impact data for a majority of the FOT flights indicates a CEP of 0.69 nautical miles.

TABLE 8.4 FOLLOW-ON OPERATIONAL TEST LAUNCH SUMMARY

Date	Launch Complex	Missile #	Result	Air Force Serial #	Code Name	Op. #	Program	Unit
24 May 66	395-C	B-91	F	64-0455	Silver Bullet	W-7955	FOT-1	381st SMW
22 Jul 66	395-B	B-95	S	64-0459	Giant Train	W-7968	ST	308th SMW
16-Sep 66	395-C	B-40	S	62-021	Black River	W-7920	FOT-2	390th SMW
24 Nov 66	395-B	B-68	S	63-7723	Bubble Girl	W-8807	FOT-3	390th SMW
17 Mar 67	395-C	B-76	S	63-7731	Gift Horse	W-7941	FOT-4	381st SMW
12 Apr 67	395-B	B-81	F	63-7736	Glamor Girl	W-7995	FOT-5	308th SMW
23 Jun 67	395-B	B-70	S	63-7725	Buggy Wheel	W-8022	FOT-6	381st SMW
11 Sep 67	395-B	B-21	S	62-0002	Glowing Bright 44	W-8038	FOT-7	381st SMW
30 Nov 67	395-B	B-69	A	63-7724	Glowing Bright 49	abort	FOT-8	308th SMW
28 Feb 68	395-B	B-88	S	64-0452	Glory Trip 04T	W-8126	FOT-9	381st SMW
2 Apr 68	395-C	B-36	S	62-0017	Glory Trip 010T	W-5576	FOT-10	395th SMS
12 Jun 68	395-C	B-82	S	63-7737	Glory Trip 08T	W-8172	FOT-11	390th SMW
21 Aug 68	395-C	B-53	S	62-12296	Glory Trip 18T	W-7537	FOT-12	381st SMW
19 Nov 68	395-C	B-3	S	61-2757	Glory Trip 026T	W-0852	FOT-13	390th SMW
20 May 69	395-B	B-83	S	63-7738	Glory Trip 039T	W-3226	FOT-14	395th SMS

S=successful; F=failure; A= abort
Titan II Missile Synopsis, Headquarters SAC 13 August 1985

Army Safeguard System Test Target Program, 1971–74

The Army's Safeguard Antiballistic Missile System Program utilized both Minuteman and Titan II missiles as targets during its research and development phase. Eight Titan II launches were scheduled with six successful flights and two that ended in aborts prior to liftoff (Table 8.5). Exercise M1-17 was an abort caused by a sequence timing problem resulting from a range safety hold after the countdown had started. Exercise M2-36 was an abort caused by faulty wiring to one of the two start cartridges on Stage I. With one abort not counted due to the range safety cause, this program had a successful launch rate of 88 percent and a 100 percent flight-to-target success rate.[15]

Army Ballistic Missile Development Agency Program, 1975

The Army Ballistic Missile Development Agency sponsored three Titan II flights during the Ballistic Missile Development Test Target Program (BMDTTP) that centered around development of sensors for use in antiballistic missile systems (Table 8.6). The first test in this program was called the Signature of Fragmented Tanks (SOFT-1). One launch was scheduled with the option of a second launch. In the SOFT program, a Titan II missile was launched along its normal trajectory, but at the point of Stage II separation, the Stage II

TABLE 8.5 SAFEGUARD SYSTEM TARGET TEST PROGRAM

Date	Launch Complex	Missile #	Result	Air Force Serial #	Code Name	Op. #	Program	Unit
26 May 71	395-C	B-69	A	63-7724	MI-17	abort	SSTTP	390th SMW
20 Jun 71	395-C	B-12	S	61-2766	MI-17	W-2709	SSTTP	390th SMW
27 Aug 71	395-C	B-100	S	65-10644	M2-1	W-0291	SSTTP	308th SMW
24 May 72	395-C	B-46	S	62-027	M2-10	W-6639	SSTTP	381st SMW
11 Oct 72	395-C	B-78	S	63-7733	M2-14	W-4006	SSTTP	390th SMW
5 Oct 73	395-C	B-69	S	63-7724	M2-27	W-8340	SSTTP	308th SMW
1 Mar 74	395-C	B-85	S	64-449	M2-31	W-7443	SSTTP	381st SMW
20 Jun 74	395-C	B-41	A	62-022	M2-36	abort	SSTTP	390th SMW

S=success; F=failure; A=abort
Titan II Missile Synopsis, Headquarters SAC, 13 August 1985.

tanks were artificially fragmented into precisely sized pieces for evaluation of the ability of sensors to discriminate among the debris and the reentry vehicle. The reentry vehicle impact of the three flights in the program are not available except that both were completely successful.[16]

Special Operations Test—Integrated Test Flight-1, 1976

By the mid-1970s it was clear that the original Titan II inertial guidance platform and associated computers would soon be unsupportable. In a project code-named RIVET HAWK, an updated guidance system was evaluated on a single test flight. The flight was successful and while the reentry vehicle did not hit the target, a computer program error was identified as the source of the error and the new guidance system was adopted fleetwide (Table 8.7).[17]

TITAN II ACCURACY

As might be expected, the overall accuracy of the Titan II Mark 6 reentry vehicle has been the subject of considerable speculation during and after its deployment. The origi-nal specification was for an accuracy of less than 1 nautical mile.[18] In the unclassified portion of a Titan II system assessment given to Congress in May 1980, the circular error probable (a circle with a radius within launch 50 percent of the reentry vehicles impact) was stated as 2.5 times better than the system requirement of 1 nautical mile, implying 0.4 nautical mile. Other nongovernmental sources give a CEP of between 0.65 and 0.80 nau-tical miles.[19] In the final report for the OT program, accuracy is described by the term cir-cular error, but no definition is given. The most accurate flight in the OT program had a circular error of 0.175 nautical miles or 1,064 feet.[20] Using the crossrange and downrange

Date	Launch Complex	Missile #	Result	Air Force Serial #	Code Name	Op. #	Program	Unit
9 Jan 75	395-C	B-27 (29)	S	62-008	ST	W-2592	SOFT-1	308th SMW
7 Aug 75	395-C	B-52	S	62-12295	DG-2	W-5006	BMDTTP	381st SMW
4 Dec 75	395-C	B-41 (18)	S	62-022	DG-4	W-5678	BMDTTP	390th SMW
27 Jun 76	395-C	B-17	S	62-2771	Rivet Hawk	W-8440	ITF-1	308th SMW

Note: Numbers in parenthesis, i.e., B-30(25) indcates a composite missile with Stage I from B-25 and Stage II from B-30. The serial number is for the Stage II component since it carried the guidance system.

S=success; F=failure; A=abort

Titan II Missile Synopsis, Headquarters SAC, 13 August 1985

errors listed in the OT final report, the CEP for Titan II for this program can be calculated as 0.789 nautical miles, which is in reasonable agreement with published values.[21]

Much has been written about the relative accuracy of Titan II when compared to the Minuteman weapon system and thus the need for the much higher yield weapon on Titan II compared to Minuteman.[22] Using actual test values for 38 of the 50 Minuteman I OT launches that represented acceptable flight criteria, the Minuteman I CEP was determined to be 1.03 nautical miles. The confidence level for these results is plus or minus 16 percent, resulting in a range of CEPs from 0.84 to 1.16 nautical miles. The available official Titan II data set is limited to 14 launches from the OT program, and thus has a confidence level of plus or minus 25 percent so the 0.78 nautical miles actually represents the middle of a range of CEPs from 0.58 nautical miles to 0.97 nautical miles. Clearly the relative accuracy of Titan II and Minuteman I was not an issue.[23]

In fairness to the Minuteman II and III programs, their CEPs were 0.26 nautical miles and 0.21 nautical miles, respectively, clearly much more accurate than Titan II. In 1979 the Minuteman III guidance system upgrade gave a CEP of 0.12 nautical miles.[24]

Non-Flight Evaluation Programs

SERVICE LIFE ANALYSIS PROGRAM, 1965–85

A total of 155 Titan II engine sets (Stage I and Stage II) were manufactured between 1961 and 1967. The Service Life Analysis Program (SLAP) engine monitoring effort was initiated in April 1965. SLAP consisted of two complete engine system tests per year to monitor the reliability of the Stage I and II engines, detect trends in component failure or degradation, and to provide the basis for extending service life and establishing necessary corrective actions.[25]

TABLE 8.7

TABLE 8.7

TITAN II OPERATIONAL TEST LAUNCHES, 1964–76

YEAR	LAUNCH ATTEMPTS[a]	NUMBER LAUNCHED	SUCCESSFUL FLIGHTS
1964	5	5	5
1965	14	14	9
1966	9	9	7
1967	5	4	3
1968	5	5	5
1969	1	1	1
1970	0	0	0
1971	3	2	2
1972	2	2	2
1973	1	1	1
1974	2	1	1
1975	3	3	3
1976	1	1	1
TOTAL	51	48	40

[a] *This number reflects attempts that culminated in engine ignition. Actual total "key turn" launch attempts could not be compiled from available data.*

Engine selection was focused on obtaining the oldest representative examples from alternating bases on an ongoing basis. The first step was to conduct an in-silo inspection and test of the engine using the missile verification procedure. Once the pretest engine condition was established, the entire missile was removed from the selected silo, and the engines were shipped via Ogden Air Logistics Center to Aerojet-General Corporation for disassembly, inspection, and hot-fire tests at their engine test facilities. After the hot-firing tests, the engines went through Ogden Air Logistics Center for disassembly, a thorough inspection, and refurbishment, and were then stored until needed in the field.

The test program began in April 1965 and continued through January 1985. Thirty-five engine sets were selected with 32 fired. Engine age, from last overhaul, ranged from 37 to 223 months for Stage I and from 37 to 224 months for Stage II. Subassembly 2 of the Stage I engine set was normally tested since it included the autogenous system used to pressurize the Stage I propellant tanks and was used to pressurize the tanks in the test stands. Five of the 35 tests used both Stage I subassemblies. The program revealed that the engine service life, while not indefinite, was being properly maintained with the procedures in place at the operational bases.

Four of the 35 tests for the Stage I, Subassembly 2, thrust chamber were less than the full 180-second test duration. All four of these were due to test stand equipment malfunctions or shortage of propellants. One test firing, on 26 January 1984, with a Stage I engine set that was 206 months old and had never been overhauled, was run successfully for 200 seconds to consume surplus propellants. Both Subassembly 1 and 2 were fired in this test with satisfactory results. Stage II engine tests were nearly as successful with only one failure due to engine hardware. An investigation revealed that the failed turbopump assembly was due to lubricant leakage resulting from a pretest pressurization problem.[26]

Reliability and Aging Surveillance Program, 1969–85

On 8 February 1967, Headquarters U.S. Air Force announced that the Titan II Weapon System would be inactivated by 30 June 1973. The original service life of the Titan II program had been 10 years, beginning with full deployment in late 1963. With 38 missiles launched and only 12 airframes remaining for the test program, not counting the spare missile at each base, consumption of the missiles in a continuing flight test program at a proposed rate of 6 missiles per year meant that one squadron from the 390th SMW, Davis-Monthan AFB, Arizona, would need to be deactivated to free up additional missiles for continuing test flights. Further refinement of this tentative decision resulted in a proposed seven-year phaseout plan, starting in 1969, leaving one squadron of missiles at the 381st SMW, McConnell AFB, Kansas, and two squadrons at the 308th SMW, Little Rock AFB, Arkansas.[27]

On 14 November 1968, the Chief of Staff, U.S. Air Force, informed the Air Force Logistics Command (AFLC) and Headquarters (HQ) SAC that the FOT program would end in May 1969. No further missile production was planned, and the remaining 12 Titan II airframes, other than spares at the bases and the missiles in the silos, had been reserved for other Department of Defense programs. AFLC and HQ SAC were requested to devise a monitoring program that could be used to predict equipment failure rates and hence reliability of the Titan II weapon system, allowing the system to remain active. There was strenuous resistance to this proposal. Neither SAC nor AFLC felt that such a ground testing program could be a viable substitute for continued testing of four to six missiles per year in the operational environment. The previous OT and FOT programs had a total of seven in-flight failures of which ground testing would have identified only one.[28]

The RASP plan was submitted to HQ SAC and AFLC and approved in July 1969. Funding for the program began in Fiscal Year 1971. RASP involved not only the missile systems but also the ground support equipment. RASP evaluations were conducted in two phases. Phase I required testing of four missiles per year optimum, with a minimum of two. Phase I included the complete series of tests and required propellant downloading and missile recycle operations. The missile guidance system was evaluated in Phase I tests. Phase IA did not require a propellant download but tested only critical airborne and support equipment for identifying aging trends.[29]

Three airframe problems were monitored during RASP evaluations. The first was the presence of minute cracks in the propellant tank welds. First observed during the leak repair process of Operation Wrap Up, the tank welds were x-rayed and compared to the older x-rays to determine if the cracks were propagating. Twenty-four percent of the missiles inspected required weld repair. One of the methods of repair was use a weld patch or a bolt patch. The bolt patch was done in the field and utilized butyl rubber as one of the components. This same butyl rubber material was used in the oxidizer tank manhole covers and pressure cap seals. Butyl rubber reacted with the oxidizer initially by swelling, ensuring an excellent seal. As the program aged, the inspections found that the rubber was deteriorating, and oxidizer leaks began to occur in both the bolt patch seals and the oxidizer tank manhole cover seals.

Through 1977 the solution was routine preventative maintenance by replacing the seals which lasted approximately 6 years. In 1978 a new seal material was ready for use in the field for the bolt patches. Kalrez was a Teflon-based seal that indicated no degradation after 55 months of accelerated exposure to oxidizer. Average service life was expected to be above 10 years. Replacement of all bolt patch seals, a total of 29 patches on 19 missiles, was completed in 1980. Oxidizer manhole cover, pressure cap, drain and fill, and pressure sensor seals required a fabricated version of the Kalrez material, and replacement process was nearly complete by the end of 1980. By the end of the program, Stage I and II oxidizer tank seal replacement averaged 7 years; Stage I fuel tank seals 14 years; and Stage II fuel tank seals 13 years.[30]

By the end of the RASP inspections in 1985, 73 missiles had received Phase I or Phase IA RASP testing. The final RASP study in 1985, which evaluated two missiles, B-75 and B-106, both at McConnell AFB, indicated that the missiles and ground support equipment were in excellent condition. While several discrepancies were discovered within the Phase IA systems studied, none was found that would have prevented successful completion of the missiles' mission.[31]

Vandenberg AFB, California, Titan II Operations

The key to the Titan II training exercise launch operations was the unique relationship between Vandenberg AFB and the operational bases. The 395th Missile Training Squadron (MTS) was activated on 1 February 1959 at Vandenberg AFB, California. Initially tasked with providing training support for missile combat crews for the Titan I weapon system, the 395th MTS was redesignated as the 395th SMS on 1 February 1964 as it assumed the same duties for the Titan II weapon system. Upon completion of the contractor Titan II flight test and development launches at Vandenberg AFB, the 395th SMS provided launch crews for the first three Titan II flights of the DASO program.[32]

In late 1964 a modification program called Operation REALISM was started in the three Titan II launch control centers at Vandenberg AFB. These modifications enabled the missile, after installation of range safety ordnance, to be placed in the "alert ready" mode up to 10 days before maintenance on the test instrumentation had to be performed. These time critical maintenance items consisted of the range safety package with the telemetry transmitter, radar tracking transponder, command receivers, flight termination ordnance, and wiring, none of which was present on the operational missiles at the wing locations. Instrumentation for range safety, telemetry, and communications located only at Vandenberg AFB were isolated as much as possible from the operational equipment found in the operational launch control centers. Part of the instrumentation equipment was placed in a locked enclosure located on Level 2 of the launch control center, rendering indicators and controls inaccessible to the launch crew. The remainder was relocated into the newly created Peace Time Launch Center at the 1st Strategic Aerospace Division Command Post. Range safety operations, as well as instrumentation communications, were still not completely isolated from the launch crews, but the improved environment made the operational training launches much more realistic.[33]

Five months after the completion of the DASO program, the Titan II OT program began. Launch operations utilized missiles from the operational bases and their maintenance and missile combat crews while the 395th SMS personnel provided support. In June 1965, HQ SAC proposed assigning Emergency War Order (EWO) targets to the Titan II launch complexes at Vandenberg AFB in between launches in the FOT program, the successor to the OT program.[34] The original plan was to put Launch Complex 395-D on continuous EWO alert through the operational life of Titan II. Launch Complexes 395-B and 395-C would alternatively be on EWO alert 10 months of the year and be available for use in the FOT program the remaining 2 months of the year. In April 1966, SAC stated that generation and maintenance of the alert posture could not be at the expense of equipment and personnel priorities that were needed to support the FOT program. These considerations indicated that only one site would be converted to EWO alert status. Delays prevented this EWO alert status in 1966. With official approval of EWO alert status complete but not yet implemented, training of the 395th SMS personnel for operational readiness capability continued in order to generate sufficient trained personnel for the complexes.[35]

On 19 January 1967, Launch Complex 395-D was placed on EWO alert status, manned by missile combat crews from the 395th SMS. HQ SAC continued to study the feasibility of committing an additional complex to EWO status while still utilizing it in the FOT program. This launch complex, 395-C, would be used in the FOT program for three months of the year and be on EWO alert status the remaining nine months. The major stumbling block for this mode of operation was the need to install the range safety equipment on the missile prior to an FOT launch. The solution was to install and test all necessary equipment and wiring, then remove range safety instrumentation and explosive destruct charges. The missile could then be installed and placed on EWO alert status. Launch Complex 395-C was placed on EWO alert status on 9 June 1967.

On 1 January 1968 the 395th SMS missile combat crew strength was increased to 13 combat ready crews. That brought the number of 24-hour alert tours each crew worked to approximately 8.5 per month, down from a peak of 10 at the beginning of EWO alert status. On 4 March 1968, Launch Complex 395-C was removed from EWO alert status in preparation for the next FOT launch. To maintain two complexes on EWO alert status, Launch Complex 395-B was reactivated and placed on alert on 19 April 1968. A proposal to place all three Vandenburg AFB Titan II launch complexes on EWO alert status, while preserving the ability to support future launches for programs such as the Army's Safeguard Antiballistic Missile Program, was turned down by the Air Force due to budget constraints. After the final FOT launch, the 395th SMS was inactivated in December 1969 and Launch Complexes 395-B and 395-D were deactivated (units are inactivated, equipment is deactivated). This involved salvaging all useful equipment, including the diesel generators, and then securing the facilities from unauthorized entry.[36]

Launch Complex 395-C was retained in "mothballed" status for probable use in the Army test programs for two reasons. It served as a communications link with the southern end of the base and was the farthest from the ocean and salt-water corrosion problems. A complication to this process, at least in terms of personnel, was the need to retain support staff for the installation of the instrumentation range safety system packages used on each Titan II missile launched from Vandenberg AFB. Nearly 800 pounds of equipment, amounting to 100 separate pieces of instrumentation, and considerable wiring had to be attached to a missile taken out of the operational fleet. Checkout alone took nearly three weeks with experienced personnel. In order to maintain Titan II launch capability but also economize on support costs, many of the 395th SMS personnel were transferred to the 394th SMS to form a combined Minuteman and Titan II squadron. Even with limited numbers of spares to support the operational Titan II missiles, plans were in progress to launch Titan II and older Minuteman I ICBMs as targets to support the Army's Safeguard Antiballistic Missile Program interceptors.[37]

The majority of the 51 Titan II launch attempts were relatively routine, at least as routine as launching a missile weighing over 300,000 pounds might be. A number of the launches have interesting stories leading up to the launch or refurbishing the launch complex. These stories are organized by the unit involved in the operation.

395th Strategic Missile Squadron, Vandenberg AFB, California

While the 395th SMS was responsible for only 5 of the 48 Titan II missiles launched during the program, the squadron also served as the support for preflight and postflight operations. On 30 November 1967 the first aborted launch of a Titan II occurred. Missile B-69 (63-7724) was transported to Vandenberg AFB from the 308th SMW, Little Rock AFB, and installed in Launch Complex 395-B. Normal launch procedures took place until the point of engine ignition. After full thrust was reached, the hold-down bolts did not fire due to a failure in a timing circuit in the control monitor group equipment chassis in the launch control center. Ordinarily when both thrust chambers reached 77 percent of thrust, each thrust chamber pressure switch closed and a signal was sent to three 1.8-second timers. After these timers timed out, a signal was sent to the missile hold-down nut circuit, triggering a FIRE-EXPLOSIVE-NUTS signal.[38] All three timers had to work, but in this case one failed, so the signal to trigger the explosive nuts was not sent. The Stage I engine fired at full thrust for 24 seconds, burning away most of the concrete in the bottom of the exhaust deflector commonly referred to as the "W," before the abort checklist was utilized to shut down the missile engine.

SSgt Carl Duggan was 395th SMS engine shop chief at this time. One of his duties during a launch was to stand by in the fall-back area with emergency equipment, ready to assist if something went wrong with the launch. He watched intently as the silo closure door opened normally and as the Stage I engine ignited as expected shortly thereafter. The missile did not rise out of the silo between the towering columns of exhaust within the normal time frame. After 24 seconds, the missile combat crew initiated the abort sequence, the engine shut down, and the silo closure door closed. Duggan and Sgt Bill Zimmerman reported to the launch control center with their protective suits and equipment. The launch crew still had a fire warning light on the Launch Control Complex Facilities Console but knew that the engine was shut down. The decision was made to send Duggan and Zimmerman to Level 2 of the launch duct to make an initial report on the status of the missile.

Donning their protective suits, Duggan and Zimmerman proceeded to the launch duct. Opening the door to Level 2 of the launch duct, Duggan leaned over to look down to the bottom of the silo and saw what appeared to be a fire at Level 8. The flame deflector was full of water, and the fire was burning on the water surface. The fire was clearly visible in the darkness of the silo with the silo closure door shut.

Duggan and Zimmerman returned to the launch control center and reported their observations. The consensus among the contractor and launch test personnel was that the engine pressure sequencing valves were leaking fuel into the flame deflector and feeding the fire. Duggan and Zimmerman proceeded to Level 7 of the launch duct to try to cap off the pressure sequencing valves. They turned on the launch duct lights, lowered the work platforms without difficulty, and began to work on the problem. With the lights now on, the fire did not seem nearly as large or intense. Zimmerman handled the fire hose while Duggan began work on capping the pressure sequence valve drain. As Duggan began work, Zimmerman was distracted when he saw the damage in the launch duct and momentarily let the water spray fall short of Duggan's hands. The heat from the fire began to melt his rubber gloves and as Duggan felt the material become tacky, he turned and yelled to Zimmerman, who directed the water back onto his gloves. They completed the remainder of the work successfully.

The team had one remaining task, that of disarming the Stage II engine and prevalve ordnance and the translation rockets. Duggan recalled the September 1964 Launch Complex 395-C translation rocket ignition incident as they rode the elevator to Level 3, and he kept thinking that here was this missile, primed for launch, and he was going out near a solid rocket motor of 5,000 pounds thrust to safe the firing circuit. They lowered the work platforms, removed an access panel, and uneventfully, if somewhat nervously, disarmed the vernier rocket squibs, the Stage II prevalve squibs, the Stage II start cartridge, and the translation rockets. A crew from the Range Safety Shop followed them and disarmed the range safety package.

SSgt Carl Duggan and Sgt Bill Zimmerman both received the Airman's Medal for their heroism in putting out the fire and safing the missile ordnance.[39] Missile B-69 was removed from the silo and refurbished at the Vandenburg AFB missile maintenance facilities.

The launch crews of the 395th SMS had a unique perspective on the Titan II program during their nearly two years on alert status at Vandenberg AFB. This perspective was to actually launch the missile that had been on alert in the launch duct during their many alert tours. Unlike the operational base launch operations which entailed moving a missile from the base in question to Vandenberg AFB, these missiles were used to test more realistically the long-term consequences of storage of the missiles in the launch duct environment.

On 1 April 1969, missile B-83 (62-7738) was removed from EWO alert status at Launch Complex 395-B and readied for the final FOT launch, code-named GLORY TRIP 039T. The

Figure 8.1. SSgt Carl Duggan receiving the Airman's Medal for Heroism after the aborted launch of Titan II B-69 (63-7724) on 30 November 1967. *Courtesy of Carl Duggan.*

war reserve W-53 warhead was removed, shipped to the Atomic Energy Commission facilities near Amarillo, Texas, to be denuclearized and configured for the MIXED MARBLE II airburst option and then returned. On 20 May 1969, Crew E-028 of the 395th SMS, composed of Capt Ted Suchecki, MCCC; 1Lt Jim Buyck, DMCCC; TSgt Jerry Meyer, BMAT; and Sgt Leroy Cubicciotti, MFT, conducted the final FOT program launch of an operational Titan II missile. Receipt of the launch order was routine, and the countdown proceeded as expected. At liftoff, Suchecki was amazed at how quiet the launch control center remained considering the incredible noise being generated a mere 250 feet away during the launch. Suchecki did not have much time to contemplate why this was true because he immediately observed both the LIFTOFF and ABORT indicators illuminate on the Launch Control Complex Facilities Console just to the right of where he had turned the launch key a minute earlier. He tried to call the 1st Strategic Aerospace Division Command Post for verification of launch but received no answer. Envisioning everyone running over to the command post windows to watch the missile as it cleared the silo, Suchecki, expecting the worst, decided to run the abort checklist instead of the post-launch checklist. This procedure started the fire deluge system, resulting in quite a bit of water ending up in the flame deflector. Suchecki's contribution to the Titan II operational crew force was the addition of a "NOTE" in the launch checklist operational technical data that, in the event of a similar situation, directed the crew to process the post-launch checklist instead! The flight of B-83 was normal, and the reentry vehicle landed in the Kwajalein impact area as planned. Further accuracy details remain classified. The success of this final flight in the FOT program was particularly significant to the Titan II program since B-83 had been emplaced in Launch Complex 395-B for slightly over a year. This was the most realistic test of the Titan II basing concept to date and was passed with complete success.[40]

Randy Welch's single most memorable Air Force experience during his tour of duty at Vandenberg AFB as part of the 394th SMS was the attempted launch of B-41 (62-0022), code name M2-36, the last of the SSTTP launches. Welch was an airman 1st class and a member of the Propellant Transfer System (PTS) shop. The PTS team was in the back of a stake-bed truck waiting for the launch. The truck was full of the PTS equipment necessary to support the self-contained rocket fuel handler's clothing outfit (RFHCO). These suits were necessary anytime there was a chance of exposure to the toxic propellants.

They chanted down the last 15 seconds of the countdown as it was broadcast over the range countdown network. Fully expecting to see a launch typical of films he had watched, Welch was surprised to see smoke and flame billowing out of only one of the two exhaust ducts. Seconds later, the missile still in the silo, flaming debris began to shoot out of that one exhaust duct and land near the silo closure door. The missile finally shut down, and an eerie black and orange haze floated above the silo while the debris continued to burn.

The radio crackled to life with "PTS move!" They drove down to the site in silence, wondering what would come next. Only two other aborts had occurred in the Titan II program in the 12 years of Titan II operations at Vandenberg AFB so this was a new experience for many of them. As they neared the site, the source of the flaming debris became clear; it was the sound suppression modules (5-foot by 4-foot boxes of aluminum with sound-absorbing material inside) that had lined the exhaust duct. They had been torn from the walls of the exhaust duct due to the prolonged exposure to the captive Stage I engine exhaust plume, flung hundreds of feet in the air, and scattered all over the site, crumpling the chain-link fence in several places as they landed.

Welch, 19 years old, and Sgt Gino Jones, 24 years old and soon to be leaving the Air Force, suited up and made ready to walk down the long cableway to see what was happening in the silo. The missile combat crew commander refused to let the PTS team through Blast Door 9 until most of the fire indicators blinked out. Welch and Jones then proceeded down the long cableway with instructions to return immediately if there were any signs of trouble. Several people had suggested the missile might be leaning against the launch duct wall or might have split open.

After closing Blast Door 9, the last protective door separating them from an explosive blast, Welch and Jones were faced with a surreal scene. Steam had filled the cableway and visibility was nil as they walked the 180 feet to the Level 2 launch duct access door. Welch remembers that they both looked at each other and opened the door, figuring they were goners. The missile was shrouded in steam but seemed intact. Breathing a sigh of relief, they returned topside to replenish their air supplies. As the full PTS crew formed topside, the decision was made to remove the range safety destruct package prior to downloading the missile propellants.[41]

The top of the Stage I fuel tank had collapsed with a 4-foot split along the Y-frame weld on the dome of the tank due to the vacuum created as fuel was sucked out. The autogenous pressurization system used for the Stage I fuel and oxidizer tanks operated from exhaust gases from the turbine on the Stage I Subassembly 2 engine but this engine had not ignited. Because the tanks could not be pressurized to replace the liquid consumed by the engines, the Stage I fuel tank dome had been partially sucked in. The Stage I oxidizer tank had not been damaged. Also, the prolonged exposure to the effects of the engine flame, which reached temperatures over 5,000 degrees Fahrenheit, had partially melted the metal cover of the Level 7 launch duct door, the access ladders leading down to the flame deflector had been melted, and one side of the deflector concrete had been severely eroded.

Sgt Al Howton, a 394th SMS maintenance technician, was at the Vandenberg AFB flight line, about 1.5 miles east of the site, at the moment of Stage I engine ignition. He too realized something was wrong when he saw smoke come out of only one of the exhaust ducts. At first the acoustic modules could be seen flying through the air as they cleared the cascade vanes at the top of the exhaust ducts. Then, all of a sudden, the modules stopped coming out and the exhaust seemed clean of debris. Inspection of the exhaust duct cascade vanes once the missile was made safe showed that the acoustical modules had plugged the vanes in the duct where the engine had fired. The pile up of debris represented an extreme hazard to any workers in the launch duct repairing the flame deflector since if any came loose they would fall approximately 145 feet down on to the workers. The decision was made to offload the missile propellants before working to clear the vanes.

Howton and several others were involved in this cleanup effort. After considerable debate, it was decided that a large lead pipe hanging from a crane hoist could be used as a battering ram to dislodge the approximately 200 acoustic modules that were lodged in the cascade vane openings. The idea worked perfectly, and with the plug removed, module replacement operations could begin. The normal method of replacement was to use a window-cleaning platform, suspended by ropes from one of the cascade vanes. Personal safety lines were required, by regulation, to be attached to the work platform. Actual operating procedure had to be changed somewhat. Figuring that if the platform attachment lines broke, they would go down with the platform, the refurbishment crews would normally attach their safety harness to the acoustic module support bolts they were working on. At this particular time, the "powers that be" insisted on the safety belts being attached to the platforms. The crews refused to operate under these conditions; they knew that this was a more dangerous method. The standoff was resolved by allowing them to work within the launch duct as they had wanted, and the vanes were safely cleared of all debris.[42]

Missile B-41 Stage I was sent to Martin Marietta Company, Denver, for examination. With the Titan II assembly line long since converted to Titan III and other members of the Titan family, the decision was made to replace the fuel tank with a slightly larger Titan III-series Stage I fuel tank. B-41 Stage I was combined with the Stage II airframe from

B-56 and returned to the operational fleet. The B-41 Stage II flew on the second to the last Titan II ICBM launch in 1975.[43]

308th Strategic Missile Wing, Little Rock AFB, Arkansas

In October 1964, a 308th SMW missile combat crew was the first Titan II missile combat crew from an operational unit to launch a Titan II missile as part of the five-flight DASO program being conducted at Vandenberg AFB. Missile B-1 (61-2755) had been received by the 395th SMS on 15 October 1962, and had been used by Martin Marietta Company for checkout and demonstration at the three Titan III training silos. In December 1963, B-1 was withdrawn from the silo demonstration program for update modifications in keeping with the changes made to the rest of the operational missiles. These included inspection and rewelding of propellant tanks and inspection of oxidizer feed lines. On 2 March 1964, the missile was accepted once again by the 395th SMS. Originally installed in Launch Complex 395-B, the missile was removed for installation of the Stage II translation rocket system, and on 2 September 1964, B-1 was installed in Launch Complex 395-C, by personnel of the 308th SMW, in readiness for Exercise BLACK WIDOW. Launch Complex 395-C was placed on alert/missile readiness on 1 October 1964.[44]

After 4 hours and 24 minutes in alert/readiness posture, the launch execution message was received. Three minutes later, after validation and decoding of the launch message, the missile combat crew composed of Maj Paul F. Moon, MCCC; Capt Cyril R. Frost, DMCCC; TSgt Esker M. Eaton, BMAT; and SSgt William E. Doss, MFT, became the first operational Titan II force crew to initiate a launch sequence using the peacetime launch procedures checklist. Approximately 40 seconds into the launch sequence, the HOLD FIRE indicator on the Launch Control Complex Facilities Console illuminated. Moon pushed RESET-SHUTDOWN, and the crew immediately entered the trouble-shooting procedures and identified the cause to be a thrust mount that was not in a fully locked condition. The launch was rescheduled for the next day, and 308th SMW ordnance personnel disconnected the missile ordnance before further investigation of the thrust mount failure took place. Inspection of the thrust mount indicated that one of the four vertical dampers had been blocked from completely locking due to improper installation of the heat-blast protective cocoon that was peculiar to the Vandenberg AFB silos. Repositioning the cocoon and cycling the lock mechanism successfully 10 times indicated the problem was resolved. Since the missile batteries had been activated and had only a 4-hour lifetime, replacement batteries needed to be installed. Upon installation and testing of the new batteries, the guidance system inertial measurement unit failed and was replaced.[45]

On 2 October 1964, Launch Complex 395-C went back in alert/missile readiness posture. The missile combat crew was composed of Maj Kenneth R. Wine, MCCC; Capt Ted L. Brown, DMCCC; CMSgt Herbert A. Hancock, BMAT; and SSgt Frank P. Sledge, MFT. Wine's crew had been able to ride an incredible string of luck in reaching the opportunity to launch this missile. Six crews had been selected earlier that year for possible participation in the DASO launch. The crews were ranked by performance and scores on their standboard evaluations.

Figure 8.2a. Missile shock isolation system vertical damper assembly as seen at the operational launch complexes. This damper system attenuated vertical motion due to a nearby nuclear blast. It also was used to lock the thrust mount into a stable position for launch. *Courtesy of the Titan Missile Museum National Historic Landmark Archives, Sahuarita, Arizona.*

The rule was that once a crew was selected, only that intact crew would go. If any member of the crew was transferred, then the crew was no longer on the short list of six. The entire crew wanted to be part of the task force but realized that they were the sixth crew on the list and while pleased to have made the selection, they were resigned to the fact that they would most likely not be going since only three crews normally made the trip.

The first faint hope that they might actually get to go came as the crews above them began to drop out. The first crew's MFT broke his leg; the second crew had a DMCCC who was concerned about the career effect of an aborted launch, so they had to drop out. The third crew dropped, too, and now the fourth, fifth, and sixth crew would be going after all.

Wine's crew had packed and was at the fall-back position watching the silo through binoculars when they saw the door open and then close soon after. They waited a while longer, figuring that the launch was being rescheduled and would occur shortly. It was then that they got the message to return to quarters and get some sleep because they would be going back on alert! Crew changeover took place, and the rest is history.

Figure 8.2b. In Titan II launch complexes at Vandenberg AFB, California, the vertical damper assemblies were protected from the Stage I engine exhaust using a "cocoon" of fireproof material to facilitate refurbishing them for the next launch. *Courtesy of Ron Hakanson.*

This time the countdown proceeded exactly as they had trained in the Missile Procedures Trainer. Wine's crew was also one of the few, if not the only one, to watch the progress of the countdown on real-time television in the launch control center. Three television monitors were positioned above the control monitor group equipment racks in front of the commander's console. The cameras for these monitors were placed below the Stage I engines, at the top of the launch duct and on the surface. As the countdown progressed, a quick look up at the monitors permitted the crew to watch the Stage I engine gimbal and the thrust mount lock down. Hancock was observed by one of the high-ranking distinguished guests momentarily watching one of the monitors. This guest was a SAC general. He asked Hancock how he could watch the monitor and still complete his task of marking the indicators as they illuminated during the countdown, his primary task. Hancock stepped back, revealing the correctly marked indicators, and then stepped forward without uttering a word.[46] Missile B-1 lifted off successfully, ending the flight with the reentry vehicle impacting within one nautical mile of the target.[47]

On 25 June 1965 missile B-19 (61-2773) was selected for the OT program with the code name NEW ROLE. The missile was removed from Launch Complex 373-6 on 9 July 1965 and shipped to Vandenberg AFB, arriving on 13 July 1965, then emplaced in Launch Complex 395-D on 28 July 1965. Alert-ready status was achieved on 13 August 1965. On 25 August 1965, the missile combat crew composed of Capt Harold D. Casleton, MCCC; 1Lt Richard W. Kalishek, DMCCC; SSgt Jerome E. Patosnak, BMAT; and A1C Dayle V. Smith, MFT, acknowledged the launch message only to have it cut off in mid-sentence. The crew members looked at one another, wondering what was going on. The command post called seconds later and told them to ignore the last message; a train coming down the coast on railroad tracks that passed close to the launch complex had caused a hold. On 26 August 1965 the launch message was again received, this time without interruption. Reentry vehicle impact was 0.421 nautical miles short and 0.310 nautical miles left for an impact error of 0.523 nautical miles.[48]

The last OT launch for the 308th SMW in 1965, Exercise SEA ROVER, took place in December. Missile B-73 (63-7728) was selected 26 October 1965 and removed from Launch Complex 374-6 on 9 November 1965. B-73 was shipped to Vandenberg AFB on 14 November 1965 and emplaced in Launch Complex 395-C on 26 November 1965. The missile was placed on at alert-readiness status 10 December 1965.

After five days on missile readiness status, on 22 December 1965 the missile combat crew composed of Maj John Radizietta, MCCC; 1Lt Thomas D. Weaver, DMCCC; SSgt Richard T. Frenier, BMAT; and A1C Samuel Cirelli, MFT, acknowledged the launch order and began the launch sequence. Liftoff was successful.[49]

At 163 seconds into flight, as Stage II engine thrust built up, the missile experienced a constant yaw-left force which remained once steady-state engine thrust was reached. Telemetry indicated no abnormal guidance signals. While telemetry indicated that the pitch/depitch and translation rockets fired, reentry vehicle separation did not occur due to the abnormal flight profile. Further data on the actual impact of the reentry vehicle/Stage II combination is not available. Analysis indicated that a misadjusted yaw actuator had prevented control by the guidance system. Further investigation revealed that the technical orders for yaw actuator adjustment were unclear. All linear actuators in the fleet were inspected, and the technical orders were clarified.[50]

On 2 October 1967, missile B-69 (63-7724) was removed from Launch Complex 374-7 and shipped to Vandenberg AFB to be the eighth missile launched in the FOT program, Exercise GLOWING BRIGHT 49. On 30 November 1967, the missile crew composed of Maj Douglas C. Cameron, MCCC; 1Lt James C. Swayze, DMCCC; SSgt Lyle R. Groth, BMAT; and TSgt Robert L. Turner, MFT, was on alert at Launch Complex 395-B. After receipt and acknowledgment of the launch order, the crew turned keys, and the countdown proceeded normally up through engine ignition. After full thrust was reached, the hold-down bolts did not fire due to a failure in a timing circuit in the control monitor group equipment chassis in the launch control center. The Stage I engine continued to fire at full thrust for 24 seconds, burning away much of the concrete in the bottom of the exhaust deflector before the abort checklist was utilized to shut down the missile engine.[51]

After a 27-month wait, the 308th SMW was again involved in the launch of a Titan II

Figure 8.3. The first launch by an operational crew, Titan II ICBM B-1 (61-2755), took place on 2 October 1964. The launch crew was from the 308th SMW, which was the first as well as the last operational wing to launch the Titan II ICBM. This was the fourth launch in the Demonstration and Shakedown Operation program, and it was successful. *Author's Collection.*

missile. The SSTTP had been completed with the abort of missile B-46 six months earlier, and now the Army was sponsoring a special test named Signatures of Fragmented Tanks (SOFT). The first test was named SOFT-1. The Army Ballistic Missile Development Agency had two primary mission objectives for SOFT-1. The first objective was to test a long-wave infrared detector package, launched in the nose of a second missile near the target area, that would observe the fragments (this was originally called the Ballistic Particle Experiment). The second objective was to complete the study of the radar signature of the fragmenting second stage behind the reentry vehicle to see if the tank pieces could hide the location of the actual reentry vehicle during the early stages of reentry into the atmosphere. Martin Marietta Company was a subcontractor to Boeing for the fragmentation program; the Army's Redstone Arsenal at Huntsville, Alabama, had developed the infrared tracking system. Missile flight operations were all Air Force by this time. Martin Marietta Company engineers had to devise a way to fragment the tank into precisely 28 pieces and place radar beacons on 3 of the pieces to facilitate designating the cloud for "skin track" radars and optical sensors.

To provide attachment points for the brackets to hold the linear-shaped charges, 250 holes were drilled in the oxidizer and fuel tanks. F. Charlie Radaz, the SOFT program manager for Martin Marietta Company, found that the hardest part of this contract was not the installation of the shaped charges, although this was close to the top of the list. The hardest part was convincing Boeing that the experience gained in maintaining the Titan II fleet indicated that the holes could be sealed properly. Once Boeing was convinced of this fact, Radaz and his staff had to prove that the brackets holding the shaped charges could be positioned at the correct angle to provide tight enough contact with the tank so that the charge would cut cleanly, resulting in the correct shape and size pieces. Different types and sizes of explosive charge were used, depending on the airframe part being cut, heavier on the engine bells and propellant tank stringers, not as heavy on the tank skin. In order to have the resulting fragment cloud give the appropriate test target, the Stage II tanks had to be vented to a given residual propellant pressure prior to fragmentation. Lockheed was responsible for flying the infrared optics experiment. At apogee a door would open in the payload fairing of the Lockheed launch vehicle. A telescope/detector package would then track the Titan II Stage II fragments. There was concern that the detector would lock onto the sun, moon, or stars, so the launch window was narrow to minimize this possibility.[52]

Missile B-27 (62-008), composed of missile B-29 (62-010) Stage I and B-27 Stage II, was selected for the SOFT test in October 1973, but after two months of work the program was put on hold due to lack of funding. Six months later the program was resumed, and on 24 October 1974 the missile was ready. On 30 October 1974 B-27 was emplaced in Launch Complex 395-C and one month later was alert ready. On 10 January 1975, the missile combat crew composed of Capt Michael W. Sayer, MCCC; 1Lt M. D. Wilderman, DMCCC; Sgt Robert B. Ribertone, BMAT; and A1C David W. Fiyak, MFT, successfully launched B-27.[53]

The Stage II tank fragmentation was successful, producing 28 fragments, but only two of the three radar transponders operated correctly. Unfortunately, a late change to the sensor package software that was to have increased observation time backfired because during the actual flight the detectors picked up a corner of the moon image, becoming saturated during the important part of the test, only coming out of saturation at the very end of the viewing window. This infrared experiment was the primary test objective; the fragmentation data was secondary. A second Stage II airframe had been modified as a follow-up test article but was not used.[54]

The fourteenth and last launch operation for the 308th SMW, as well as the final launch of the Titan II ICBM program, was given the name Project RIVET HAWK. RIVET HAWK was the first flight test on a Titan II ICBM of a new inertial measurement unit, the heart of the missile's inertial guidance system (see Chapter 3 for details on the guidance system). Missile B-17 (62-2771) was selected and shipped to Vandenberg AFB, arriving on 12 December 1975. By 21 June 1976 B-17 was flight ready.

On 27 June 1976, the missile combat crew composed of Capt Roger B. Graves, MCCC; 1Lt Gregory M. Gillum, DMCCC; SSgt David W. Boehm, BMAT; and SSgt Kenneth R. Savage, MFT, began the launch procedures.[55] Key turn took place and within seconds a GUIDANCE HOLD occurred due to an INERTIAL GUIDANCE SYSTEM NO-GO signal. The shock produced during prevalve opening was sensed by the inertial measurement unit, triggering the hold. Robert Popp, the Delco representative for the RIVET HAWK program, remembers that about 30 minutes prior to launch, one of the Martin Marietta Company engineers had asked him if the hammer effect caused by the sudden flowing of propellants due to a pre-valve opening would be a problem. The new software had retained both MEMORY and BLAST DETECT modes, so the Delco team felt that the countdown could be resumed by pressing RESET if the hold did occur. Popp remembers thinking, as the launch window continued to slip by, that here was the chance to find out! The inertial guidance system was returned to the READY mode, the countdown recycled, and after downrange checks, the countdown resumed 18 minutes later. The flight to target was successful, but the reentry vehicle impacted approximately 1.46 nautical miles long and 0.36 nautical miles cross range.[56]

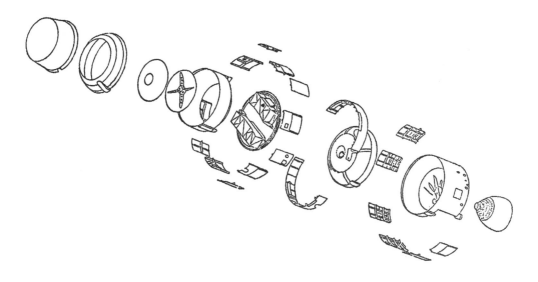

Figure 8.4. The Signature of Fragmented Tanks experiment required precise cuts in many of the Stage II structures of Titan II ICBM B-27 (62-008) as shown in this illustration. The resulting debris cloud was used to test the ability of the Safeguard Anti-Ballistic Missile radar system to discriminate between the debris cloud and the reentry vehicle. *Courtesy of F. Charlie Radaz.*

As one might imagine, this was more than a little disconcerting. Review of the telemetry from the guidance system, as well as extensive computer modeling, revealed an error in the software. The unique feature of the new USGS inertial measurement unit was the rotating X-Y platform. This feature mitigated a source of error in the X-Y plane that had to be accounted for in a nonrotating system. The newer computer in the system allowed the continuously changing X-Y instrument outputs to be monitored for updating the platform alignment. In the USGS equipment used on Titan III, the platform rotated at one revolution per minute. For deployment in the operational Titan II fleet, the decision was made to slow the platform down to one-quarter revolution per minute due to a failure rate with the one-revolution-per-minute system that was unacceptable for the Titan II program. It seems that Titan II USGS programmers failed to provide a program path for the updating of the instrument coefficients after one minute; rather, it was after one *revolution* or four minutes. The resulting uncompensated instrument errors actually grew exponentially, and after four minutes were unacceptably large. This was not known at the time but, by a quirk of fate, the instrument error compensation values at the time of launch were four minutes old, causing the resulting impact error. Post-launch review of the guidance software clearly revealed the cause of the error. The fix, which did not require another launch, was to refresh the instrument compensation factors after 90 degrees of rotation, or with a one-quarter-revolution-per-minute system, once a minute as before.[57] With only four spare Titan II missiles remaining in the inventory, including one each at the three operational Titan II wings, and Western Test Range support equipment unavailable in time for a second launch before the USGS purchase decision date of 1 October 1976, the decision was made to proceed with the USGS modification.[58]

LtCol Charles Simpson had the dubious honor of addressing a committee in the Pentagon tasked with the decision to purchase the USGS system for Titan II. Simpson was chairman of the Ballistic Missile Evaluation Division, Headquarters Strategic Air Command. He was part of a group that evaluated the results of missile launches, determining the accuracy and reliability of the various ICBM systems. He recalls many discussions among his group about the Delco engineers' explanation for the large miss distance. They finally agreed that the error was accounted for, and Simpson was now before a committee of one- and two-star generals. After some discussion and explanation by the experts on the impact error and its cause, the committee asked for an opinion from Simpson, the lone Evaluation Division representative. Simpson remembers briefly reiterating once more the causes and the fix and then discussing in detail how this new guidance system was an absolute must if Titan II was to be kept in the inventory. Simpson joked that with the big warhead that Titan II carried, it would still rattle some windows—even at that miss distance. The committee members laughed, and agreed to purchase the USGS system.[59]

381st Strategic Missile Wing, McConnell AFB, Kansas

Originally, the fourth Demonstration and Shakedown Operation launch, Exercise HIGH RIDER, was to have been missile B-32 (62-013), launched by a crew from the 381st SMW, McConnell AFB, Kansas. On 21 September 1964 final work was being conducted prior to placing the missile on alert-readiness status at Launch Complex 395-D. Dick Rector, a civilian working for the Navy flight safety office at the Western Test Range, was working the night shift as a flight termination crew member. He was inspecting all of the flight ordnance for the range safety destruct system. This was to be a no-notice launch, so the missile would sit in the silo for what was termed the "cold soak" period, for a week or more, before launch. This simulated the missile being on alert in an operational silo. The destruct system was installed in both stages during this cold soak period. If at any time prior to launch anyone had to go into the launch duct, one of the flight termination crew members had to be present to make sure the flight termination equipment was not compromised.

The flight termination system destruct packages were located in the between-tanks areas on both stages. Rector was down on Level 5 to do a final check along with A2C D. Hockman, TSgt Paul Seashore, in charge of all airborne ordnance, and two other personnel from the 381st SMW Task Force. SSgt Charles Smith, 395th SMS, and SSgt Sam Ancheta, a BMAT from the 381st SMW Missile Handling Shop, and two additional personnel were in the launch duct on Level 3 working on the guidance system.[60]

In the launch control center, the missile combat crew composed of Maj Robert Arnold, MCCC; Capt L. Miller, DMCCC; Master Sergeant Hensley, BMAT; and Sergeant Weinholdt, MFT, were on duty. Arnold had just finished talking with a maintenance technician who had returned from the launch duct. Working from a guidance system checklist, Arnold directed Hensley to bring the guidance system back to READY. Hensley began the process by pressing the STANDBY button on the Missile Guidance Alignment-Checkout Group panel. The system advanced from STANDBY to ALIGN 1. A fraction of a second later, a shock was felt from the direction of the silo, and Blast Door 9, the blast door closest to the silo, slammed shut with a bell-like resonance.[61]

On Level 5, the work platform segments were down, and Rector and his team had just walked out onto work platform Segment C directly in front of the launch duct door. Rector had turned to latch the door closed when he heard a high-pitched scream as the Stage II translation rocket on the opposite side of the missile ignited. The next thing he knew, the launch duct door, which swung out into the silo equipment area, flew back open, bending the hinges, and the exiting air pressure threw him into the shock isolation platform supporting the massive launch duct air-conditioning unit outside of the launch duct. The platform was mounted 12 inches off the floor, and Rector's left leg slid underneath, tearing muscles and breaking his leg at the knee. His head hit the corner of the sled, fracturing his safety helmet, blackening his left eye, and breaking a tooth in the process. Seashore was unhurt. Rector recalls not really feeling the pain at the time since his first concern was to get away from an accident that might be spreading throughout the missile complex. The elevator was not available, so they started up the access ladders. With Hockman helping from above, and Seashore from below, they were able to get Rector up three ladders to Level 2. Hockman had sprained his ankle so it was somewhat slow going, but they finally brought Rector into the launch control center. Arnold took one look at Rector and organized his crew to render first aid, then notified the base medical personnel.

Smith and Ancheta were not as fortunate since they had been working on the Level 3 platforms much closer to the base of the translation rockets. Luckily, they had both just finished closing up the guidance bay door and were approximately 10 feet away and partially shielded by the circumference of the missile from the nozzle of the 5-foot-long, 5,000-pound-thrust translation rocket. The heat and intense light from the rocket exhaust burned both of Ancheta's hands and partially flash-blinded him. Smith was flash-blinded and also suffered burns to his arms. Rector ended up requiring 38 stitches but recovered fully; Ancheta and Smith were hospitalized for their burns but recovered fully. On 21 September 1964, Seashore and Hockman were awarded the Airman's Medal for heroism involving risk of life in assisting Rector to safety. Rector's hardhat was retained and used at the Pacific Missile Test Center, Point Mugu, California, as an example of why one wears a hardhat![62]

Damage to the missile airframe consisted of a large hole, approximately 25 inches in diameter, one foot directly above the rocket nozzle.[63] Fortunately, the blast from the translation rocket firing did not set off the Stage II destruct charges, and the flame had not reached the oxidizer tank skin. Martin Marietta Company immediately began to investigate the incident. Since a nearly identical event had taken place at McConnell AFB earlier in the year (11 February 1964), with no injuries because no one was in the launch duct at the time, this problem was no longer a random event. Three possibilities seemed probable sources for the failure: connector contamination, missile wiring faults, or electrical shorts within the connector.[64]

The first step was to get the connector back to Denver for a thorough analysis. F. Charlie Radaz, a resident Martin Marietta Company engineer at Vandenberg AFB, flew back to Denver that night, hand carrying the translation rocket initiator connector. Charles Carnahan, a Titan II program manager at Vandenberg AFB, stressed the need for a thorough multistep plan to open up the connector and make sure they did not miss anything. Everyone wanted to open it up right away, but Carnahan insisted that it be done slowly and correctly. Once separated into the two major pieces, gold flakes from the connector surfaces could be seen lying among the pins. In the 11 February 1964 incident, the cause had been found to be small metal shavings left in the connector body during the manufacturing process. Radaz reported the presence of gold flakes in the connector to Carnahan by phone, and Carnahan asked him a string of detailed questions concerning the color of the flakes, their exact location, the current that would have had to flow through them, et cetera. The Denver engineers felt that the flakes were the shorting source, but Carnahan said that his team just did not think this was possible. The Denver engineers put the connector into a temperature chamber set at 60 degrees Fahrenheit, the temperature of the launch duct. By changing the temperature several degrees up or down, they could create a temporary short circuit. This did not correlate with the gold flakes. The connector was then x-rayed, and the problem was revealed.[65]

After soldering the wires within the connector, final assembly required pouring an insulated potting compound into the voids between the wires. It turns out that a worker had apparently used a screwdriver to tamp the potting compound around the wires, inadvertently bending over several bare wires almost into contact with the metal casing of the connector. With small temperature changes, the exposed wires contacted the metal casing, causing a transient short circuit. When the guidance system was brought to the READY mode, there was a surge of current that was sufficient to short the connector, which then caused the initiator to fire. The Titan II fleet was immediately inspected, and several more plugs from this same lot were found and replaced. In addition, as a result of the incident, portions of the missile wiring were inspected, and all igniters for translation and vernier rockets were inspected and replaced as necessary.[66]

Stage II of missile B-32 was removed from the silo, and repairs were made to the missile at the Martin Marietta Company facilities at Vandenberg AFB. B-32 Stage II was reinstalled at Launch Complex 395-D. On 4 November 1964, the missile combat crew composed of Capt Thomas A. Grant, MCCC; Capt William Reinken, DMCCC; A1C J. G. Montgomery, BMAT; and MSgt Irwin L. Glore, MFT, successfully launched B-32. The surface range was 4,135 nautical miles to the Kwajalein Tango target area. The reentry vehicle impacted was 0.838 nautical miles long and 0.497 nautical miles right cross-range, generating an impact error of 0.974 nautical miles. On 17 November 1964, Headquarters Strategic Air Command announced that the DASO program had been highly successful, with all five reentry vehicles impacting within one-half of the distance established by the special operational requirement.[67]

Exercise GOLD FISH was the fifth launch in the OT program and the third for the 381st SMW. Missile B-22 (62-003) was removed from Launch Complex 533-2 on 30 March 1965 and shipped, via truck convoy since the C-133 fleet was grounded, to Vandenberg AFB on

Figure 8.5. On 21 September 1964 the 5,000-pound-thrust translation rocket located between launch duct quadrants 2 and 3 on Stage II of Titan II ICBM B-32 (62-013) inadvertently fired during a guidance system power up at Launch Complex 395-D, Vandenberg AFB, California. After an extensive investigation, a short circuit was found in the igniter connector. The short was caused by substandard manufacturing technique. The white wire to the left of center in this photograph was in transient contact with the metal casing of the connector and was sensitive to temperature fluctuations. *Courtesy F. Charlie Radaz.*

22 April 1965. On 11 May 1965 the missile was installed in Launch Complex 395-C and 17 days later was placed on alert-ready status. On 14 June 1965 the missile combat crew composed of Maj Henry A. Curtin, MCCC; Capt Robert S. Luckett, DMCCC; SSgt Eugene R. Myer, BMAT; and SSgt Bill J. Lamb, MFT, acknowledged receipt of the launch execution message.[68] Before key turn could take place, a range safety hold was called. The launch execution message was resent, and B-22 lifted off successfully. Approximately two seconds into vernier engine operation at the end of Stage II powered flight, vernier engine number one shut down prematurely. This prevented reentry vehicle separation and decoy deployment. The reentry vehicle–Stage II combination impacted approximately 7 nautical miles short of the target area. Engineering Change Proposal 512100 was issued fleetwide for installation of a redesigned vernier nozzle, and the problem did not reoccur.[69]

Exercise BOLD GUY was the tenth launch in the OT program at Vandenberg AFB. Missile B-58 (62-12301) was removed from Launch Complex 532-1 on 28 July 1965, prepared for shipment, and airlifted via C-133, arriving at Vandenberg AFB on 7 August 1965. B-58 was inspected and installed in Launch Complex 395-B on 23 August 1965. The missile was placed on alert-ready status on 11 September 1965. On 21 September 1965 the missile combat crew composed of Capt Thomas E. Macomber, MCCC; 1Lt Robert W. Bloodworth Jr., DMCCC; SSgt Lester T. Pratt, BMAT; and A1C Jackson G. Shelton III, MFT, acknowledged the launch execution message and turned keys. The launch was successful. At initiation of the staging sequence by closure of the Stage I thrust chamber pressure switch, telemetry indicated that the guidance computer was issuing erratic but discrete signals, including decoy eject, reentry vehicle release, ignite pitch rocket number one, ignite pitch rocket number two, ignite translation rockets, and sustainer engine cutoff. Only the sustainer engine cutoff signal was acted upon, resulting in the premature shutdown of the Stage II engine. Stage II and the reentry vehicle impacted well short of the target area.[70]

Monitoring of telemetry as the flight continued showed that the computer continued to issue appropriate commands during the remainder of the flight, indicating that the computer was functioning properly. Review of the telemetry revealed that heat generated in the Stage I engine compartment in the area where the Stage I control wiring harnesses were located had caused the wiring insulation to fail, causing an electrical short. The computer was affected by the resulting external noise on the 28 volts DC ground return line. A Time Compliance Technical Order, TCTO-21M-LGM25C-661, was issued, calling for the wrapping of Stage I engine electrical control harnesses, thrust chamber pressure switches, and the pressure sequencing valve with reflective tape. Engineering Change Proposal 11934 for improved guidance computer grounding was initiated, and the problem did not reoccur.[71]

Exercise CROSS FIRE was the thirteenth launch in the OT program. On 24 September 1965, maintenance personnel began work at Launch Complex 533-2, preparing missile B-4 (61-2758) for removal and shipment to Vandenberg AFB. The missile left McConnell AFB on 12 October 1965, arriving at Vandenberg AFB the same day. Installation in Launch Complex 395-B took place on 28 October 1965, and the missile assumed alert-ready status on 20 November 1965. On 30 November 1965 the missile combat crew composed of Capt Milo L. Carlson, MCCC; 2Lt James G. Ellis, DMCC; TSgt James A. Collins, BMAT; and A1C Gerald A. Humes, MFT, successfully launched B-4.[72]

After 63 seconds of Stage II powered flight, a drop in fuel discharge pressure took place, followed almost immediately by hard-over actuator movements and a 6,700-pound drop in thrust (the Stage II engine normally developed 100,000 pounds of thrust). The uncontrolled actuator movement resulted in a pitch altitude that could not be controlled by the guidance computer, and the mission failed. Telemetry analysis indicated that the guidance and control systems had not failed; rather, they had attempted to correct for an external sideways force that had developed on the missile. The primary cause was a rupture either in the fuel return manifold on the thrust chamber bells or one or more of the coolant tubes of the regenerative cooling system. The sideways thrust that developed subsequently overpowered the pitch actuator. Aerojet-General conducted an inspection of the regenerative cooling system before and after engine firing to evaluate possible weakened areas. This failure was apparently a one-time event, because no physical changes in the engine design were made and the problem did not recur.[73]

The first launch exercise in the Follow-on Operational Test program was Exercise SILVER BULLET. Missile B-91 (64-455) was removed from Launch Complex 532-8 on 4 March 1966 and transported to Vandenberg AFB via C-133 aircraft, arriving on 23 March 1966. The missile was emplaced in Launch Complex 395-C on 20 April 1966. The missile was placed on alert readiness on 7 May 1966. After five days of alert-readiness posture, the missile autopilot had to be replaced, and B-91 was returned to alert readiness on 14 May 1966. On 24 May 1966, the missile combat crew composed of Capt Marion C. Fasler, MCCC; 1Lt James L. Stapp, DMCCC; SSgt David Struthers, BMAT; and TSgt Thomas Bloominger, MFT, successfully launched B-91.[74]

Stage I and II flight through vernier engine cutoff was normal. Telemetry indicated that the inertial guidance system issued the reentry vehicle release command followed by the pitch/depitch and translation rockets on and off commands. The reentry vehicle release signal was given but no separation occurred. According to Kwajalein ground optic tracking equipment, Stage II appeared to break up at 175,000 feet, and what appeared to be the reentry vehicle was seen emerging from the tank debris at approximately 148,000 feet. The reentry vehicle impacted 23 nautical miles short of the target area.[75]

390th Strategic Missile Wing, Davis-Monthan AFB, Arizona

The 390th SMW began participation in the operational flight program when preparations for Exercise CARD DECK were started with the removal of missile B-54 (61-12297) from Launch Complex 570-3 on 11 March 1965. It was shipped four days later by C-133 to Vandenberg AFB. Installation into Launch Complex 395-D took place on 1 April 1965, and B-54 achieved initial alert readiness on 20 April 1965 and remained so until launch.

On 30 April 1965, the missile combat crew composed of Maj Norton M. Hewitt, MCCC; 1Lt Joel J. Lefave, DMCCC; MSgt Miguel L. Dezarraga, BMAT; and A1C Ronald V. Szabat, MFT, received the launch execution message. The countdown was uneventful, and B-54 was launched successfully.[76]

After approximately 100 seconds of powered flight, the Stage I engine Subassembly 1 shut down prematurely. Telemetry indicated that the subassembly's turbopump had malfunctioned due to bearing breakdown or seizing from insufficient lubrication. As a result, B-54 did not reach the target area. In order to monitor the possible recurrence of this malfunction more carefully, additional telemetry measurements of the next two flights were requested. Additionally a time compliance technical order change was issued to the operational fleet for new servicing and inspection procedures for the turbopump assembly gearboxes. This malfunction did not occur again during the 12 years of Titan II ICBM launch exercises.[77]

In October 1970, Headquarters SAC selected missile B-69 (63-7724) for use in the first Titan II launch in support of SSTP. B-69 had been the first operational missile to experience an aborted launch (30 November 1967) and had been refurbished and stored at the 394th SMS, Vandenberg AFB, during the intervening years. Only one Titan II launch site was operational at Vandenberg AFB, Launch Complex 395-C, at this time.

On 17 February 1971, a task force under the command of LtCol Ed D. Bailey left Davis-Monthan AFB to begin preparations for the upcoming launch, code named MI-17. B-69 was removed from storage on 23 February 1971. Numerous small problems due to prolonged storage had to be fixed, and the missile was installed on 9 March 1971. Problems continued to plague the missile as the initial flight control response test failed. An autopilot, rate gyros, and two actuators were cannibalized from B-12, the other missile in storage at the missile assembly and maintenance facilities at Vandenberg AFB. The retest was successfully completed on 24 March 1971. Fuel and oxidizer loading was completed by 26 March 1971. On 26 May 1971, the missile combat crew composed of 1Lt Ed Moran, MCCC; 1Lt Bruce J. Stensvad, DMCCC; SSgt Ray Hersey, BMAT; and Sgt William C. Shaff, MFT, was on alert at Launch Complex 395-C. A GREEN DOT launch exercise message was received and decoded. On Moran's count, the two officers turned their launch keys. Meanwhile, in the flight test program command post, less than one second after initiation of the launch sequence, the Army representative pulled off his headphones and yelled, "Hold! Hold!" Capt Ted Suchecki, the Air Force test conductor in the command post, pushed LAUNCH DISABLE, and the countdown was halted just after it had begun. Discussions immediately began as to the ability to recycle and re-initiate the countdown. F. Charles Radaz, the Martin Marietta Company engineer for the Safeguard test series, asked for and received time to evaluate the telemetry data to see at what point the missile systems were poised in the launch sequence.[78]

The Army's need to be able to call a hold after key turn due to any downrange problems with the Safeguard equipment had been a point of heated discussion among the Army, the Air Force, and the Martin Marietta Company. Both the Air Force and Martin Marietta staff were emphatic that once key turn took place the only hold that might occur was one generated by the launch sequence itself. The Army, realizing that they had relatively few Titan II airframes to expend on these tests, wanted to be able to hold and, if necessary, recycle the launch as range conditions dictated. The Army finally prevailed, with unforeseen consequences.[79]

In the launch control center at Launch Complex 395-C, the countdown had begun as normal. On the Launch Control Complex Facilities Console, the LAUNCH ENABLE indicator illuminated as expected and then, just as the ACTIVATE BATTERIES signal was issued by the launch control set, the system was shut down so rapidly by the LAUNCH DISABLE signal sent from the flight test command center that the ACTIVATE BATTERIES indicator on Control Monitor Group 1 (this instrumentation gave a visual indication of the count-down sequence) never illuminated and thus was not observed by Hersey as he monitored the progress of the countdown. The LAUNCH DISABLE indicator illuminated on Moran's console simultaneously with Suchecki's actions in the command post and the launch sequence went into SHUT DOWN mode.

Radaz's major concern was whether or not both the first- and second-stage propellant prevalves had opened. If the FOT program instrumentation had been in place it would have indicated prevalve status, but this was no longer the case. Radaz listened on the sup-port net as his team reviewed the sparse available data and as the range people tried to clear the range. Little further information was given to Moran's crew. They remained at their consoles for at least an hour. Moran decided to run the pre-launch checklist and made sure that the launch control center was now operating on internal air supply just in case something happened on the missile during the wait. They also checked to be sure that Blast Doors 8 and 9, forming the blast lock between the launch control center and the launch duct, had their pins fully extended.

Back at the command post, the "range ready" message was finally received. Radaz had reviewed all available data on the status of the prevalves but simply could not give an answer as to the feasibility of continuing on to a launch. He did not know, short of going out to Level 7 of the silo and looking, just what the status of the Stage I prevalves were. This could have been done but would have taken a significant amount of time. Radaz rec-ommended a visual check, but the Air Force representatives felt that the technical data allowed a second try without the need for a time-consuming check.

Moran received another launch time and waited, listening for the final check of the range. He and Stensvad again turned their launch keys on his count, and the sequence pro-ceeded normally. At the completion of the launch sequence both the LIFTOFF and ABORT lights illuminated simultaneously. Meanwhile, Hersey had been watching the Control Monitor Group 1 indicators and saw the ENGINE START SWITCH CLOSED light illumi-nated. He began to count off the seconds to liftoff. When he reached four seconds, he knew something was wrong because liftoff did not take place. Liftoff would have been indicated by the flashing red FLASHER MONITOR light on the Control Monitor Group 1 panel as well as the white LIFTOFF indicator on the Launch Control Complex Facilities Console which he could not see at the moment. At six seconds an abort would normally be indi-cated.

Meanwhile, Radaz had several readouts so that he could clearly see the sequence of engine ignition events as the countdown proceeded. Radaz watched as the turbopump speed monitor pegged in overspeed, and with a sinking heart, he realized immediately what had happened.

Moran, unsure of whether liftoff had occurred, covered his microphone and turned to the pad chief, who was on a direct phone link to a mess hall from which launches were frequently observed. The pad chief relayed back to Moran, "Lots of smoke, no missile yet." Moran did not want to push RESET-SHUTDOWN for fear the missile might be slowly climbing out of the silo with the umbilicals not yet released. He waited 10 to 15 seconds and then pushed RESET-SHUTDOWN. The crew quickly went through the abort checklist, receiving no further information from the command post. Moran remembers another reason for his hesitation was that with RESET-SHUTDOWN initiated, the silo closure door returned to its closed position. With a fully fueled missile in the launch duct in an unknown state, Moran felt that leaving the silo closure open was better for crew safety, but this was not an option. The crew remained on duty overnight as the situation was assessed.[80]

What had happened? Under normal conditions, it took approximately one second to complete the ACTIVATE BATTERY signaling sequence. The relay responsible for carrying this signal also signals the Stage II prevalves to open. It also commands a Stage I prevalve motor-driven switch that takes 0.1 second to close, initiating the prevalve opening sequence on Stage I. Unbeknownst to anyone, on the first launch attempt key turn, this motor-driven switch had operated for only 0.08 seconds prior to the sequence being shut down by the LAUNCH DISABLE command, so the Stage I prevalves were *not* open and propellants could not flow to the turbopump assemblies.

As a consequence, the launch logic was now poised in an unforeseen configuration. When the countdown was restarted, ACTIVATE BATTERIES was already sensed by the launch logic since they had been activated in the earlier countdown, so the OPEN STAGE I PREVALVE signal was not repeated. Unfortunately, this meant that the Stage I prevalve motor-driven switch was not commanded to close, therefore it did not initiate opening of the Stage I prevalves.[81] The remainder of the countdown proceeded normally and both engine start cartridges fired, spinning up the turbines that drove the fuel and oxidizer turbopump assemblies for both thrust chambers. However, since the Stage I prevalves had not opened, the turbopumps were dry and quickly accelerated past their design limits, shattering and spewing shrapnel in an expanding disk pattern, cleanly cutting off both turbine exhaust pipes and spraying the acoustical modules with turbopump debris. Fortunately, the sound attenuation modules absorbed the debris, preventing it from puncturing the Stage I fuel tank and possibly collapsing the fully loaded missile.[82]

The majority of the 390th SMW task force had returned to Davis-Monthan after the preparations to launch B-69 were complete since there was a shortage of missile maintenance personnel at the base. The skeleton team that remained, and the four missile combat crews that had stood alert, pitched in with the personnel of the 394th SMS and the 3901st SMES to emergency offload the propellants from B-69 on 27–28 May 1971. On 29 May 1971, Headquarters SAC informed the 1st Strategic Aerospace Division that missile B-12 (61-2766) was to be prepared for launch as soon as possible and that the 390th SMW would provide the task force manpower. From 27 to 30 May 1971, the 1st Strategic Aerospace Division Launch Analysis Group investigated the damage to the launch duct. On 30 May 1971, B-69 was removed and the amazingly minor damage to the launch duct was repaired.[83]

On 1 June 1971, missile B-12 was installed. This missile installation had a tense moment during the handling of Stage I. Sgt Ed Ryan, a missile maintenance technician with the 394th SMS, watched in the early morning light as Stage I was hoisted by the crane of the transportation cradle and was being positioned for lowering into the launch duct. All was normal until a strong gust of wind blew through the launch complex and the crane cab began to

Figure 8.6. On 26 May 1971, Titan II B-69 (63-7724) experienced its second ground abort. The Stage I start cartridges ignited and spun up the turbines which shattered due to a lack of propellant in the turbopump assembly. At the top is a normal Stage I engine set showing the Subassembly I turbine location indicated by the white arrow with the exhaust stack below. The Subassembly 2 exhaust stack is behind and slightly to the right. The bottom photograph shows the aftereffect of the ground abort on Subassembly I. The start cartridge is seen on the left; the turbine housing, minus the sheared-off exhaust stack, is in the center of the photograph. Shrapnel impact marks can be seen on the Subassembly 1 thrust chamber. *Upper photograph, from the author's collection. Lower photograph, courtesy of Bill Shaff.*

spin. The crane operator froze. Ryan's eyes got wider and wider as he watched Stage I head in slow motion straight for a fully loaded fuel holding trailer. The downwind tag line holders had dropped their lines and were headed for the protection of the oxidizer hardstand, which, at Vandenberg AFB, was below grade level. All of a sudden Ryan and SSgt R. Belk, who were on the upwind tag line, were being taken for a ride as the Stage I airframe continued to rotate toward the fuel trailer. SSgt Elisia Tatum, a missile equipment handling

Figure 8.7. B-69 Stage I oxidizer prevalve showing the shaft indicator which should have been aligned with the downward pointing arrow. A visual inspection would have shown that the valves were not open, but at the time such an inspection would have been highly dangerous. *Courtesy of Robert Dreyling.*

operator, jumped into the cab and dropped the cab slew lock, stopping Stage I just short of collision. The rest of the installation went smoothly, and B-12 was installed in record time.[84] The entire installation, propellant loading, and checkout took only 8.5 days from ABORT to READY GREEN, a record that stood until the end of the system. On 8 June 1971, just 12 days after the aborted launch attempt, a new missile was ready for launch.[85]

On 19 June 1971 two successive launch attempts resulted in APS (Accessory Power System) HOLD indicators due to faulty motor-driven switches during the inertial guidance system power transfer process. The switches were replaced, and on 20 June 1971, B-12 was successfully launched by the missile combat crew composed of Capt John R. Ransome, MCCC; 1Lt Charles W. Schubert Jr., DMCCC; TSgt George V. Fetchik, BMAT; and SSgt Johnson Randolph, MFT. A unique feature with this launch was conducted with the guidance system in MEMORY MODE, simulating the effect of a nearby miss from a nuclear weapon disrupting the beam of light that normally kept the guidance system aligned. All Titan II-related mission requirements were met for this first test in the SSTTP.[86]

B-69 was shipped to Denver. After a thorough inspection, a new engine was installed and the missile was sent back to Vandenberg AFB for storage. Understandably, however, missile checklist procedures were modified to avoid a recurrence of this situation.[87]

Summary

Fifty-one Titan II operational missile launch training exercises spanned a 12-year period from 1964 to 1976 (see Table 8.7). The 308th SMW successfully launched 16 Titan II missiles over a 12-year period, with 2 failing to complete the flight to target and one aborted launch. The 381st SMW successfully launched 17 Titan II missiles with 3 failing to reach the target. They had no aborted launches. The 390th SMW launched 13 Titan II missiles with one failure to reach the target area and two aborted launches.

Despite the concerns of SAC and AFLC, the bench test concept encompassed by the RASP was able to successfully monitor and provide preventative maintenance in support of 12 launch attempts over a seven-year period. With one launch aborted due to range-related problems, the 10 for 11 successful launches and flights to target was a record illustrating that Titan II was still more than able to perform its mission.

Titan II operations were not, however, without their dangers, and in some instances, tragedies. With such a successful training launch record, what was the reason for the all too prevalent appelation of Titan II as "an accident waiting to happen"? A total of five fatal accidents in 24 years is the reason. Careful consideration of each accident reveals how unwarranted this description is for the Titan II program, which was in reality less dangerous than equivalent industrial conditions.

IX

FATAL ACCIDENTS
IN THE TITAN II PROGRAM

Though Titan II was a reliable and effective weapon system, Titan II was not forgiving—it did require the close attention of its operations and maintenance crews. During its 24-year period of operation, five fatal accidents occurred involving Titan II missiles or launch complexes. These events claimed the lives of 58 men, 53 of whom perished in a single accident during silo modification. In each case, human error was the cause of the fatality, not a hidden flaw of the missile, its subsystems, or the launch complex.

The accident investigation process in the Air Force has two major components. The first is the Mishap Investigation Board (earlier known as the Safety Investigation Board). This panel is tasked with finding out the actual cause of an accident and with proposing fixes to prevent reoccurence. Testimony taken by this type of investigation is privileged and cannot be used against individuals who might later be found at fault. The findings of the Mishap Investigation Board are considered classified and remain within the Air Force safety community. The second type of investigation is that carried out by the Accident Investigation Board. This board provides a publicly releasable report of the circumstances surrounding the event and usually includes a cause for the accident. The findings are a resource for use by the Air Force for disciplinary action.[1]

Launch Complex 373-4, 308th Strategic Missile Wing,
Little Rock AFB, Arkansas

After several years of operation, modifications of the launch complexes to facilitate maintenance and increase reliability, as well as hardness against nearby nuclear blasts, became necessary. The first large modification program was Project YARD FENCE. On 9 August 1965, Titan II ICBM Launch Complex 373-4, near the town of Searcy, Arkansas, was undergoing Project YARD FENCE modifications.[2] As was the case with previous work conditions at the other operational bases during the YARD FENCE program, the missile, B-25 (62-006), remained in the launch duct fully loaded with propellants but with the Mark 6 reentry vehicle removed. The contractor had over 50 local workmen on site. Work was in progress simultaneously throughout the silo and on the surface of the site with major emphasis on flushing Hydraulic System 2 (HS-2), the system that operated the launch duct work platforms and silo blast valves. A hydraulic reservoir and pump were located on the surface with pressure and return lines fed through a reopened 8-inch-diameter construction access port in the silo closure door apron. The silo closure door was closed. The silo equipment area hardening modifications required oxy-acetylene cutting torches as well as electric arc welding equipment on Levels 2, 3, 5, 6, and 7 of the silo equipment area. Work within the launch duct included painting the access hatches for the silo closure T-lock wells on Level 1 and installation of perforated steel covers on acoustic modules on and just above Level 7. The missile combat crew on duty was composed of Capt David A. Yount, MCCC; 1Lt James C. Markey, DMCCC; MSgt Ronald O. Wallace, BMAT; A1C Donald E. Hastings, MFT; and A2C Bennie L. Williams, a spare BMAT.

At 1309, laborer Gary Lay was standing with a group of 12 coworkers near the emergency escape ladder that connected silo equipment area Level 2 to Level 3. Lay felt a rush of warm air on his back and turned to see flames directly behind him near the water chiller shock isolation platform. The lights went out, and the emergency lighting system came on as the large group of workers made a rush for the ladder. Lay decided to try to go through the fire area and out to the cableway. Hurbert A. Sanders, a painter working in the launch duct on Level 1, smelled smoke and left the launch duct. Just as the silo lighting went out, he found the access ladder to silo equipment area Level 2 and descended. There he met Lay, and they both ran down to the launch control center with acrid, choking smoke billowing behind them.

In the launch control center, the first indication of a problem came with the illumination of the FIRE DIESEL AREA indicator on the Launch Control Complex Facilities Console in the launch control center. As the klaxons sounded throughout the complex, Yount began the silo fire emergency checklist and ordered the evacuation of the silo area. Personnel on the surface noticed smoke coming out of the hose access area. As the surface warning control lights and siren activated, all power was lost in the complex. At 1311 Lay and Saunders, the only two survivors from within the silo equipment and launch duct areas, entered the launch control center. Lay had numerous burns on his hands and face, while Saunders was suffering from smoke inhalation.

Yount called the wing command post requesting a Missile Potential Hazard Team (MPHT) be formed. At 1320, Col Charles P. Cosgrove, deputy wing commander, 308th SMW, alerted the base hospital and requested that ambulances be sent to Launch Complex 373-4. At 1316, Wallace and Hastings donned air packs and proceeded to Level 2 of the silo. They reported extreme heat and smoke but did not see flames and returned to the launch control center. At 1407, Col Charles P. Sullivan, the 308th SMW commander, requested a physical count of personnel on the site and was told that 53 workers were missing.

Firefighters and equipment were dispatched from the main base at 1411 by helicopter, followed by medical vehicles. From 1440 to 1630 additional support equipment was dispatched. At 1800 the missile combat crew attempted to open the silo closure door to permit ventilation of the launch duct, but the door failed to open. At 1915 no fire was visible in the silo equipment area on Level 2, but smoke and fumes were still too dense to con - duct further rescue operations in the silo. One hour later, a rescue team was able to reach Level 5. Finding no survivors and with their air supplies running out, they returned for air pack replenishment and proceeded to Level 8 at 2030. At 2240 the SAC Disaster Control Center reported 53 fatalities, with two hospitalized survivors. The last casualty was removed from the silo at 0530 on 10 August 1965. The 53 fatalities were distributed throughout the silo. Twelve were on Level 2, 24 on Level 3, 8 on Level 4, 1 on Level 5, and 4 each on Levels 6 and 7.

At 1010, 10 August 1965, members of the Air Force Aerospace Safety Missile Accident Investigation Team arrived at the complex. Although the investigation team did not take over control of the site until 2000, team members participated in the penetration of the silo during a preliminary investigation. This first site penetration team noted a slight soot residue distributed partially down the cableway from the launch control center to Level 2 of the silo. Footing on the silo equipment area flooring was difficult due to a widespread film of what later was determined to be hydraulic fluid. The floor plan of the silo launch duct area is divided into four sections or quadrants with Quadrant I and II meeting at the cableway entrance to the silo equipment area. Proceeding past the Quadrant II portion of the exhaust duct and into Quadrant III, the investigators found the first evidence of substantial fire damage. Burned electrical cables and fire debris on the floor surface was noted. They found only mild soot deposits in Quadrant IV.

Descending in the elevator to Level 3, light sooting was noted as the team moved from Quadrant I to Quadrant IV, past the diesel generator. The floor was more heavily covered with hydraulic fluid in a layer approximately 1/4-inch thick. At the far end of the diesel generator, the aluminum partitions between Motor Control Center 1 and the diesel generator were burned out and melted in several places. Once past these partitions, and now in Quadrant III, massive fire damage to the Motor Control Center 1 was apparent with meter dial faces broken or melted; a wooden work table heavily and uniformly charred and two areas of light concrete spalling (cracking and flaking) were evident. In the area of the pipe race, evidence of extreme heat was readily apparent. Hydraulic fluid was present on nearly all surfaces. Burned electrical cabling insulation covered the floor. Moving past the Motor Control Center to Quadrant II, damage again was present only as a general sooting on all surfaces. The team returned to the surface.[3]

The Air Force Logistics Command Missile Engineer Technician Team had by now arrived. The results of the preliminary investigation showed that the missile was undamaged and the Missile Engineer Technician Team, working with the Missile and Launch Complex Group Team investigators, prepared for propellant offload and removal of the missile from the site while at the same time not compromising the investigation efforts. A new missile combat crew reentered the launch control center at 0825. At 1230 the missile combat crew was directed by the wing command post to turn over records to the investigators. By 1258 the silo was reported clear of carbon monoxide except for a reading of 25 parts per million (ppm) concentration in the collimator room which housed the Azimuth Alignment Set used to align the guidance system and which was diagonally opposite the launch duct from the site of the fire. At 1700 missile tank venting began and was terminated at 1853. The Missile Potential Hazard condition was terminated at 2000 on 10 August 1965.[4]

The results from the first inspection and statements from the survivors, rescue personnel, and medical personnel responsible for locating and removing the casualties strongly pointed to an intense fire of short duration on Levels 2 and 3 in the silo equipment area. The Accident Investigation Team split into subgroups to evaluate the possible major contributory subsystems of the complex as well as to be sure the missile, still fully loaded with propellants, was in a safe condition. Since complex power was down, the launch duct air-conditioning system—three 25-ton water chillers—was inoperative. The launch duct air temperature was above 70 degrees Fahrenheit, the boiling point of the oxidizer onboard the missile. The thermal mass of nearly 53 tons of fuel and 100 tons of oxidizer was sufficient to prevent tank rupture due to expansion from increased temperatures but not for very long. Oxidizer tank pressures increased from 9 to 13 psi to 60 to 70 psi before venting took place. That the missile tanks had not ruptured was testimony to the sturdy construction of the missile airframe. A second, more thorough inspection was made by the Explosive Material and Fire Pattern Group as well as by the Facility Pneudraulic Team (compressed air and hydraulic systems). The focal point of the flame pattern was found to be located at Quadrant III, Levels 2 and 3. They discovered a fresh, unfinished weld near a ruptured hydraulic line in Quadrant III, Level 2. Inspection of the ruptured hydraulic hose revealed that the rupture had been caused by a weakening of the stainless-steel braiding covering the hose due to possible contact with the electrode of an arc welding tool. This area became the focal point for the source of fuel and site of ignition. The Life Sciences Group began an intensive survey of the casualties, emphasizing location and analysis of cause of death. Conclusive evidence was found by this group that clearly placed a welder in the area of suspected flame origin.

Project YARD FENCE modifications included the flushing of Hydraulic System 2, located on Level 6 of the silo. The flushing system had been operating at 500 pounds per

Figure 9.1a. The scene of the accident at Launch Complex 374-4, 308th SMW, Little Rock AFB, Arkansas. The welding rod as it was found by the investigators. The actual nick in the braided line in the right upper corner of the photograph is not visible. The silo wall is to the left and air-conditioning ductwork is just out of view to the right. *Courtesy of the Air Force Historical Research Agency, Maxwell Air Force Base, Montgomery, Alabama.*

square inch pressure with a flow of 110 gallons per minute through a pair of hoses leading from the surface hydraulic reservoir and pump. At the time of the accident, these hoses were attached to the Hydraulic System 2 panel on Level 2 Quadrant IV. These hoses ran within 14 inches of a welding operator who was attaching a triangular stiffener plate to the existing web stiffener on the Spring Can S-3 support for the Motor Control Center 1 platform. The contractor personnel locator board showed the welder to be on Level 3. The location of the weld was in an extremely awkward position that was really only accessible working from Level 2, kneeling on the floor, leaning through the guardrails, and reaching around the hydraulic lines to the stiffener plate. A hardhat located on Level 2 at the welding operation site confirmed that the welder had been on Level 2.

The accidental contact of the welding rod to the hose caused the failure of the exterior metal braiding. Thus weakened, the braiding no longer prevented the interior Teflon hose from rupturing, spraying, and effectively atomizing the fluid into a mist that permeated to Levels 2 and 3. The residual heat from the just-welded fixture or the heat from the electrode touching the metal braiding was significantly higher than the 200-degree Fahrenheit flash point of the fluid and served as the ignition source.

Tragically, a set of oxyacetylene cutting-torch tanks was located on Level 2 in Quadrant III as well. While the valves on the torch were closed, the tank valves were open. Both tank lines were burned through by the initial flames, and the subsequent release of oxygen and

Figure 9.1b. Close-up of the ruptured section of hose showing the melted stainless-steel braiding. Without the restraint of the metal sheath, the inner Teflon hose ruptured, spraying high-pressure flammable hydraulic fluid as a mist that was ignited by the still-hot weld. *Courtesy of the Air Force Historical Research Agency, Maxwell Air Force Base, Montgomery, Alabama.*

acetylene contributed to the intensity of the fire. The resulting flames instantaneously consumed the oxygen on Levels 2 and 3. Between this lack of oxygen, and the tremendous levels of carbon monoxide and toxic fumes from the burning hydraulic fluid, the majority of the casualties in the silo were due to either asphyxiation or the inhalation of high concentrations of aldehydes and other combustion byproducts with death occurring within two to five minutes. Workers on Level 7, approximately 75 feet below the site of the fire, were most probably overcome by toxic combustion fumes.[5]

The final report attributed the primary cause of the accident to be "that a welder (contract employed civilian) caused a flexible high pressure line containing flammable hydraulic fluid to rupture by accidentally striking it with a welding rod."[6] Combined with a lighting system that was not strong enough to penetrate the thick smoke from the fire and an escape system that was never meant to accommodate 53 people in an emergency, the resulting tragedy unfolded. Each of the workers had been issued a face mask for use if a rocket propellant spill occurred, but these masks offered no protection from the fumes of a fire.

The Accident Investigation Team report detailed a number of additional causes and contributing factors to the magnitude of the loss of life. Inadequate ventilation was the result of partial work completed on the silo blast valves that impeded airflow. The silo elevator did not have an independent source of power. The Level 2 collimator room partition blocked access to the cableway on Level 2, forcing exit in only one direction, in this case back through the intense heat of the oxyacetylene-fueled fire. Safety procedures for each of the tasks were well identified, but the combination of several tasks on one level at the same time made for incompatible work safety conditions. The intensity of the fire generated a large concentration and volume of toxic combustion byproducts.

The report ended with 26 individual findings and changes to the launch complex facilities and Project YARD FENCE protocols. The report noted that SAC was ill-equipped to manage the safety aspects of Project YARD FENCE. While the missile combat crew commander on duty during maintenance operations was responsible for verifying that safety conditions were adhered to, this was not possible with 53 workers carrying out a multitude of different projects in the silo. A strong recommendation was made that SAC have primary responsibility for the missile, but the contractor would have primary safety and quality-control responsibility at the site. Greater efforts by the contractors would thus be necessary to ensure that all personnel were thoroughly educated in the hazards of combined tasks that individually appeared safe.

A lack of safety discipline by the contractor contributed to the accident. The hydraulic flushing lines were haphazardly draped; welding blanks and standby fire extinguishers were lacking; cigarettes and lighters and other prohibited articles were found on the persons of the workers or within the work areas. Hand tools in the launch duct were not tied to the personnel or support beams to prevent them from falling and striking the missile.

Base firefighting personnel were not familiar with the layout of the launch complex. Critical medical information was lost when no record was made of the injuries of the first 11 casualties removed from the accident area. No record was made of the body locations within the silo by name.

Titan II missile complex ventilation systems were never designed to provide sufficient capacity to support prolonged painting or welding operations. Compounding this was the fact that fresh air intake was prevented due to dismantlement of sections of the inlet and exhaust ducts. The collimator room was constructed so as to block exit from Quadrant IV should Quadrant III be obstructed. A kick-out panel was later installed to permit two exit paths.[7]

Ironically, the hydraulic fluid was not listed as flammable under ordinary conditions. A study was undertaken to replace the flammable hydraulic fluids then used in the complex with aircraft-quality noncombustible fluids. The cost for replacing the fluid was excessive, and the fluid was not replaced. It is worth noting that there was never another hydraulic fluid–fueled fire in the Titan II program.[8]

On 13 August 1965 all but minor above-ground work with Project YARD FENCE at the 308th SMW was halted at the five launch complexes that had work in progress. The sites were brought back to operational status. Work did not resume on Project YARD FENCE at the 308th SMW until 1 December 1965. A similar work delay occurred at both the 381st SMW and the 390th SMW.[9]

In January 1966, SSgt Robert L. Cunningham, A1C Joseph D. Rollings, A1C William A. Hand, and A2C Donald D. Trojanovich were awarded the Airman's Medal for their heroism "involving voluntary risk of life at Titan II Missile Complex 373-4." Staff Sergeant Cunningham had aided in the recovery of the victims; Airman 1st Class Rollings and Airman 2nd Class Trojanovich entered the launch duct to perform a propellant tank decompression; and Airman 1st Class Hand had entered the complex in an attempt to rescue any survivors.[10]

On 8 September 1966, 13 months after the tragic fire that had claimed 53 lives, operational control of Launch Complex 373-4 was returned to the 308th SMW, and on 29 September 1966, 373-4 was returned to alert status.[11]

On 9 August 1986 more than 400 people attended a ceremony to dedicate a memorial to the 53 civilian workers killed 21 years earlier in the flash fire at Launch Complex 373-4. The 7-foot-tall granite monument, engraved with the names of the 53 victims, was paid for by private funds raised by Mrs. J. Turley, whose father, Willis Bailey, was one of those who perished in the fire. The monument was placed near the entrance to Little Rock AFB.[12]

Launch Complex 373-5, 308th Strategic Missile Wing, Little Rock AFB, Arkansas

Any underground facility of the size and type of a Titan II launch complex has its share of dangerous areas. Perhaps one of the most dangerous was the launch duct. With a depth of 145 feet, working off the lowered work platforms required attention to detail as well as normal safety precautions.

Launch Complex 373-5 was fairly quiet on Sunday, 24 January 1968. MSgt R. Eugene Bugge and Airman 1st Class Natasii came out to perform a general cleanup of the launch duct prior to an upcoming inspection. They arrived on the complex and met with Capt Nathan Hartman, MCCC, for the required maintenance briefing. Hartman read the safety briefing, emphasizing the need for the two-point safety harness if they did not lower all of the work platforms as they worked on each level. The gap between the platforms was so small that if all platforms were down the harnesses were not required. Afterward, Hartman, Airman 1st Class Jackson, and Technical Sergeant Shrage went out to clean up on Level 6, while Bugge and Natasii began work on Level 2 by lowering the work platforms. Two DMCCCs were on the site that day, 2d Lieutenant Lind and 2Lt Donald J. Jacobowitz. They remained in the launch control center, maintaining the requirement for at least one officer and a total of at least two crew members manning the consoles on Level 2 of the launch control center.

Bugge and Natasii planned to work their way down each level within the launch duct, checking and cleaning up as they descended level to level, alongside the missile. Bugge wanted to check one of the Level 2 platforms for normal operation after he noticed hydraulic fluid on the grid of the platform. He lowered the platform segment again and stepped forward to take a closer look, not wearing the safety harness as previously instructed. Unfortunately, the part of the lowered platform he stepped on was wet from the cleanup procedure. He lost his footing and fell backward, off the platform, dropping nearly 80 feet and landing on his back across the 23,000-pound thrust mount that supported the missile in the shock isolation system.

Jacobowitz remembers hearing Natasii calling excitedly on the wire maintenance network system, reporting that Bugge had fallen down the launch duct. Jacobowitz looked across the console at Lind and asked Natasii to repeat what he had just said, which he did. Jacobowitz asked if Bugge had fallen all the way down and was temporarily reassured when Natasii replied that no, Bugge had not fallen down to the flame deflector, he was lying across the thrust mount.

Lind, seeing that Jacobowitz had turned as white as a sheet, asked what was wrong. Jacobowitz filled him in and then emphasized that they first had to call in a SAC Two-Man Policy violation, since Natasii had been in the launch duct alone once Bugge had fallen and was apparently now unconscious or dead. Jacobowitz used the voice signaling system to inform Hartman of the situation. After filling him in, Jacobowitz asked that he return to the launch control center and take command. Jacobowitz would then return to the launch duct and attend to Bugge. Hartman replied that he would first lower all the work platforms on Level 7 and look at Bugge, then pick up Natasii and return to the launch control center. Hartman got the Level 7 platforms lowered and saw that Bugge was not moving. He returned to the launch control center and after dispatching Jacobowitz and Jackson to the launch duct, called in the Two-Man Policy violation to the wing command post and awaited an update from Jacobowitz.

After finding a ladder to climb up to the thrust mount, Jacobowitz and Jackson proceeded as fast as they could to the silo and descended to Level 7 where they entered the launch duct. Jacobowitz was the taller of the two, but the ladder was still just a little too short. He remembers hoping that Jackson had a good grip on the ladder as he jumped up to the thrust mount and turned to face Bugge. He had fallen across the thrust mount with his head toward the missile and legs dangling off the other side. He was not breathing and Jacobowitz was unable to find a pulse on his wrist, but he did find a slight pulse, or so he thought, on his neck. Jacobowitz began the old-style artificial respiration as best he could, while awaiting further medical assistance. When the medical team arrived, the doctor joined Jacobowitz on the thrust mount, quickly examined Bugge and told Jacobowitz to stop, Bugge was dead.

After Bugge's body was removed, Hartman was asked if he wanted his crew relieved. He polled the crew, and they all agreed there was no need to bother another crew. Jacobowitz bumped into Hartman 15 years later at the Air War College at Maxwell AFB, Alabama, and after a brief discussion of the accident, Jacobowitz remembers that Hartman expressed his thanks and pride that the crew had remained on duty for the full alert.[13]

Launch Complex 374-7, 308th Strategic Missile Wing, Little Rock AFB, Arkansas

"Routine" maintenance in the Titan II program is somewhat of a misnomer. Routine tasks had to be thought out to ensure that the complexity of the launch facility did not hinder the maintenance operation at hand.

On 8 October 1976 a four-man maintenance team was dispatched to Launch Complex 374-7 to clean up hydraulic fluid that had sprayed onto the missile the day before from a ruptured work platform hydraulic hose. This work involved wiping down the missile with rags soaked in Freon 113. As the rags were used up, they were dropped down into the flame deflector for collection when the task was complete.

When they were finished wiping down the missile, several team members descended into the flame deflector from Level 8 to clean the puddled hydraulic fluid. They were ordered out of the area by the missile combat crew commander when several of the team members became dizzy. They had left the Freon-soaked rags on the floor of the flame deflector, and the decision was made to go back and retrieve them. Sgt Mark A. Davis was the first to go back down to retrieve the rags. He held his breath as he bent over to pick up the rags because the odor of Freon was quite strong. He became ill and returned to Level 8. Sgt Larry South then descended into the "W" and resumed retrieving the rags. A1C Larry E. Woods noticed South was having difficulty as he bent over to pick up the rags and relayed this information to the missile combat crew commander in the launch control center. The commander ordered everyone out of the launch duct. South was able to climb the ladder to the Level 8 launch duct access door but collapsed in the silo equipment area at approximately 1500. South was declared dead on arrival at the base hospital.

The highly volatile Freon 113 was heavier than air. The residue left on the rags, as well as all the fumes generated in wiping down the missile, had accumulated in the lowest area of the launch duct, the bottom of the "W." As more rags were dropped, the layer of displaced air became thick enough so that when South began to collect the rags, he was breathing an anoxic (minimal oxygen) atmosphere. There was meager air circulation in the area of the flame deflector, only 60 cubic feet per minute were removed from the area by Exhaust Fan 104, and the duct was 8 feet above the floor. South had entered an area that was most likely blanketed with perhaps an 8-foot-thick layer of predominately Freon 113 fumes and, by bending over to pick up the rags, had repeatedly inhaled the concentrated fumes.[14]

Capt Bill Howard, the Technical Engineering Division engineer assigned by the Missile Potential Hazard Team to investigate the cause of the fatality, arrived at the launch complex only to find that the missile combat crew had been ordered to purge the launch duct. Howard donned his CHEMOX unit anyway, as ordered, and descended into the "W" carry - ing an oxygen meter and portable vapor detector (PVD) unit to check for propellant fumes. With the launch duct exhaust system running, he was fairly sure that no residual gases would remain, but he still had to run the check. As he suspected, oxygen levels were back to normal, and the portable vapor detector had given no propellant fume indication. Now the supposition had to begin.[15]

Just as he was biting into his sandwich during a later-than-normal lunch, Jimmy R. McFadden, chief missile facilities engineer for the 314th Civil Engineering Squadron at Little Rock AFB, was called by the 308th wing commander. He had called to inform McFadden that a message had been composed describing an accident and that he needed the concurrence of the missile facility engineer. McFadden listened as he read him the text of the message and then quickly told him that he could not concur. The message stated that the flame deflector was inadequately designed to sustain life, causing the loss of one life and nearly two. McFadden explained that it was not a design flaw; there had never been a design requirement for any greater air turnover in the bottom of the flame deflector. There was, in fact, more than sufficient air, even though stationary, to support occupation for a reasonable amount of time. McFadden was thanked for his advice, and the conversation ended.[16]

Needless to say, this was a case of small oversights compiling into a fatal accident. Both Howard and McFadden had not known of the habit of dropping the rags to the bottom of the silo for later retrieval or of the amount of Freon 113, nearly 2.5 gallons, that was routinely used. If they had, they would have required entry onto the floor of the flame deflector be considered only after the area had been purged for several minutes.[17] After the accident the maintenance protocols for use of Freon 113 were modified to include a log of the amount in use, kept by the missile crew commander, and the declaration of off-limit areas until air quality was checked.[18]

Launch Complex 533-7, 381st Strategic Missile Wing, McConnell AFB, Kansas

Without a doubt the most dangerous part of the Titan II system from the maintenance standpoint were the propellants. Both were highly toxic. The oxidizer, nitrogen tetroxide, was highly corrosive when dissolved in water. Therefore, all but the most minute of leaks of either fuel or oxidizer had to be taken seriously. With this considerable potential for catastrophic accident, large propellant spills were rare in the Titan II program. Ironically, the two largest oxidizer spills took place 14 years apart at the same site, Launch Complex 533-7, 381st SMW, McConnell AFB, Kansas.

The first took place on 12 November 1964 when approximately 1,800 gallons of nitrogen tetroxide were spilled into the flame deflector during a propellant transfer operation. Subsequent investigation showed that two valves in the propellant transfer system had been improperly positioned. This resulted in the oxidizer being routed to the vernier tanks on the holding trailer, then overflowing and entering the silo through a 2-inch nitrogen vent line which emptied into the flame deflector. Damage to the launch duct and missile was extensive, but there were no injuries to any of the support personnel or the missile combat crew on duty at the time. Missile B-23 (62-004, actually a combination of missile B-30 Stage I and missile B-23 Stage II) was removed, refurbished at the Martin Marietta Denver facilities, and returned to the 381st SMW in August 1966. The launch complex was returned to alert status in mid-January 1965.[19]

The second spill took place 14 years later on 24 August 1978. Launch Complex 533-7, located just south of the small town of Rock, Kansas, had been chosen for both a Reliability and Aging Surveillance Program evaluation of the launch complex and missile airframe and a Service Life Analysis Program evaluation of the Stage I and Stage II engines. Both of these tests were part of an ongoing analysis of the Titan II program designed to replace the launching of test missiles. The missile combat crew composed of 1Lt Keith E. Matthews, MCCC; 2Lt Charles B. Frost, DMCCC; A1C Danford M. Wong, BMAT; and SrA Glen H. Wessel, MFT, was on duty monitoring the missile recycle operation. At 1200 a two-man Propellant Transfer System team was on Level 7 in the launch duct, finishing loading the Stage I oxidizer tank of missile B-57 (62-12300). Airmen 1st Class Erby Hepstall and Carl Malinger were engaged in the final steps of disconnecting the Stage I oxidizer propellant transfer hose from the oxidizer airborne quick-disconnect fitting at the bottom of the Stage I fuel tank. The Stage I oxidizer tank was located above the Stage I fuel tank with the oxidizer feed lines coming through the fuel tank to the engine attachment point. At 1220 Hepstall and Malinger had nearly completely unthreaded the transfer line fitting when the airborne (missile side) quick-disconnect poppet valve failed to seal. The 13,220 gallons of nitrogen tetroxide that had just been loaded into the Stage I oxidizer tank began to gush out, forcing them to drop the oxidizer fill line.[20] The dark red clouds of vapor quickly cut visibility to zero inside the launch duct. The airmen screamed into the Radio-Type Maintenance Network (RTMN, a radio communication system used by the launch control center and topside propellant transfer control trailer staff), "The poppet did not seat, it won't stop, let's get out of here." Although Malinger and Hepstall were in the midst of a cloud of toxic oxidizer fumes, they were both wearing a rocket fuel handler's clothing outfit (RFHCO) and were in no immediate danger at this point.

Matthews was on Level 1 of the launch control center when the klaxon sounded, indicating oxidizer vapors present in the launch duct. He descended partway down the stairs to Level 2 where Frost advised him that the PTS team had just disconnected the transfer line. Matthews commented that a little vapor was normal and returned back to Level 1. The klaxon sounded again, and Frost heard a fragment of the message from the PTS team. Matthews quickly returned to the launch control center, and both officers tried without success to use the RTMN to establish contact with the PTS team.

Simultaneously, 2Lt Graham B. Sorenson, the site maintenance officer, was monitoring the operation from the control trailer on the surface. Upon hearing the screams of the team members, Sorenson looked out the trailer window and saw clouds of vapor billowing out of the silo exhaust vent. Sorenson immediately left the control trailer to find SSgt Robert J. Thomas, the PTS team chief, who had left the control trailer to assist in repositioning equipment for the start of the fuel upload. Sorenson and Thomas ran back to the control trailer to find out what was happening. When Thomas heard that the poppet was not seated, he quickly left the trailer and entered the access portal, intent on rescuing his team members. A1C Mirl R. Linthicum, a PTS team chief trainee, radioed to the PTS backup team to have a RFHCO suit ready for Thomas since he was going down to Level 7 to attempt to stop the flow of oxidizer. Sorenson radioed the security police at the end of the launch complex access road and advised them to stand by for evacuation of civilians in the area.

Before Thomas made it through Blast Doors 6 and 7, Hepstall had managed to find his way out of the launch duct on Level 7 and taken the elevator to Level 2 where he proceeded down the cableway to the backup PTS team position near the silo side of Blast Door 9. His helmet visor was clouded from direct exposure to liquid oxidizer, but he was uninjured. When Thomas entered the area, Hepstall informed his team chief of the situation and said that he was going back down to find Malinger. Thomas suited up in RFHCO, Hepstall changed his helmet, and they both left to find and assist Malinger. Airman Gary L. Christopher left the PTS position to return to the surface to retrieve additional RFHCO suits and air packs. A1C Francis A. Cousino, Liquid Fuels System inspector and evaluator, had come downstairs to the blast lock just as Hepstall returned for the second time, coughing badly as he stumbled and fell to the ground. Hepstall told Cousino that Malinger was trapped on Level 7 and that Thomas had descended in the silo elevator to get him. Sorenson and 2Lt Richard I. Bacon Jr., site maintenance officer trainee, arrived at the blast lock just as Hepstall finished changing his RFHCO and returned down the cableway. By this time oxidizer vapors had arrived at Blast Door 9 and were beginning to fill the area between Blast Door 8, which led to the launch control center and was shut, and Blast Door 9. Sorenson directed everyone in the room to evacuate topside.

Meanwhile, Matthews had tried to open Blast Door 8 in order to investigate what was happening. The RTMN was not working properly since too many people were trying to use it at once. Blast Door 8 would not open because of the interlock with Blast Door 9 that prevented both blast doors being opened at once. Matthews smelled oxidizer and told Wessel to set up a portable vapor detector to determine how much oxidizer vapor was leaking into the launch control center. A reading of 3 parts per million was close to the maximum allowable concentration. Wessel shut the blast dampers through which the vapors were coming, and Matthews directed him to open the escape hatch on Level 3 of the launch control center. Suddenly the portable vapor detector alarm sounded and then went silent as the concentration went past the full-scale reading. The missile combat crew donned CHEMOX masks and prepared to evacuate.

Hepstall managed to find Malinger and Thomas. Thomas had tried to directly stop the flow of oxidizer and was now injured; Malinger and Hepstall had pulled Thomas into the silo elevator and returned to Level 2. Either Hepstall or Malinger began calling for help on the RTMN. They reached Blast Door 9, which was closed but not locked. Opening Blast

Door 9, Malinger or Hepstall called on the emergency phone for someone to unlock Blast Door 8 and help them.

Frost was still on Level 2 of the launch control center when he heard either Hepstall or Malinger call on the RTMN and asked that Blast Door 8 be opened. Frost replied that Blast Door 8 was locked and that they should evacuate up the access portal stairs. As he continued this conversation, the emergency phone in the launch control center rang. Frost answered and again stated that Blast Door 8 was locked. Much to his amazement, Blast Door 8 swung open, and Hepstall stumbled into the launch control center, followed by Malinger and a cloud of oxidizer vapor. Both had their helmets off. Malinger was screaming that Thomas was dead. Matthews donned a fresh CHEMOX unit and went through the open Blast Doors 8 and 9 to pull Thomas into the area between the two blast doors. Matthews, Malinger, and Frost together pulled Thomas into the launch control center and realized he was still alive as he moved his head from side to side. Matthews returned to Blast Door 8 and noted that he could see but a few feet in front of him due to the dense red-black oxidizer vapors. He closed but did not lock Blast Door 8.

Frost received a call from the wing command post and advised them that "the locks are on the safe and the keys are in it; we have one man possibly down and we're evacuating now." Frost turned around and saw that Thomas's helmet had been removed and his suit unzipped. Either Malinger or Hepstall had tried to administer mouth-to-mouth resuscitation.

At this point the personnel in the launch control center began to evacuate. Malinger and Matthews helped Hepstall down the stairs to Level 3 and the escape hatch. Matthews returned to Level 2 and determined that Thomas was dead. The crew and remaining PTS team members reached the surface amidst clouds of oxidizer vapor following them up the air intake that doubled as the escape route. Above-ground personnel assisted Malinger and Hepstall to a nearby water hydrant to wash off the remaining oxidizer residue but found little water pressure. A section of the fence was removed and they evacuated the site, with the aid of security police, to a nearby farmhouse where both PTS team members were washed down with water after their suits were removed. Frost contacted the wing command post and was instructed to take Hepstall and Malinger to a hospital in nearby Winfield. Evacuation of nearby civilians began, and a protective cordon isolating the launch complex was set up.

Sorenson and Bacon came to the farmhouse shortly after 1300. Linthicum and A1C John G. Korzenko volunteered to suit up and try to retrieve Thomas's body. The four returned to the complex, and Linthicum and Korzenko attempted to descend to the launch control center through the air intake shaft. Korzenko quickly returned to the surface since he smelled oxidizer vapor in his suit, and Linthicum was right behind him since he found it too dark to see. At the direction of the wing commander no further retrieval attempts were made until additional PTS personnel and equipment reached the site.

The oxidizer fumes formed a cloud about 1 mile long and 1/2 mile wide and reached an altitude of approximately 1,000 feet. The town of Rock, Kansas, population approximately 200, located 2 1/2 miles north of the site, was evacuated at 1345 without incident. Only one civilian was treated for inhalation of oxidizer fumes.

Approximately one hour later, at 1400 hours, a team of nine additional PTS personnel arrived with RFHCOs and air packs: TSgt John C. Mock Jr., SSgt Robert A. Sanders, A1C Scot A. Jaeger, Michael L. Greenwell, Middland R. Jackson, Rodney W. Larson, Terry J. Watke, and Gregory W. Anderson, and SrA James C. Romig. Mock, Jackson, and Anderson tried to penetrate through the launch control center air intake but were unsuccessful. They then forced their way through the access portal entrapment area. Between the heat from the oxidizer vapor reacting with the materials in the access portal and the physical exertion of using a small crowbar to break through the entrapment area, the team had to resurface for new air packs. Mock, Jackson, and Greenwell made their way down the access portal stairwell with barely any light to see by. The oxidizer vapors were so dense that the flourescent lights, normally more than sufficient illumination, cast hardly any usable light. They finally reached the blast lock area and were able to feel their way to the launch control center where they found Thomas's body. They carried him to the access portal elevator, which was inoperative. A fifth member then joined the rotation of teams as they operated in pairs to carry Thomas up the 55 steps of the access portal, one flight at a time. At approximately 1600, two hours after they arrived on the scene, Thomas's body was on the surface and transported to the base hospital.

Two fatalities resulted from the accident, and 25 personnel were slightly injured. Thomas was declared dead on arrival at the hospital. The autopsy revealed that he had died from acute pulmonary edema due to inhalation of high concentrations of oxidizer vapor. Subsequent investigation showed that Thomas had apparently tried to stem the flow of oxidizer from the tank with his glove. The high velocity stream of fluid penetrated the glove/cuff interface, instantaneously filling his suit with a dense cloud of vapors.

Hepstall was transported to Wesley Hospital, Winfield, Kansas. On 3 September 1978 he died due to lung and renal failure resulting from inhalation of concentrated oxidizer vapors. This was apparently due to either direct entrance via a 7-inch gash in the left leg of his RFHCO or due to vapor inhalation while attempting to give Thomas mouth-to-mouth resuscitation.

Malinger was transported to the William Newton Memorial Hospital, Winfield, Kansas, and then to USAF Medical Center, Scott AFB, Illinois, on 25 August 1978. Malinger developed severe complications but eventually made a gradual recovery, although he suffered permanent damage to his vocal cords and lungs and paralysis in his left arm. The accident left him 100 percent disabled.[21]

The final report of the investigation into the accident was issued by Col Ben G. Scallorn on 10 October 1978. The primary cause was failure to follow recommended procedures. The actual cause of the accident was the lodging of a Teflon "O" ring from the bottom of the oxidizer filter unit in the poppet valve mechanism, jamming it open. When Hepstall and Malinger began to disconnect the oxidizer transfer line, they had quickly unscrewed the quick-disconnect rather than follow the technical order specification that the quick-disconnect should be slowly unscrewed and if any leak was seen, screw the disconnect back to the fully connected position so that the tank could be unloaded and the quick-disconnect replaced. The filter unit had been removed during the propellant download several weeks earlier, and the lower "O" ring was inadvertently left in place. Technical orders called for the replacement of the filter prior to oxidizer upload, but this was not

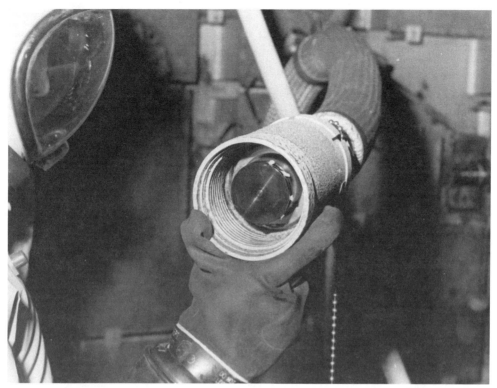

Figure 9.2. Ground half of the oxidizer propellant transfer line quick-disconnect showing the Teflon O-ring pieces jamming the poppet valve into the partially open position. *Courtesy of Mark Clark.*

done. The flow of oxidizer during the upload dislodged the ring into the flowing stream. Adherence to applicable technical orders would have prevented this primary cause from occurring.

Thomas died due to an unforeseen flaw in the design of the RFHCO. The attachment points for the gloves and boots were not designed to withstand direct impingement of a high-pressure stream of oxidizer or fuel. The subsequent failure of the left glove/cuff on Thomas's suit allowed direct penetration of the suit environment by liquid oxidizer. A detailed inspection of the suit and later tests with the helmet visors showed that the suits offered only marginal protection to prolonged exposure to liquid oxidizer.[22]

A contributing cause to the confusion immediately after the spill began was the inadequacy of the RTMN system. While it had functioned reasonably well during the normal PTS procedures, in the developing emergency the attempt by up to six people to use the system at the same time, combined with the override capability in the launch control center, only magnified the confusion.

Three significant equipment changes were made as a result of the investigation. The RFHCOs were modified with a visor made out of high-impact material much more resist - ant to clouding during prolonged exposure to liquid oxidizer. The glove and boot/cuff interface was redesigned to protect against liquid propellant impingement failures.[23] The

RTMN upgrade, identified in 1976 but slow to be implemented, was accelerated as a direct result of this accident. A major flaw in the system was removed: no longer could a constantly keyed microphone render the net useless.[24]

On 24 August 1978, Jim Sturdivant, the resident Martin Marietta Company technical representative at Little Rock AFB, called company headquarters in Denver and reported the accident at Launch Complex 533-7. Shortly thereafter, Gil Selsor, Martin Marietta Company representative at McConnell AFB, called and gave more details. A four-member team left for Wichita that afternoon: Jim Greichen, Tom Fujiyoshi, John McDonald, and Keith Wanklyn. Wanklyn and Greichen arrived in the early evening and traveled immediately to the site while daylight remained. The sight of the black-red fumes still billowing out in a steady stream from the exhaust vents was something they both vividly remembered 20 years later. They realized that the situation could have been much worse since the airflow from the silo was limited to the silo exhaust vents, air intake for the launch control center, and the access portal. The silo closure door was still closed.[25]

On 25 August 1978, after conferring all night on the possible options, the Martin Marietta Company team recommended dilution of the oxidizer in the flame deflector with water to create a 5 to 7 percent nitric acid solution. While the reaction of the nitrogen tetroxide with water would generate a large amount of heat and a cloud of fumes, this alternative was better than the problems that might be caused by trying to neutralize the oxidizer. The first step along this path was to get the silo closure door open so that the heat and fumes could be released into the atmosphere rather than damaging the silo and missile even further. At 1930 the first of many site penetrations took place by a PTS team from the 308th SMW, Little Rock AFB, which had been called upon as the primary PTS response team for cleaning up the accident. The use of a PTS team from another wing was standard procedure when a fatality had occurred and morale might be down. Propellant tank pressure readings were taken in the launch control center, but the conditions underground prevented further work. Blast Door 9 was closed at this time, greatly simplifying control of the above-ground environment, because fumes could now only be coming from the silo area and not the access portal. An emergency silo closure door-opening procedure was used by the PTS team since launch complex facility power was unavailable.

Capt Craig Allen, chief of missile quality control, 381st Civil Engineering Squadron, joined the recovery team on 26 August 1978. Their first setback was that the 100,000-gallon hardened water supply located in the silo, adjacent to the launch duct, was unavailable because of a previous test procedure that rendered it unaccessible. The complex had undergone a combined systems test procedure the day before. The test had not been completed successfully, and the manual deluge valve on Level 7, at the bottom of the hard tank and upstream of the automatic valve, was left in the closed position. This was a routine procedure, but in this case rendered the hard water contents unavailable to dilute the oxidizer in the flame deflector. The 100,000-gallon water tank on the surface could not be directly used in the silo. This meant that over 80,000 gallons of water had to be trucked in by tankers from fire hydrants in the city of Winfield approximately 20 miles away. The original plan was to offload multiple tankers at once by using their built-in pumps, but as the transfer pumps in the tankers failed one by one, the tankers were gravity drained, one at a time, into the "soft" water reservoir topside. This reservoir was full at the beginning of the effort and was replenished by the convoy of tankers. From there a Winfield City Fire Department pumper truck, which had responded to the scene under a mutual aid agreement with the McConnell AFB Fire Department, drafted the water from the reservoir and pumped it into the silo. After water flow began, clouds of dark red vapors alternated with white clouds of steam due to the heat of the reaction.

Martin Marietta Company personnel monitoring the tank pressures became concerned when the pressure in the Stage II oxidizer tank, still fully loaded with 3,000 gallons of oxidizer, began to rise. To cool this stage and prevent the situation from becoming worse, another pumper truck from McConnell AFB with a deluge nozzle was positioned adjacent to the launch duct opening next to the cooling tower pit to spray an arc of water onto the Stage II oxidizer tank. Water flow continued for several hours in accordance with the instructions from Headquarters SAC to keep pumping till "the red cloud goes away."

On 27 August 1978 samples were taken of the diluted oxidizer that had accumulated in the flame deflector. The concentration was still too high, and an additional 20,000 gallons of water was added. The PTS team was able to reach the launch duct for the first time and found the missile was leaning over and touching the extended work platforms used during the earlier Stage II oxidizer upload. The decision was made to download Stage II to relieve the load on the empty and buckled Stage I oxidizer tank for fear that it would collapse and rupture the Stage I fuel tank. Draining the Stage II oxidizer tank represented a unique challenge to both the Air Force personnel and Martin Marietta Company staff since the pumps normally used for this operation were on Level 8 and unavailable and the hydraulic system for raising and lowering the work platforms was inoperative. The rest of the day was spent generating a solution using nitrogen pressure to download as much oxidizer as possible and using K-bottles of compressed air to operate the platforms.[26]

On 28 August the Stage II oxidizer was successfully offloaded. The missile creaked and groaned but straightened out somewhat, and the buckled area was reduced considerably. Now that the missile was as safe as it could be under the circumstances, the next problem was removal of the nearly 100,000 gallons of nitric acid solution that had filled the flame deflector. Not only was the physical removal a problem to consider, equally important was locating a place to dispose of the highly corrosive solution. Kansas Industrial Environmental Services was able to provide 5,600-gallon acid transportation tankers as well as a site for disposal.[27]

The next day the plans for pumping out the acid solution were finalized and the equipment prepared. Four tanker trucks would be used in a continuous process to haul the solution to the disposal site. With an estimated 100,000 to 120,000 gallons of water and oxidizer mixed together, this would prove to be a time-consuming process if the pumps held up under the extremely corrosive conditions in the silo. On 30 and 31 August 1978 the first pumping system was assembled and the cabling and hoses were protected as much as possible. This was an all-day process since corrosion-resistant steel fittings had to be used wherever possible to prevent the acid from dissolving the apparatus. All of the above-ground equipment was positioned, and dry runs of the procedure were utilized to troubleshoot possible problems. On 1 September the pump was lowered 145 feet down the north

Figure 9.3. Titan II ICBM B-57 (62-12300) showing the partially collapsed Stage I oxidizer tank. The Stage I fuel tank was fully loaded, as were the Stage II oxidizer and fuel tanks. *Courtesy of Mark Clark.*

exhaust duct and turned on. Allen and Wanklyn remember feeling that finally the problems were behind them and now they just had to get the liquid to the surface for disposal. After 15 to 20 minutes flow stopped, and the pump was pulled to the surface. The hose fittings had dissolved, and most of what had been pumped out had siphoned back into the silo. So much for the easy part.

The next day the hose was repaired with corrosion-resistant clamps and the pump was inspected only to find that the impeller had dissolved. A search for a more resistant pump was immediately begun. Eventually a Grundfus pump was located in Houston. An all stainless-steel pump, it proved to be the solution, and by 7 September 1978 the pumping operation was running smoothly. All that remained was a decision about whether or not to refurbish the launch complex.[28]

As soon as the investigation teams were finished on site, evaluation teams from Headquarters SAC and the Ogden Air Logistics Center (ALC) at Hill AFB, Utah, began the task of determining whether the site could or should be rebuilt. In December 1978, SAC designated the rehabilitation project at Launch Complex 533-7 as Project PACER DOWN. Capt Craig Allen was assigned as officer in charge. This unique program involved an emergency Military Construction Program administered by SAC and a high-priority depot repair program run by Air Force Logistics Command to bring the site back on line. Phase I consisted of safing the complex and removing the damaged equipment. Phase II included the refurbishment of the silo by general contractors and equipment and airframe refurbishment by Martin Marietta at Denver. Phase III was equipment installation, the emplacement of the missile, and return to alert. Phase I work continued on the seemingly insurmountable job of preventive cleanup as cabling and equipment still in the silo were decontaminated and a damage assessment and possible repair evaluation were conducted.[29]

Maj Jerald Bozeman was the field-grade officer on the first propellant transfer operation at the 381st SMW in late November after the accident. While the sector maintenance officer was coordinating the process, Bozeman had overall on-site maintenance responsibility for the propellant transfer operation. The 3901st Strategic Missile Evaluation Squadron and Martin Marietta Company representatives worked side by side with the propellant transfer team personnel as they wrote and rewrote the technical manuals governing propellant transfer while the operation was taking place so that there were no ambiguities or any room for error. Launch Complex 532-9 was next to Lake Cheney, and temperature inversions were a common occurrence in late November and early December, often delaying the operation for several days since propellant could not be transferred if an inversion layer was present. The entire process lasted until the first week of the new year, and the resulting changes to the technical manuals provided the basis for virtually error-free propellant transfer operations for the remaining years of the Titan II program.[30]

On 5 September 1980, Headquarters SAC approved $4.17 million for the repair of Launch Complex 533-7.[31] In October 1980, the General Accounting Office awarded the Project PACER DOWN contract to Mayfair Construction Company. Mayfair was to rebuild the structure of the silo while Martin Marietta Company installed the missile and launch equipment. Work was delayed further by the inability to coordinate a work schedule that permitted both Mayfair and Martin Marietta personnel to work at the site. These issues were resolved, and on 28 April 1981, Headquarters SAC approved a joint-occupancy schedule with work to begin 1 August 1981, nearly three years after the accident with a proposed completion date of 8 January 1982.[32]

On 2 October 1981, the Defense Department announced plans to retire the Titan II system. A stop-work order for Project PACER DOWN was issued by Headquarters SAC on 5 November 1981. Contract termination negotiations with Mayfair Construction Company and Martin Marietta Company were completed and by December 1981, Project PACER DOWN was terminated.[33] Launch Complex 533-7 was disposed of in the same manner as the rest of the 381st SMW sites during the Titan II deactivation program.

Launch Complex 374-7, 308th Strategic Missile Wing, Little Rock AFB, Arkansas

The most highly publicized accident involving the Titan II program took place in September 1980, approximately one year prior to the deactivation decision. Earlier that year, the review of the Titan II program instigated by congressional concerns from the 1978 oxidizer spill near Rock, Kansas, had been released.[34] The Titan II program was given a clean bill of health, with specific recommendations on upgrades and modifications. There was no call for dismantlement. Four months after the release of the report, the picture changed dramatically.

From 2 September to 14 September 1980, missile B-25 (62-006), installed at Complex 374-7, had undergone a Stage II oxidizer download to permit replacement of the Stage II oxidizer tank manhole cover seal.[35] Stage II oxidizer was loaded on 12 September 1980. The complex was returned to alert status 14 September 1980.

Following the upload of oxidizer on 12 September, the Stage II oxidizer tank pressure began to decrease. This was expected as the liquid oxidizer absorbed the nitrogen pressurization gas until the liquid became saturated. On 16 September 1980, a propellant transfer system team was sent to the complex to repressurize the Stage II oxidizer tank. Following this repressurization, the Stage II oxidizer tank continued to lose pressure, and at 2345 the missile was declared "not ready" due to the tank pressure again decreasing to below acceptable limits.

On 18 September 1980, a highly experienced wing instructor crew was on duty at Launch Complex 374-7: Capt Michael T. Mazzaro, MCCC; 1Lt Allan Childers, DMCCC; SSgt Rodney L. Holder, BMAT; and SSgt Ronald O. Full, MFT. First Lt. Miguel A. Serrano was a student DMCCC. Originally assigned to Launch Complex 374-5, the crew had been reassigned to 374-7 due to major maintenance work that was going to be conducted at 374-5. With the delay of reassignment and a faulty alternator in the crew vehicle, the day had not started out on a particularly positive note.[36]

On the morning of 18 September 1980, a pneudraulics team, a missile handling team, and a PTS team were dispatched to Launch Complex 374-7. The pneudraulics team was to fix the HS-2 equipment, the hydraulic system for the launch duct work platforms, which was suffering intermittent failures, while the missile handling and PTS teams were moving the above-ground PTS support equipment to Launch Complex 373-2. HS-2 repair was completed at 1640, and the PTS team, which had been waiting for several hours, was ready to repressurize the Stage II oxidizer tank. After a maintenance briefing in the launch control center, the eight-man PTS team departed to begin the pressurization procedure. Two technicians in RFHCOs were on the surface at the oxidizer hardstand, with an environmental unit technician. The team chief and four technicians were underground on Level 2 of the launch control center with two technicians in RFHCOs, ready to proceed, and two partially suited and ready as backup.

At 1827 the suit environmental control units were activated and a two-man PTS team, composed of SrA David F. Powell and Airman Jeffery L. Plumb, departed down the cableway toward the launch duct. Recently updated technical orders required Powell and Plumb to use a torque wrench and socket to remove the oxidizer vent pressure cap from the Stage II oxidizer tank. The team had brought these tools with them from the base but had left them in the truck. The previous procedure had been to use a ratchet and socket that were stored in the equipment area on the silo side of Blast Door 9. Powell picked up the ratchet and socket, not checking to see if they were securely connected by the spring loaded retaining pin, from the decontamination area on the silo side of Blast Door 9. They reported to the launch control center when they reached the launch duct.

The team chief in the launch control center read the procedure checklist to Powell and Plumb over the radio. They completed the first three steps of the repressurization procedure. As the team chief read the caution statement prior to the next step, Powell picked up the ratchet with the socket seemingly securely attached. As he swung it up into operating position, the 8.75-pound socket separated from the ratchet at waist-high level, fell onto the Level 2 platform, bounced once onto the rubber boot between the platform edge and the missile airframe, and before either technician could grab it, pushed through the boot, and fell approximately 80 feet. The socket hit on the thrust mount ring, then bounced upward and toward the missile, puncturing the Stage I fuel tank skin. Both technicians watched as a stream of white liquid poured out of the missile. There was now quite obviously a fuel leak in Stage I. In 35 to 40 seconds, a noticeable cloud of Aerozine 50 vapor had reached Level 5 of the launch duct, approximately 30 feet below them.

At 1836 Powell notified the missile combat crew of a possible fuel leak, mentioning nothing about dropping the socket. Almost simultaneously, a warning klaxon sounded, and Mazzaro had the indications of a fuel leak on the Launch Control Complex Facilities Console (LCCFC). Mazzaro directed the PTS team chief to instruct Powell and Plumb to return to the decontamination area only after they were sure they had not reconfigured any equipment. Mazzaro then turned to the Fuel Vapor in Silo checklist and began attacking the problem at hand. Powell and Plumb left the launch duct with their tools, less one socket, and locked the Launch Duct Level 2 access door. Mazzaro contacted the wing command post, informing them of a fuel leak and possible fire in the launch duct.[37]

At 1800 on 18 September 1980, Col John T. Moser, 308th SMW commander, was at home getting ready for an evening out with the base commander and other unit commanders stationed at Little Rock AFB. Before they left the house, Moser received a phone call from the wing command post that something had happened at Launch Complex 374-7. Moser immediately canceled his plans for the evening. As he drove to the wing command post, he radioed ahead and activated the Disaster Preparedness Response Team that would set up the hazard corridors and start the evacuation process. Upon arrival at the wing command post, Moser initiated the checklists detailing the next sequence of actions. Headquarters SAC was immediately notified, but there was no plan in place to notify the local or state governments because of the "We will neither confirm nor deny the presence of nuclear weapons" mentality of the time. Moser assigned a staff member to record conversations and actions in the command post at the outset of the accident.[38]

Back at Launch Complex 374-7, the PTS team backup personnel in the decontamination area reported that the Vapor Detector Annunciator Panel on the launch control center side of Blast Door 9 indicated 40 ppm fuel vapor. Mazzaro watched as the FUEL VAPOR LAUNCH DUCT, FIRE IN ENGINE, FIRE LAUNCH DUCT, LAUNCH DUCT, and ENGINE SPRAY indicators illuminated. The missile combat crew immediately entered the Fire Automatic Corrective Actions checklist. LAUNCH DUCT TEMPERATURE HIGH and LAUNCH DUCT AIR CONDITIONING OFF indicators illuminated at this point, supporting the prior indications.

Mazzaro continued his checklist, closing Blast Valve 5, isolating the launch control center from outside air in case of above-ground air contamination. At 1835 Childers directed the topside PTS team members to evacuate approximately 2,000 feet up the launch complex access road. They removed their RFHCOs, loaded the equipment into the transport truck, and contacted Childers to request that the access gate be opened. They then drove upwind and stopped, remaining in contact with the launch control center via the RTMN. During this period, at 1838, VAPOR SILO EQUIPMENT AREA, VAPOR FUEL PUMP ROOM, and OXIDIZER VAPOR LAUNCH DUCT had illuminated on the LCCFC. While Mazzaro continued the Fire Automatic Corrective Actions checklist, referencing other checklists for all flashing indicators as needed, the underground PTS team members had exited the silo and, with the backup team members, reentered Blast Lock Area 201, closed and locked Blast Door 9, closed Blast Damper 2 at the direction of Mazzaro, and proceeded to the launch control center. They left the large ratchet handle in the blast lock area with two RFHCOs and environmental control units. Approximately 10 minutes after the socket had been dropped, the Stage I fuel tank pressure had dropped from 11.6 psi to 9.7 psi. Mazzaro reported the readings to the wing command post.

At 1848 Moser directed the Missile Potential Hazard Team to form, and at 1854 Powell informed the wing command post that he had dropped the socket. Stage I fuel pressure was dropping at approximately 0.38 psi every five minutes, equal to approximately 10 gallons of flammable fuel. The intact propellant tank pressures continued to rise at a slow rate while the water spray was in operation. Moser directed that the Missile Potential Hazard Network (MPHN) communications system be activated. This was an open conference line that connected Headquarters SAC, Martin Marietta Company, Headquarters Eighth Air Force, and the Ogden Air Logistics Center, permitting constant expert input toward solving the developing crisis. At 1940 LtGen L. R. Leavitt Jr., Vice Commander in Chief, Strategic Air Command, joined the conference.

At 1930 the three-member PTS team topside reported that a steady stream of green vapor or smoke was coming out of the silo exhaust shaft, shortly thereafter turning to white. Several minutes later, a security police team arrived from the base and established an entry control point. At 1949, slightly over one hour from the time the fuel tank puncture occurred, the water spray ceased operation as the 100,000-gallon supply tank was depleted. Due to a malfunction documented three weeks earlier but not yet resolved, the above-ground 100,000-gallon soft water tank could not supply additional water. Stage I fuel tank pressure continued to drop, while the intact tank pressures began a rapid climb as the cooling effect of the deluge water was removed.

At 1956 the MPHT directed Mazzaro to remove power from the missile. At 2000 Stage I fuel tank pressure was reading -0.5 psi, indicating that the tank was now either empty, considered unlikely, or the pressure was below the accurate calibration level of the sensor. Fifteen minutes later, the MPHT directed the PTS technicians to retrieve the RFHCO equipment from Blast Lock 201 and to prepare to return to the silo to vent the Stage I fuel tank. When the PTS team opened Blast Door 8 they reported smoke and vapors in Blast Lock Area 201 and immediately re-closed the door. At 2014 the MPHT directed Mazzaro to turn off the launch duct air conditioner.

At 2057 Moser directed all personnel to evacuate the launch control center after all classified documents were secured. Mazzaro directed Holder and Full to take gas masks and descend to Level 3 to open the emergency escape hatch. Mazzaro and Childers remained on Level 2 of the launch control center and secured the targeting tapes and Emergency War Order materials in the Emergency War Order safe, locked it, and attempted to lock the administrative classified documents safe. They were unable to do so because there were too many documents. Before leaving the launch control center, they set the DIESEL POWER TRANSFER SWITCH to HAND, at the direction of the MPHT, thus preventing automatic operation of the diesel generator should commercial power fail. Two portable vapor detectors (PVDs) were left operating in the launch control center: one was in the short cableway just behind Blast Door 8, the other was behind the alternate launch office console. The two officers descended to Level 3 and the escape hatch door. The emergency escape hatch was opened, and all personnel climbed the 55-foot ladder to the surface where they proceeded to the fence breakaway panel and left the site, joining the two security police and three topside PTS team members 2,000 feet upwind of the complex. It was now 2043, slightly more than two hours after the socket punctured the missile fuel tank.

At 2057 Moser's on-site representative, Col James L. Morris, the deputy commander for maintenance, and Sgt Jeff K. Kennedy, a PTS technician, arrived via helicopter and joined the PTS personnel on the access road. Morris stressed that there would be no penetration of the complex without explicit orders. Kennedy and Powell, accompanied by Mazzaro and Childers, went down the road to the silo to evaluate the extent of the fuel vapor venting. They observed a large quantity of white vapor that they first took to be smoke, but which turned out to be fuel vapors, belching in puffs out of the silo exhaust vent. No fumes could be seen at the air intake for the launch control center. Kennedy decided to return to the launch control center via this route to obtain tank pressure readouts. Powell accompanied him only partway but was ready to come to his assistance if Kennedy did not return in several minutes. Since this action did not involve opening the blast lock doors, Kennedy did not consider this a penetration of the complex. Upon returning to the surface, he returned to the fall-back position and reported to Morris that the Stage I fuel tank pressure was now slightly negative and that Stage I oxidizer was at 29.6 psi, Stage II fuel was at 41.0 psi, and Stage II oxidizer was at 41.6 psi. Morris made it clear to Kennedy and Powell that no further unauthorized penetrations underground would take place.

At 2150 security police arrived to provide area security and assist in local area resident evacuation. At this time the 308th SMW MPHT and the MPHN members were developing a plan to reenter the launch control center to further monitor tank pressures. Communications and decontamination equipment were being positioned while the penetration plan was further refined. At 0106, 19 September 1980, the final plan for penetrating to the launch control center was relayed to Morris. The plan called for one RFHCO-equipped person to sample the silo exhaust shaft with a portable vapor detector and the following actions were to be carried out by the primary and backup PTS teams. The primary team would try to break through the entrapment area with a crowbar and open Blast Door 6 with a portable hydraulic pumping unit. They would monitor vapor levels with the portable vapor detector and return to the surface if readings reached 250 ppm fuel vapor. If conditions permitted, they would then take a reading with the PVD between Blast Doors 6 and 7. If possible they would take readings through the breathing holes in Blast Door 7. If readings were above 250 ppm, they were not to open Blast Door 7. If readings were below 250 ppm, they were to open Blast Door 7, observe the vapor detector annunciator panel (VDAP), and return to the surface. A second team would then open Blast Door 8, enter the launch control center and record the tank pressure readings and levels of vapors in the launch control center, and report to the surface.

At 0150 two volunteer PTS teams were formed. The primary team, which would take exhaust shaft readings and then attempt to enter Blast Lock Area 202, was composed of SrA David Livingston, SrA John G. Devlin, and SrA Rex W. Hukle. The second team, consisting of SSgt Stephen L. Riva, SrA James R. Sandaker, and A1C Joseph P. Tallman, would be partially suited in RFHCO, ready to respond if needed. Both teams were transported to the complex fence where Morris, Capt George H. Short (Chief, Field Maintenance Branch), and TSgt Michael A. Hanson (PTS team chief) disassembled a portion of the fence near the entry gate. Hanson was monitoring the radio communications with the PTS team, while Morris remained in contact with the command post.

At 0205, Livingston proceeded to the exhaust shaft with a PVD. The instrument's meter pegged at maximum scale, indicating at least 250 ppm fuel vapor concentration. Livingston placed his gloved hand over the exhaust shaft and reported that he could feel that the area was extremely hot. He returned to the fence staging area. Meanwhile, Devlin and Hukle had proceeded to the access portal, broken through the entrapment area doors, and descended to Blast Door 6. After 15 to 20 minutes of unsuccessful efforts to manually pump open the blast door lock pins, they were recalled to the surface since their environmental control units had only 30 minutes of air remaining.

While the air supply packs were being replaced, Hanson reviewed the use of the emergency hand pump unit with Devlin, Hukle, and MSgt Ronald W. Christal, a pneudraulics technician. Kennedy entered the conversation, stating that he had experience using the emergency pump. Hanson assigned Livingston and Kennedy to reattempt blast lock entry. They reached Blast Door 6 and found the fuel concentration to be 0.5 ppm. They successfully opened Blast Door 6 and found that the fuel concentration was now 180 ppm. After opening the breather holes in Blast Door 7, they took a reading and found the air on the other side of Blast Door 7 in the 180 to 190 ppm range, still below their back-out limit. Blast Door 7 was opened, and they entered Blast Lock Area 201 between Blast Doors 8 and 9. A heavy fuel fog caused the PVD to peg at 250 ppm. Moving close to the VDAP located to the left of Blast Door 9, Kennedy reported that all the readings were at their maximum.

The team exited the area, leaving Blast Doors 6 and 7 open. Testimony differs regarding the next action. Hanson recalled that he told Kennedy and Livingston to stand by at the upper switch for Exhaust Fan 105, located in the outside stairwell of the access portal. Kennedy recalls that they were told to turn the fan on, which Livingston did. Both men returned to the surface to sit on the concrete ledge of the access portal, awaiting further instructions. Twenty or 30 seconds later, at 0300, 19 September 1980, the silo exploded.

Back at the wing command post, the room became stunningly silent as the radio suddenly went dead. Moser looked around the room and back to the radio, thinking that the nuclear warhead had detonated. It seemed like an eternity before a call from Morris came in from the site, indicating that the explosion had been due to propellants and not the warhead.

Kennedy was beginning to sit down on the concrete curb at the access portal and Livingston was standing at the top of the stairs when the explosion occurred. Kennedy was blown approximately 150 feet by the explosion into the complex fencing on the southwest corner of the complex. He tried to stand up but was unable to do so because his leg was broken. As he made his way to the surface gate for help, he could hear Livingston calling for help. Livingston was found 20 to 30 feet from the access portal. When found, he had removed his RFHCO but was unconscious.

Devlin, wearing an RFHCO but no helmet, was standing next to the surface gate. Hukle was sitting on the tailgate of Morris's truck. Christal and Hanson were standing beside the truck on the driver's side, and Morris had just sat down in the driver's side to use the radio when the silo exploded. Devlin was blown approximately 50 feet, suffering a broken heel and serious burns. Hukle moved up to the truck cab for further protection but suffered serious burns and a serious leg injury. Christal and Hanson were blown approximately 50 feet by the blast and suffered burns but were able to run back up the road for help. Morris took cover in the truck and called the command post to report the explosion. As he left the truck, he heard Hukle ask for help and moved him to safety. He then found Devlin and likewise moved him to safety.

All of the injured, except Kennedy and Livingston, were evacuated from the immediate area within 20 minutes of the explosion. Kennedy was able to radio for help from Morris's truck and was rescued at 0327. Livingston was found at 0356 and rescued approximately 10 minutes later. All injured personnel were evacuated to local hospitals.

TSgt Donald V. Green and TSgt Jimmy Roberts, two security police, arrived at the site at 0148 on 19 September 1980, having escorted an all-terrain forklift being driven to the site. Before returning to the base, they asked TSgt Thomas A. Brocksmith, the on-duty missile security police chief, if they could help in any way. Brocksmith was literally swamped and asked if they would ensure that no civilians were inside the hazard corridor and in the process ensure all posted security police knew how to use their gas masks.

Green and Roberts headed north while listening to the radio net and the conversations

Figure 9.4. Aerial view of Launch Complex 374-7 nearly directly overhead. In the upper left-hand corner of the photograph, the silo closure door can be seen where it came to a stop. This photo was taken several days after the explosion, as evidenced by the tank truck on the oxidizer hardstand and the crane located between the silo closure door wheel tracks. *Courtesy of Bill Shaff.*

between Morris and Kennedy. The road they were on did not completely encircle the launch complex, so after they had gone as far to the north and west as the road would allow, they backtracked. When they reached the southern side they proceeded on a dirt road which took them to the southwest and west side of the site. They were west of the complex when the explosion took place. Green remembers it was deafening. The force of the blast shoved the truck into a roadside ditch. Green immediately turned the truck around and raced back to the paved road and the complex. They tried to use the radio, but all they heard was static. Green was beginning to think the worst since he could not contact Brocksmith at the on-site command post. Finally they made contact, and evacuation north of the complex was begun. Green and Roberts were assigned to evacuate everyone residing beside the highway near the complex south to the town of Damascus.

As Green and Roberts neared the outskirts of Damascus, they heard a cry for help on the radio. It was Kennedy. They turned around and sped back to the launch complex. Unfortunately, they blew the engine in the truck. Stalled beside a farmhouse with a Cadillac parked in the front yard, Roberts tried to hot-wire the car to no avail. An Air Force truck sped by headed toward Damascus without slowing down. It was transporting the first group of injured back to local hospitals and the base for treatment. They stopped the next Air Force truck and, using their police authority, commandeered the vehicle with the driver accompanying them to help if he could.

When they arrived at the complex access road, Morris informed them Kennedy had been removed from the site but Livingston had not been located. Morris relayed the information that Kennedy had last seen him near the access portal. Green and Roberts volunteered to return to the complex to search for Livingston. Donning gas masks and wondering about the possibility of radioactivity from the warhead, Green and Roberts drove down the access road, only to find it blocked by a large cylindrical object which they thought was the warhead (it was the compressed air reservoir for the silo closure door hydraulic system). They could not drive around it because of debris on both sides of the road. They climbed out of the truck, jumped over another cylindrical object (which did turn out to be the warhead), and walked to the entrance gate area, searching for Livingston.

Launch Complex 374-7 was covered in smoke and dust with several small fires, all combining for low visibility. A nearby truck with a "light-all" unit mounted in the back was still running, and Green tried to redirect the lights, which were still on, to illuminate the search area but could not. He entered the truck from the passenger side to use it to crash through the fence and then transport Livingston when they found him. The truck would only move a couple of feet. Getting out on the driver's side he could see why. The tires were burnt and flat. By this time Roberts had headed south, and Green found a gap in the fence and began searching to the north. He called out for Roberts, but the mask so muffled his voice that it did not carry far. He thought of removing the mask but the smoldering debris and a tank on fire with a leaking liquid convinced him to keep it on.

As he continued the search he nearly stumbled into the 50-foot-wide crater which was the remains of the launch duct and silo equipment area. Fearing that Roberts might have fallen into the crater, he began to search the perimeter looking for both Roberts and Livingston. Green was lost and disoriented when he heard Maj Joseph A. Kinderman, security police commander, calling him on the truck radio. Green had left the vehicle radio on, using the external speaker which was directed toward the site. He followed the sound since he could not see the truck, and as he reached the vehicle another rescue team looking for Livingston passed him going into the site. They were surprised to see Green and asked if he had found Livingston. He replied that he had not but that Roberts might have; he was still on the site to the best of Green's knowledge. He showed them the area where he had last seen Roberts and was then taken away for medical attention. Green remembers that he had just been given a shot to calm him down when Roberts appeared and said that he had located Livingston and the job was done.[39]

Senior Airman Livingston died following surgery on 19 September 1980. Sergeant Hanson was released from the hospital on 22 September 1980. Senior Airman Devlin was released from the hospital on 1 October 1980. Sergeant Kennedy was released from the hospital on 3 October 1980. Master Sergeant Christal was released from the hospital on 6 October 1980. Senior Airman Hukle was released from the hospital on 7 October 1980. In all, 21 were injured as a result of the explosion or during the rescue efforts.

Investigation

At approximately 2000 on 18 September 1980, Col Richard A. Sandercock, Vice Wing Commander, 381st SMW, received a phone call from the Eighth Air Force missile safety officer informing him that a serious situation had developed at the 308th SMW, Little Rock AFB. Sandercock was informed that fuel was leaking in an uncontrolled manner from a missile at Launch Complex 374-7 and that he would be heading up the Eighth Air Force Mishap Investigation Board. His task would be to determine the cause of the mishap, make recommendations to prevent it reoccuring, and submit a formal report. Concerned that his limited technical experience with the Titan II program would prove a hindrance, Sandercock said as much, but it was clear that he was the choice. He had spent several years in missile and nuclear safety at the unit, command, and air staff level so he was familiar with accident and mishap investigation board procedures, techniques, and management. Testimony to Sandercock's team from individuals involved in the mishap could not be used for legal proceedings so this made his task somewhat easier.[40]

Informed that he was to report to Base Operations at 2200 for transportation to Little Rock via a T-39, Sandercock did so only to find that the flight had been diverted to Ogden AFB to pick up technical experts from the Air Force Logistics Center and to get them down to Little Rock as soon as possible. He returned to quarters to get a few hours sleep, anticipating the next 24 to 48 hours would provide little opportunity for rest. He was back at Base Operations prior to 0700 the next morning. "Good Morning America" was on the television. When David Hartman announced that a Titan II ICBM with a nuclear warhead had exploded at 0300 near Damascus, Arkansas, Sandercock experienced a "heart-grabbing shock" thinking of what that might mean. Fifteen minutes later he was airborne on the way to Little Rock for what ended up to be a six-week mishap investigation involving experts from all parts of the Air Force, a team that reached 40 people in size.

Sandercock arrived at mid-morning and was immediately briefed on the actions taken by the 308th SMW Accident Investigation Board. He made sure that all communications logs, maintenance records, medical records, et cetera, were being impounded and witness lists compiled. Sandercock next met with Moser, expressed his regrets for the loss of life, and explained his charter and authority. Even though Moser had been under a great deal of strain over the past 18 hours, Sandercock was impressed with his composure and gracious assistance.[41]

At approximately 1200 on 19 September 1980, just nine hours after the explosion, Sandercock and Moser flew out to Launch Complex 374-7 in a helicopter. The scene was surrealistic. The silo closure door was blown off, landing in a lightly forested area approximately 750 feet from the silo. The silo itself was an almost unrecognizable tumble of twisted rebar and concrete. The head works had been ripped from the launch duct and silo equipment area substructure. The large steel cascade vanes used to deflect the engine exhaust had been blasted clear of the silo, coming to rest inverted. Huge chunks of concrete had miraculously missed a number of parked trucks. The charred earth, burned vehicles, and severely damaged equipment painted a grim picture in the light drizzle that had developed as they flew around the site. The earlier concern as to the location of the reentry vehicle containing the warhead had been resolved at daybreak, and a SAC recovery team was in the process of preparing it for removal. Nearby they could see the national media presence at the end of the access road to the site with numerous satellite dishes, large parabolic dish microphones, and several cherry-pickers with platforms extended as everyone tried to get a clear view of the surface of the site and whatever activity was taking place.[42]

Returning to base, Sandercock found that the 308th SMW operations people were extremely interested in returning to the launch control center and recovering the classified documents that were still in the safe on Level 2. While sensitive to this issue, he was also concerned about disturbing the configuration of the equipment in the launch control center before the environment and console settings could be photographed and documented for use in the investigation.

LtCol Ronald Gray, 381st Missile Maintenance Squadron commander and part of Sandercock's mishap investigation team, was responsible for leading the team down through the blast locks and into the launch control center to record the evidence and recover the classified materials. While the operations staff was ready to walk right down and get the materials, he knew that it was not a simple matter to suit them up in RFCHO and stroll down to Level 2. He waited until the fuel vapor levels had subsided sufficiently so that chemical breathing masks could be used. His team took a very structured approach to recording all that was necessary, while at the same time escorting two operations personnel to the safe and permitting the sensitive documents to be recovered. One sight in the launch control center astounded everyone. The launch control center was so well isolated from the shock of the explosion that a half-full glass of Coke was still on Mazzaro's console just as he had left it prior to the explosion.[43]

On 20 September 1980, the second day of the investigation, it became clear that an important part of the investigation would be the tape-recorded conversations of the MPHT. Sandercock called the SAC senior controller, introduced himself as the Mishap Investigation Board president, and said that the tapes needed to be impounded as part of the investigation. There was momentary silence and then he was told that someone would get back to him shortly. Within 30 minutes, the SAC director of command and control called back and said that the tapes would be on a T-39 flight leaving for Little Rock AFB that night. Sandercock spent the next day listening to the tapes and made sure that a complete transcript of the tapes was made part of the record.

Missile Accident Investigation Board Evaluation

On 25 September 1980, Col Lloyd K. Houchin, 351st SMW (Minuteman), vice wing commander and director of operations, was appointed president of the Missile Accident Investigation Board. The difference between the Mishap Investigation Board and the Missile Accident Investigation Board was that Houchin's team would be taking testimony

Figure 9.5. Famous "Deep Throat" photograph of the remains of the launch duct and silo structure of Launch Complex 374-7. The buckled south exhaust duct wall can be seen on the left of the picture underneath the Level 1 work platform segment that has been thrown backwards. The north exhaust duct, seen on the right-hand side of the photograph, was even more severely buckled. The cascade vane exhaust deflectors were ripped out of the top of the silo and flung out into the countryside. The thrust mount and lower dome of the Stage I fuel tank are also visible on Level 7 of the launch duct. *Courtesy of Mark Clark.*

that could be used in a court of law. Houchin's group's interviews took place after Sandercock's team had completed their interview process.

The Missile Accident Investigation Board evaluated the response of the various commands to the unfolding situation as well as the liability for the cause of the accident. At 1945 18 September 1980, Headquarters SAC had directed that all actions would only be taken with its approval. At this point, major immediate concerns were the potential for Stage I fuel tank collapse with a resulting rupture of Stage I oxidizer and the probable fire/explosion; the potential of an 11,000-gallon fuel spill in the launch duct; and the probability of a toxic vapor release. Consideration was given to opening the silo closure door to vent the fuel vapors and reduce the increasing temperature that was caused by either a fire, which was unknown, or by the mixing of the fuel and deluge water. If there was no fire, use of the silo purge system to vent the silo and launch duct might increase the potential for ignition to take place. Leaving the silo closure door closed would help contain the results of an explosion should it take place. Danger to the surrounding populace eliminated this venting option. The MPHT recommended evacuation of the surrounding area to a distance of 2,500 feet. Evacuation of the missile combat crew and PTS personnel prevented direct monitoring of launch duct and silo equipment area conditions. Telephone continuity did permit remote monitoring for the sound of PVD or klaxon alarms, either of which would have permitted a lower limit of fuel concentration to be estimated.

With the additional information from Kennedy's return to the launch control center, the probability of a Stage I collapse was reduced, and attention was focused on the explosive potential of the accumulating fuel vapors. Removal of power would prevent operation of monitoring equipment in the launch control center and might cause ignition as electrical equipment switched off. While the Martin Marietta Company representatives recommended that the complex be left alone for the next several hours, the rest of the conference recommended a limited site penetration to establish criteria for removing the potential for explosions. After the second penetration, the silo exploded.

The MPHT was properly activated and contacted the Eighth Air Force as soon as it became apparent that a fuel leak had occurred. All agencies were on a conference call; Headquarters SAC, Eighth Air Force, Ogden ALC, Martin Marietta Company, and 308th SMW were working toward a solution when the explosion occurred. A major problem in resolving this crisis was that the command and control capabilities were severely hampered by marginal communications, incomplete or incorrect information, and misunderstanding of complex equipment configuration and operation. Efforts in response to propellant hazards in the Titan II system had been focused on the oxidizer because of the spill two years earlier at Launch Complex 533-7, at Rock, Kansas. Consequently, little attention had been given to a large fuel spill at the time of this accident.[44]

The conclusion from these results is clear. The near disaster at Launch Complex 374-7 was not a result of a creaky, obsolete missile system just waiting to cause an explosion. Instead, human error was the cause. That the silo contained the explosion in the manner that it did and that the W-53 nuclear weapon had not undergone an explosion due to its conventional explosives was testimony to the design considerations for both the silo and the warhead.

Figure 9.6. This view looking south shows that the box girder at the right side of the photograph over the launch duct Level 1 access door was the only one not blown free of the headworks. *Courtesy of Mark Clark.*

Most Probable Explosion Scenario

The missile airframe had no preexisting factors that contributed to the accident. The 8.75-pound socket impacted the Stage I fuel tank causing a puncture estimated to be 0.25 square inches in area, or 0.5 inch by 0.5 inch in size. The leak was located between Quadrants I and II, 16 to 60 inches above the thrust mount. The initial leak rate was calculated to be 10 gallons per minute based on tank pressure readings. Total loss of fuel was estimated to be approximately 3,700 gallons between 2145 and 2336. The missile was still able to support itself at this time.

The fuel sprayed out of the tank due to normal tank pressure, vaporizing rapidly. Decomposition and oxidation of the Aerozine 50 was accelerated by contaminants, metallic oxides, and acoustical batting on the launch duct wall, which had a large surface area that promoted catalytic degradation, probably resulting in numerous small fires. These events heated the launch duct, causing pressure increases in the other three missile propellant tanks. Water sprays activated and caused the tank pressures to decrease, but once the 100,000 gallons of available water were exhausted, the tank pressures began to rise again.

Figure 9.7. Looking down to launch duct Level 7, quadrants 3 and 4. The launch duct wall acoustical modules are amazingly intact just above Level 7. The large cylindrical object lying across the thrust mount is the Stage I oxidizer feed line that passed through the Stage I fuel tank. The two large vertical cylinder/spring combination devices are two of the four vertical shock isolation springs. *Courtesy of Mark Clark.*

Fuel vapor concentrations reached explosive levels in the launch duct. Penetration into Blast Lock Area 201 and subsequent withdrawal leaving Blast Doors 6 and 7 open, along with turning on the access portal Exhaust Fan 105, accelerated the migration of explosive vapor concentrations from the launch duct into the silo equipment area. Numerous electrical motors and other potential ignition sources were present in the silo equipment area. Ignition was probably electrical and a vapor fire propagated into the launch duct, resulting in an explosion as the confined fuel vapors ignited. The explosion and resulting overpressure ruptured the Stage I oxidizer aft dome, dumping the oxidizer into the ruptured Stage I fuel tank. The resulting hypergolic reaction generated the major in-silo explosion. The explosion blew the 740-ton silo closure door several hundred feet into the air. The door landed 625 feet away and slid through a small grove of trees, coming to rest 750 feet west of the launch duct.

This same explosion pushed the Stage II engines into the Stage II fuel tank as Stage II was being thrown clear of the silo. The Stage II oxidizer tank ruptured, allowing mixing of the Stage II propellants, and Stage II exploded above the silo, jettisoning the Mark 6 reentry vehicle containing the W-53 warhead. Stage I fragments were found within 300 feet of the silo; Stage II fragments were found much farther away, within 2,300 feet, supporting these conclusions.

Large fragments from the silo structure were thrown considerable distances from the launch duct. Pieces of the silo cap structure that supported the silo closure door, weighing from 122 to 320 tons, were from 290 to 490 feet away to the east. The north flame deflector, weighing 22 tons, was thrown 1,400 feet to the north, while the south flame deflector was broken into two pieces, and thrown 1,733 feet to the south.[45]

The W-53 warhead was found near the helicopter pad access road damaged but basically intact. No radioactive contamination was found around the surface of the site nor in proximity to the warhead itself. After a thorough examination of the external damage, the warhead was separated into two major components at an appropriate bolt circle, and the two sections were packaged in separate jet engine containers for transport on 21 September 1980. The damaged warhead components were returned to the Department of Energy facilities in Amarillo, Texas, on 23 September 1980.[46]

Aftermath

The total replacement cost of Launch Complex 374-7 was estimated at $225,322,670. Demolition and cleanup were estimated at $20,000,000 based on the Rock, Kansas, oxidizer spill accident in August 1978. The missile airframe replacement costs were estimated at $129,975,000, based on the fixed costs to reopen the Titan II assembly line and the variable costs to produce a given number of airframe and airborne systems assemblies. Silo support equipment was estimated at $4,150,000 and rebuilding the launch duct, silo closure door, and topside facilities was estimated at $91,000,000.[47]

Cleanup at Launch Complex 374-7 began three weeks after the accident. From 6 through 11 October 1980, personnel worked 12-hour days to remove debris that was scattered as much as half a mile from the launch complex. Approximately 400 acres of farm and woodland were scoured for pieces of metal and concrete from as small as an acorn to 30 tons. Cost of the cleanup was $48,120 and took 6,240 man-hours. In addition to the above-ground debris, approximately 100,000 gallons of contaminated water remained in the silo. Bio-environmental engineers from the 308th SMW sampled 106 surrounding water wells to see if the fuel-contaminated water had penetrated the aquifer but found no contamination. From 10 to 27 October 1980, the 314th Civil Engineering Squadron pumped out the majority of the water and then neutralized the remainder.[48]

On 22 May 1981 a ceremony was held at Little Rock AFB where the Secretary of the Air Force, Verne Orr, awarded Airman's Medals for Heroism to SrA John G. Devlin, TSgt Donald V. Green, SrA Rex W. Hukle, Sgt Jeff Kennedy, SrA David L. Livingston (posthumous), and TSgt Jimmy E. Roberts. Secretary Orr awarded Air Medals to Col James L. Morris, TSgt D. G. Rossborough, and SSgt Silas L. Spann Jr., for their efforts to administer first aid at great danger to themselves. Technical Sergeant Roberts was the only one not to receive his medal at this time because he had transferred to Davis-Monthan AFB.[49] On

Figure 9.8. View looking southwest along the access road 50 yards northeast of the entry gate. The truck in the foreground is the vehicle Colonel Morris drove down to the entry gate with the volunteer Propellant Transfer System team members prior to the site inspection by Sergeant Kennedy. *Courtesy of Don Green.*

10 November 1983 in a ceremony at the 308th Missile Inspection and Maintenance Squadron facilities on Little Rock AFB, the building was redesignated as the Livingston Building in honor of SrA David L. Livingston.[50]

In June 1981, a planning conference was held with the 308th SMW concerning how best to make safe and seal the remains of Launch Complex 374-7 in a manner that would preserve the structural integrity of the site if restoration were to be considered feasible at a later date. Concerns over the Strategic Arms Limitation Talks II agreement and losing the complex as a potential launch facility if it were abandoned further muddied the waters. There would not be any attempt to recover any of the equipment within the silo because of the extremely hazardous conditions of the unstable walls. The final decision was to seal the site with soil, gravel, and small concrete debris, thus allowing access at a later date.[51] This was the final decision on Launch Complex 374-7 as the decision in late 1981 to deactivate the system rendered no further action necessary.

Congressional Investigation

On 25 September 1980, Secretary of the Air Force Hans Mark announced that an independent committee would be established to conduct an in-depth review of the Titan II weapon system. This committee would be independent of the ongoing Air Force investigations. Gen Bennie L. Davis, commander of the Air Training Command, would head the review group. Its focus would be the safety and supportability of the Titan II system. Extensive use was made of the report submitted to the Senate and House Armed Services Committees seven months earlier on the physical condition and maintainability of the Titan II system, as well as the reports available from the accident investigation.[52]

On 6 January 1981, the committee report was released to the public. This second full-scale review of the Titan II program within seven months agreed in principle with most of what the May 1980 report had found. There was no question that Titan II remained a reliable system. The Air Force had kept up with modernization as funding had allowed. The two major ongoing validation studies, the Reliability and Aging Surveillance Program, which monitored the missile systems and ground support equipment, and Service Life Analysis Program, which involved selecting a missile and exhaustively checking its flight worthiness as well as firing the Stage I and II engines, were effectively monitoring the Titan II program.

Potential hazards were obviously the missile warhead and the propellants. The missile warhead was considered by the committee to be the "most forgiving and least hazardous of the hazard sources associated with the Titan."[53] The worst plausible accident would be the detonation of the high explosives within the warhead, spreading radioactive debris. Since the Damascus accident could easily be construed as an example of this scenario, clearly the warhead design had been more than adequate. The design safety review did, however, recommend further attention to bringing the warhead design up to date with modern design criteria.[54]

The propellants were considered to be a much more serious potential hazard with the oxidizer's volatility and extreme corrosiveness making it the most dangerous. Evaluation of the true potential hazard required review of the entire system's subsystems, such as design safety, currency of support equipment, operations, maintenance, accident prevention, and crisis management. Modifications to the missile structure to make it less vulnerable to the type of accident that had occurred at Damascus were not feasible. Instead, procedural changes during maintenance operations needed to be implemented.

The currency of the missile launch and support equipment was reviewed. The committee did agree that a rigorous and successful inspection process had identified and corrected problems by replacing the old guidance system, updating the propellant transfer system mobile equipment, and replacing the oxidizer system seals with a new material highly resistant to the oxidizer corrosiveness. The committee felt that funding was limiting the continued modification of missile support equipment in a timely manner.

Missile crews were clearly proficient in the operation of the system. One glaring deficiency was the need for an improved missile procedures trainer that would provide expanded multiple hazard training capability. Improvements to the launch control center to ensure that missile combat crews could safely remain inside and direct recovery efforts were suggested by the committee.

The maintenance system remained fully capable of safely supporting the Titan II system. The problem was that the age and experience of many in the maintenance force were reaching a critical point since retention of veteran maintenance personnel was decreasing. Hazardous duty pay and updated trainers were recommended.

Accident prevention within the system was compared to equivalent industrial sector operations and found to be similar. Throughout the life of the system, only seven accidents due to material failure had occurred. The primary cause for the major accidents had been human error, and this was seen as an increasing trend in the accident data—something that had to be rectified immediately.[55]

The key to both the May 1980 and December 1980 studies was crisis management. While the system was found to be capable of handling plausible accident scenarios, major improvements were necessary. Interaction with local authorities to mitigate exposure to propellant vapors needed to be improved. The Rock, Kansas, oxidizer spill had so focused the crisis management efforts on the oxidizer that consideration of fuel spills, a flammable material rather than primarily corrosive, had been neglected. Solutions to fuel spills were not the same as for oxidizer. Containment procedures within the site boundaries had to be improved through such methods as covering the spill to prevent volatilization and positioning scrubbing or burn units over exhaust air shafts to burn the contaminated air stream prior to release into the atmosphere. Crisis management exercises needed to become much more realistic as well as be held more often at the missile complexes. Joint planning with the Air Force and state and local authorities needed to be done on a site-by-site basis and funds needed to be spent to ensure rapid warning of nearby populace.

The most strident recommendation of the committee's report was the need for improved communication and interaction with the public:

> First, there is a need for expeditious, authoritative statements locally. *Second, these public statements should confirm the presence or absence of nuclear weapons in the accident and frankly discuss safety features and potential hazards.* [author's emphasis] Third, procedures should be refined to provide timely notification and an ongoing flow of information to interested members of Congress and other federal agencies. And fourth, interagency agreements should be established to coordinate government-wide public affairs responsibilities, procedures and activities in nuclear accident situations.[56]

There is no doubt that the 1978 and 1980 accidents played a significant role in the decision to deactivate the Titan II program. Was the potential of an accident with Titan II growing? Was it any more or less significant a possibility than one of the strategic bombers crashing with its nuclear weapons on board? Titan II's time had come and was rapidly passing by, these last two accidents simply served to usher the conclusion along. However, the cost of the recommended improvements from the investigation board's reports as well as that of the review committee sealed the aging missile's fate. To the Department of Defense and the Strategic Air Command planners, money spent on implementing these recommendations could be better spent on newer systems.

X

THE END OF THE TITAN II ERA

With the two high-profile accidents of August 1978 and September 1980 still fresh in the minds of many, the report on the Titan II program released after the Damascus explosion was of little solace. During Senate Committee on Armed Services hearings held on 18 February 1981, Gen. Richard H. Ellis, USAF, Commander in Chief, Strategic Air Command (SAC), was asked by Senator John Warner, about the role of Titan II in the current strategic targeting plan. Ellis replied, "Titan's [deleted] warhead is designed for and effective against wide area, soft targets. Titan targets are specifically constructed high-yield aggregate Designated Ground Zeros (DGZ's) which contain more than one primary DGZ. The 52 Titan weapons presently cover [deleted]. Each of these primary DGZ's could require [deleted] to provide damage levels equivalent to those acheived by Titan."[1] When Warner asked about the status of Titan II and the ability of the SAC to continue to support the system safely, Ellis replied, "Although the system is aging, it has not become unreliable or inherently unsafe. With adequate maintenance and adherence to established safety precautions, the Titan system can be kept safe and operable."[2] Ellis concluded that to remove Titan II from the nation's war plan would not only require a complete reallocation of available forces to required targets, it would also be seen by the Soviet Union as a lessening of the United States resolve.

On 24 September 1981, an official of the Reagan administration first stated publicly that the Titan II system would be retired.[3] On 2 October 1981, President Ronald Reagan announced the initiation of his Strategic Forces Modernization Program. Titan II was not specifically mentioned during the subsequent press conference but in the background statement issued from the White House, one sentence spelled the demise of the program, "All aging Titan missiles were to be deactivated as soon as possible."[4]

On 11 November 1981, Congressman Dan Glickman, Kansas, released to the public correspondance with Secretary of Defense Caspar W. Weinberger that outlined a Titan II deactivation plan which would retire one missile a month as long as weather, equipment, and personnel workload permitted. No decision as to which strategic missile wing would be first had been made, nor had a start date been determined.[5]

Congressional hearings began immediately on the strategic modernization plan. Titan II was almost a side show as Congress debated the wisdom of vast new funding initiatives and new weapon systems. The termination of the Titan II program was initially discussed from the viewpoint of using the soon-to-be-empty Titan II silos in yet another basing mode option for the MX system. The Air Force was advocating the temporary use of the Titan II silos for the initial deployment of MX until a final basing plan was formulated. While this seemed to some a contradiction of the SALT II Treaty, the advocates of this option pointed out that the silos would not have to be enlarged in diameter so they were not really being modified in a way that was a treaty violation.[6]

On 26 February 1982, Gen Bennie Davis, USAF, Commander in Chief, Strategic Air Command, explained the rationale for deactivation of the Titan II missiles in testimony before the Senate Committee on Armed Services. General Davis described Titan II as

> designed at a time when the United States possessed a marked nuclear superiority, and the assured ability to destroy a large portion of Soviet society was viewed as the most credible deterrent to both nuclear and conventional war. The Titan II's massive yield and poor accuracy tied in well with a strategy of massive retaliation—accuracy and discrimination were neither necessary nor possible. The Titan II was, and remains, a highly effective weapon for use against a collection of targets that can all be damaged by a single, high yield weapon.[7]

It is interesting to note the change in attitude toward Titan II in the one year between Gen Richard Ellis's congressional testimony and that of General Davis.

The safety record in Titan was better than a comparable chemical industry facilities. Both major accidents had been a direct result of human error, not a failure in the missile or its support equipment. If the cost of the remaining proposed safety upgrades, as well as the yearly support costs, were eliminated with program deactivation, an estimated $500 million would be saved over a six-year period. The Air Force felt that this money would be better spent on more modern systems. Current land-based and submarine-based ballistic missile systems were much more accurate than Titan II. Smaller yield weapons on the more accurate systems could provide a more precise attack which was more in line with the current SIOP-6 options. The Titan II silo hardening (300 psi) against nuclear blast effects was significantly below those of Minuteman III (2000 psi), thus leaving Titan II vulnerable to the newer Soviet missiles systems.[8]

Considerable discussion of these issues took place during the subsequent days of the committee hearings. Concern was expressed that the estimated savings of $500 million was a drop in the bucket compared to the projected $180 billion cost for the modernization program. Some argued that the United States should keep the Titan II missiles active until their replacement was fully activated. The Air Force countered that with unlimited budgets, yes, this was true. However, funding was not unlimited, and the Air Force felt that $500 million was a good start on a variety of key aspects of the modernization program.

Others were concerned that a lag between Titan II deactivation and activation of any replacement system would result in a diminution of American strategic nuclear strength since 25 percent of the land-based and 11 percent of the overall offensive megatonnage would be lost, adding to the window of vulnerability which was the whole point of the strategic modernization program. The Air Force felt that since Titan II deactivation was programmed over several years, loss in target coverage would be gradual and compensated by deployment of the newer systems as the modernization program reached fruition. Senators Robert Dole, Kansas, and Dennis Deconcinni, Arizona, both from states with Titan II wings, testified that while the Air Force had made significant progress in responding to the recommended safety measures subsequent to the two major accidents, retiring the system was the ultimate solution to the safety problem.[9]

390th Strategic Missile Wing, Davis-Monthan AFB, Arizona

On 2 February 1982, Headquarters U.S. Air Force announced that the 390th Strategic Missile Wing (SMW) would be the first Titan II wing to be deactivated with all activities to be completed by February 1985. The 390th SMW had been chosen for two reasons. A large number of missiles would soon need extensive maintenance due to corrosion problems. Since the maintenance required draining the propellants and removing the missiles from the silos, the timing was ideal. An additional reason was that 3 of the 18 launch complexes were on the verge of being inactivated due to housing developments encroaching the safety zones of the sites. (see Chapter 5, Tables 5.6–5.8 for deactivation dates for all three Titan II wings).

The deactivation program was code named RIVET CAP. The official start date for deactivation was October 1982. The major concerns expressed at the first Titan II Deactivation Conference held at the Ogden Air Logistics Center 13–14 April 1982 was the storage and disposal of Titan II propellants. The storage and transportation capacity could accommodate only one inactivation per month at best. With a planned 45 days/complex deactivation schedule, one of the most critical factors was the timely disposal of propellants as they were removed from each missile. This was to be a prophetic concern. In addition, the commercial transport trailers for both propellants were in substandard condition and not considered reliable for the task at hand. Thorough inspections by the San Antonio Air Logistics Center transportation manager and the 3901st Strategic Missile Evaluation Squadron Propellant Transfer System inspectors was the starting point for bringing the commercial units to common specifications by the time the third site was deactivated. The major modification made to the launch complexes was raising the fuel hardstand overhead water deluge system to accommodate the height of the commercial fuel transport trailers.[10]

RIVET CAP was organized into three phases: deactivation, caretaker status, and dismantlement. Once the missile was off alert, Phase I began. The reentry vehicle was removed, propellants were downloaded, and the missile was removed to the base to be readied for shipment and storage at the Military Aircraft Storage and Distribution Center facility at Norton AFB, California. The 18 reentry vehicles were inspected and segregated into three groups. Nine reentry vehicles, minus their W-53 warheads, were shipped for storage at the 3096th Aviation Storage Depot, Nellis AFB, Nevada. Two were shipped, without their warheads, to the Directorate of Special Weapons, Kelly AFB, Texas, for replacement of expired components and then one each was shipped to the remaining Titan II bases as spares. The remaining 7 reentry vehicles were disposed of at Kelly AFB. Phase I also included salvaging parts that could be used at the other wings prior to their dismantlement.

Propellant disposition turned out to be the single most difficult part of the deactivation program. Before the commercial trailers could leave a launch complex, the drivers and trailers had to pass an array of safety tests. The drivers had to sit through safety classes from both the Environmental Protection Agency and the Department of Transportation. The truck routes were carefully planned and strictly adhered to. Propellant storage locations ranged from Aerojet storage facilities in California to the eastern and western test ranges at Cape Canaveral, Florida, and Vandenberg AFB, California, respectively.

Figure 10.1. On 5 July 1984 the reentry vehicle was removed from Titan II ICBM B-75 (63-7730) at Launch Complex 533-8, marking the beginning of the deactivation process for the 381st SMW. *Courtesy of the Air Force Historical Research Agency, Maxwell Air Force Base, Montgomery, Alabama.*

In 1983 the storage of propellants became the single largest problem holding back the deactivation process. The storage facilities at Vandenberg AFB had been the logical point to dispose of the propellants from the 390th SMW. Unfortunately, they were still being overhauled and not ready for use at the time. Propellants were stored in railroad tank cars and some oxidizer was sold to industry while some fuel was incinerated. When the Aerojet storage facility developed leaks in the transfer piping, the logistics of propellant storage became a critical problem. Fortunately other industrial users were found, as well as tanker car storage locations, and finally the Vandenberg AFB and Aerojet storage capability came back on line.

With the recent accident at Damascus on everyone's mind during the entire process of deactivation, several new hazard control systems were utilized during propellant transfer. The first was the portable foam vapor suppression system, a trailer-mounted unit that could generate foam to cover a spill in the launch duct or on the surface with a chemical foam that prevented vapors from escaping. The unit was completely self-contained, including a generator so that complete power loss at the site would not interfere with its use.

The second major piece of equipment was the portable vapor scrubbing system. This consisted of a packed tower that was connected to the silo air exhaust shaft, reagent trailers for the scrubber chemicals, and bladders to collect the tower effluent. The packed tower system used water to trap fuel vapors and a sodium hydroxide solution to trap oxidizer vapors in the event of a spill contained in the launch duct.

The fire water recirculation modification was installed at the 390th SMW launch complexes in November 1982, and installed at the two remaining Titan II wings well before their deactivation process began. Previous to this point, the 100,000 gallons of water in the hard water tank within the silo could only be used once as it collected in the launch deflector area after use. A recirculation system allowed it to be pumped from the flame deflector to the spray nozzles if necessary.

Launch Complex 570-9 was the first site for Phase I work, having been removed from alert on 7 July 1982 as part of the Service Life Analysis Program for engine analysis and the Reliability and Aging Surveillance Program for the missile and operating ground equipment. The decision was made to use this complex to test the deactivation time lines and procedures prior to the full-scale deactivation program implementation. The first site for the official deactivation process was Launch Complex 571-6, beginning with its removal from strategic alert on 29 September 1982.[11]

Phase I deactivation was almost identical to normal missile removal. The reentry vehicle was removed and transported back to the base. The explosive ordnance for prevalve opening, engine ignition, and stage separation was removed and Circuit Breaker 103, which provided power to the launch system, was permanently tagged in the "off" position. With this the missile was reported to SAC as being permanently off alert for deactivation.

The propellants were downloaded and removed from the launch complex, then the missile itself was removed from the silo and returned to the base to be prepared for shipment and storage. After the reentry vehicle had been removed, the launch documents and keys, along with all other classified materials, were removed and returned to the Plans and Intelligence staff. This meant that the crew, while still listening to the Primary Alerting System, no longer had to respond to message traffic and could concentrate on the business of deactivating the site.

Propellant download alerts were a major effort. The crews normally had to report to the wing command post for a special briefing at around 0500 or so, then head straight out to the site. The PTS team would have already spent a couple of days emplacing equipment and preparing the site and would be busily setting up as the alert crew arrived. Downloads were an all-day operation, especially if the weather did not cooperate. Conditions on site had to fit within specific parameters to start propellant flow. Waiting for the correct weather often took considerable time.

Figure 10.2. The diesel generators were recovered from Level 3 of the silo by using plastic explosives to open the 1.5-foot-thick launch duct wall clad with 0.5-inch-thick steel plate. This photo shows the results of the recovery effort at one of the abandoned Titan II launch complexes at Vandenberg AFB, California. *Courtesy of Fred Epler.*

Once all the propellant was removed, the task of removing the missile began. A crane was placed on site to pull the missile out, then transfer the stages to the waiting flatbed trucks. Once on the trucks, the stages were covered with tarpaulins and tied down for the trip back to the base.

Now that the missile was gone, a majority of circuit breakers in the launch control center could be permanently turned off and tagged. Tagging meant hanging a blue Air Force danger tag on the breaker, indicating that it should not be turned back on. Slowly but surely, the remaining major and minor systems were turned off and removed, finally eliminating the need to do much at all in the silo. Eventually, the hydraulic systems that operated the blast doors were shut down, which required the crews to bring a portable hydraulic unit with them to open and close the doors when they came on site.

As deactivation progressed, the need for crews to remain on site 24 hours a day was eliminated. The blast doors were all locked, the elevator shaft and access point was secured with heavy locks, and the front gate was chained and locked. At this point, crews came out to supervise maintenance activities and to provide a fire watch since the automatic sprinkler system had been shut down. The silo was emptied of items the Air Force could use elsewhere, and the propellant lines were all purged of any remaining fuel or oxidizer residue and fumes. Teams removed equipment for use at the sites that still remained active at the 381st SMW and 308th SMW.

Phase II began when the complex was turned over to Base Civil Engineering and a

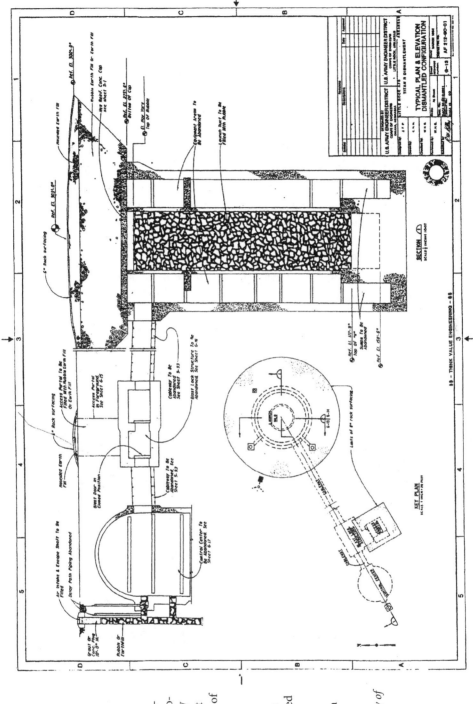

Figure 10.3. Drawing illustrating the deactivation demolition process at the 308th SMW sites. The launch duct had the upper 25 feet of headworks destroyed with explosives. The debris was pushed into the silo and then capped with reinforced concrete and a thick layer of dirt. The air shafts were filled with grout, and the blast lock doors were tack welded shut. *Courtesy of the Titan Missile Museum National Historic Landmark Archives, Sahuarita, Arizona.*

civilian contractor, who removed miles of wiring and copper tubing and salvaged specified equipment for the Air Force, including the diesel generators, warning sirens, air compressors, and silo closure door hydraulic system to name just the major items.

Phase III began when actual silo demolition took place. The Air Force did not want to spend any more time or money than necessary for the dismantlement process, yet, at the same time, the SALT I and II protocols would continue to count the Titan II silos if they were not sufficiently destroyed. The major verification item was the silo headworks, the massive concrete structure that supported the silo closure door. This structure had to be destroyed to a depth of 25 feet below the surface in a manner that Soviet reconnaissance satellites could confirm. Jack hammers and cutting torches were used to prepare the headworks area for the application of 2,800 pounds of high explosive. After the demolition, the site was left exposed for a six-month period so the Soviets could have several satellite photos of the demolished launch complex to confirm that no new weapon system was being installed. After the observation period, the resulting rubble was bulldozed into the launch duct. The silo air intake and exhaust shafts were plugged with grout, the launch control center air intake was plugged with grout, all blast doors were tack-welded shut, and the long cableway was dug up and removed. After satellite observation for 30 days, and the acceptance of the launch complex as destroyed, a reinforced concrete cap was fabricated to span the silo walls and the excavation was back-filled to a level surface. The bottom 20 feet of the access portal was filled with a slurry of dirt. The surface was smoothed to a natural contour and the operation moved to the next site.[12]

Easily lost in and amongst all of the work at the sites and parts being inventoried and dispersed to the other wings was the plight of the crew members and maintenance

Figure 10.4. Launch Complex 570-8, 390th SMW. The silo closure door has been dismantled, the cooling towers have been removed, and the site is now ready for the demolition of the headworks. Rubber tires were lashed together and placed on the surface of the area to be demolished in an attempt to contain the debris. *Courtesy of the Titan Missile Museum National Historic Landmark Archives, Sahuarita, Arizona.*

THE END OF THE TITAN II ERA

Figure 10.5. The detonation sent the tires sailing in all directions. *Courtesy of the Titan Missile Museum National Historic Landmark Archives, Sahuarita, Arizona.*

Figure 10.6. The site from approximately the same view as Figure 10.4. The similarity to the destruction after the Damascus accident is striking. *Courtesy of the Titan Missile Museum National Historic Landmark Archives, Sahuarita, Arizona.*

personnel who had elected to stay and be part of the deactivation process rather than take attractive assignments elsewhere. One can imagine that with an entire Titan II wing deactivating there were only so many positions at the other Titan II wings to be filled, plus a few positions at the logistics centers. Col John Chambers, the 390th SMW commander at the time, requested and received assurances from Headquarters SAC and the Air Force Personnel Center, Kelly AFB, Texas, that each reassignment would be treated individually. While many were reassigned to a second Titan II wing that was undergoing deactivation, none had to participate in all three deactivation programs.

On 2 December 1983, three days shy of the twentieth anniversary of its turnover to SAC as an operational Titan II missile squadron, the 571st SMS was inactivated. The deactivation process was now running smoothly and ahead of schedule. On 31 July 1984, 20 years and seven months after assuming operational alert status, the 570th SMS was inactivated, ending the Titan II era at Davis-Monthan AFB.[13] One launch complex and one task remained. Launch Complex 571-7 had been maintained intact for possible preservation as a museum.

381st Strategic Missile Wing, McConnell AFB, Kansas

In September 1983 the 381st SMW learned of a change in the deactivation schedule. Originally chosen as the last Titan II wing for deactivation, to take place from April 1986 to September 1987, the decision to base the new B-1B bomber at McConnell AFB beginning in 1986 meant that the 381st SMW would switch places with the 308th SMW. Deployment of the new bomber to McConnell AFB simultaneously with the deactivation of the Titan II wing would have been a severe logistical strain on the base support facilities.[14]

On 2 July 1984, Launch Complex 533-8, the first on alert for the 381st SMW, became the first complex to be deactivated. At the time of deactivation, Launch Complex 533-8 housed missile B-75 (63-7730). B-75 had been in place at Launch Complex 533-8 since 1969. After staying on alert for five years, B-75 had a Stage II download in support of a Stage II engine ablative skirt repair. After four additional years of uninterrupted alert status, Stage I and II oxidizer prevalve seals had to be replaced on a biannual basis through to the date of deactivation. Considering the fact that the operational requirement for the Titan II missile was a one-year time span for a propellant load without maintenance or download, this was an excellent testimonial to the design and integrity of the missile.[15]

Before the deactivation program began in earnest, the missile combat crews of the 381st SMW had to receive extensive training in deactivation operations. In early 1984, only a handful of crews were certified as "major-maintenance qualified," meaning they could monitor all operations dealing with missile propellant download, missile removal, and the other lengthy operations involved in such activities. This had been a direct result of the Damascus explosion in 1980. Only major-maintenance–qualified missile combat crews could pull alert duty at sites undergoing deactivation activities. This training was intensive and quite technical as each crew member had to be well-versed in the intricacies of these operations. Safety was paramount, especially as the deactivation of each site was trumpeted in the local media, and there were often news crews and civilians gawking at what was taking place topside at the complex.

Once major-maintenance qualified, crews could pull alert at any site, regardless of deactivation or regular alert status. This became a significant event as the deactivation program progressed because there were some crews that did not have this qualification. If someone was going to be leaving in a few months, especially if they were a standboard or instructor crew member and did not pull all that many alerts to begin with, they were usually not given the extra training.

On 1 November 1985, the 533d Strategic Missile Squadron was inactivated. Nine months later, on 8 August 1986, the 532d Strategic Missile Squadron and the 381st Strategic Missile Wing were inactivated, ending nearly 23 years of vigilant duty with the Titan II ICBM program.[16]

308th Strategic Missile Wing, Little Rock AFB, Arkansas

On 1 February 1984, Col Jack A. Leach, 308th SMW commander, held a press conference to confirm that the 308th SMW would be the last Titan II wing to be inactivated. This meant that the missile deactivation process would start in October 1985 and be completed by September 1987. In the meantime, all 17 missiles had to be kept in optimum readiness.

With deactivation scheduled to start in one year, Richard N. Holbert, president of the Arkansas Aviation Historical Society, began an effort to create a museum similar to that being pursued by the 390th SMW. Initial attempts in 1983 had been met with rejection. This time Holbert enlisted the assistance of Congressman Ed Bethune. Bethune urged the Secretary of the Air Force to give the request his personal attention. On 5 September 1984, Bethune announced that Headquarters SAC had officially agreed to transfer one launch complex in Arkansas, as well as one each in Arizona and Kansas, to an eligible organization once all the complexes had been deactivated. Choice of site would be left to the organization, and the Air Force was not going to be involved in paying for conversion to a museum. The criteria for allowing the complex to become a museum would be to block the silo closure door halfway open, erect a permanent barrier to keep it from closing or opening any further, and provide a clear cover over the half-opened launch duct for satellite viewing.

Holbert announced plans to have legislation drafted that would transfer the sites to the Department of the Interior for operation as National Parks or Monuments. Three 308th SMW sites were mentioned, 373-3 near Heber Springs, 374-9 near Quitman, and 374-1 near Blackwell. The Little Rock Community Council and the Greater Little Rock Chamber of Commerce endorsed these efforts, but the cost of conversion to a museum facility proved too expensive. Only the 390th SMW succeeded in preserving a site, Launch Complex 571-7, for a museum.[17]

On 19 April 1985, Headquarters SAC announced a major revision in the 308th SMW deactivation plans. The process would start six months earlier; in fact, immediately. Launch Complex 374-8 was to begin the deactivation process on 24 April 1985. By starting six months early, the hope was that much of the work could be completed before harsh winter weather set in. Weather conditions unique to Arkansas, such as lack of sufficient winds for propellant transfer operations to take place safely, would increase the possibility of weather-induced delays. One factor still needed careful management. This was the transport and storage of the propellants drained from the missiles. Storage facilities were anticipated to be available in time for this accelerated schedule, but this was not certain.

On 22 April 1985, Col John E. Chambers, 308th SMW commander, held a news conference announcing the acceleration of the deactivation process beginning on 24 April 1985. The scheduled end was still September 1987. It was estimated that a period of 30 to 45 days and a cost of $1,000,000 per site would be required. All other sites would remain active until one day before the scheduled phase-out date for that particular site. Launch Complex 374-8 was selected as the first site because the missile's oxidizer had already been offloaded due to a routine oxidizer seal change.[18]

The 374th Strategic Missile Squadron was formally inactivated on 15 August 1986 just 17 days shy of 24 years of service. On 14 July 1987 at Launch Complex 373-8, the missile combat crew composed of Capt J. Neil Couch, MCCC; Capt Steven W. Martin, DMCCC; TSgt James P. Ross, BMAT; and TSgt Michael W. Lee, MFT, completed the last 24-hour alert tour of the 308th SMW and the Titan II program. Deactivation activity had begun on 5 May 1987 as Launch Complex 373-8 was taken off of alert. Captain Couch made the last Titan II alert log entry:

> This concludes the last twenty-four hour alert in Titan II history. For the last twenty-four years, the men and women of the 308th, 381st and 390th Strategic Missile Wings had endured the elements and boredom to stand as the guardians of peace for the free world. We now close another chapter in military history as the mighty Titan; the bastion of peace; the dinosaur of ICBMs now fades away. This final crew changeover is complete.[19]

The 308th SMW was formally inactivated on 18 August 1987 after 25 years 4 months and 18 days of service. So ended the Titan II ICBM program.

Born amidst the stark realities of the cold war, with the need for a heavy lift, rapid response strategic misisle system, the 54 Titan II missiles at the three operational bases stood alert for 24 years, 14 years past the original design specification. Upgrades and modifications kept the launch complexes capable of supporting the missile and the launch crew. The missile, once deployed, was hardly modified, except for the guidance system, which was more a matter of a lack of spare parts than that of diminished accuracy. Considering that the program consisted of only 54 out of a total of 1,054 strategic missiles, Titan II must have had a major contribution to offer for those 24 years.

Perhaps Col Richard Sandercock, Vice Wing Commander, 381st SMW in 1980, best summed up the feelings of all those that were part of the Titan II system,

> The Titan II *was not* an old, leaky, creaky accident waiting to happen as the media and politicians would have you believe. It was instead a bright and shiny symbol of power that our enemies respected and those of us that were in the Titan II business were very proud of. The Titan II ICBM system was awesome in its complexity. It required your full attention every day. It was a true adventure that we all shared in and will always remember.[20]

EPILOGUE

By their very nature, weapons systems that are retired are rarely assimilated in large number for civilian use. On the other hand, cargo aircraft frequently have their civilian counterparts, such as the Boeing KC-135 and the Boeing 707. Such was the case with the Titan II ICBM and the Titan II Space Launch Vehicle (SLV). The Titan II ICBM was built to accurately place a reentry vehicle into the proper trajectory, so why not do the same with a satellite or space probe?

As early as 1972, Martin Marietta Company submitted a review of the Titan II systems available at Vandenberg AFB, California, with the goal of utilizing at least one of the training silos as a dedicated launch facility for satellites and space vehicles using the Titan II ICBM airframe. Martin Marietta detailed the refurbishment of Launch Complex 395-C to meet space launch vehicle demands once the Titan II ICBM launch program was complete. The existing silo closure door would be replaced with a portable system and the thrust mount/shock isolation system would be replaced with the static launch ring thrust mount in use with Titan III. Since 24-hour crew duty was not necessary, the blast valves and associated equipment to protect against the effects of a nuclear blast would be removed. Total estimated cost was two million dollars. The proposal was not accepted.[1]

Titan II Space Launch Vehicle Program

Nine years later the idea of using the Titan II ICBM airframes as space launch vehicles was again considered. This time no thought was given to using the in-silo launch facilities of Launch Complex 395-C. At the time of the Titan II program deactivation decision, 55 Titan II airframes remained, 52 in silos and one at each of the strategic missile wings as a spare. Unlike aircraft that would have been retired and stored out in the weather, the Titan II airframes would be stored indoors at Norton Air Force Base, California. In January 1986, Martin Marietta Company received a $45.2 million letter contract from the Air Force (F04701-85-C-0085), to convert up to 13 Titan II ICBM airframes into SLVs.[2] The actual contract for $483.7 million was finalized in September 1986 for eight conversions with five options. In August 1987, the Air Force exercised the option clause for an additional 5 missiles and in November 1987 added 1 more for a total of 14 missiles designated for conversion.[3] Nearly a quarter of a century after fabrication, these Titan II airframes and engine sets were found to be in excellent condition due to the efforts of the missile designers, fabricators, and the Air Force personnel that maintained the missiles.

The concept behind the Titan II SLV program was to minimize conversion expense by utilizing as many of the Titan II ICBM subsystems as possible and by upgrading where necessary using readily available Titan III space launch vehicle components. Figure E.1 shows the members of the Titan family. Titan I and Titan II were ICBMs. A common misconception made within the Titan family of space launch vehicles is that the Titan II airframe was directly incorporated into the later designs. This is incorrect. Airframe fabrication techniques were

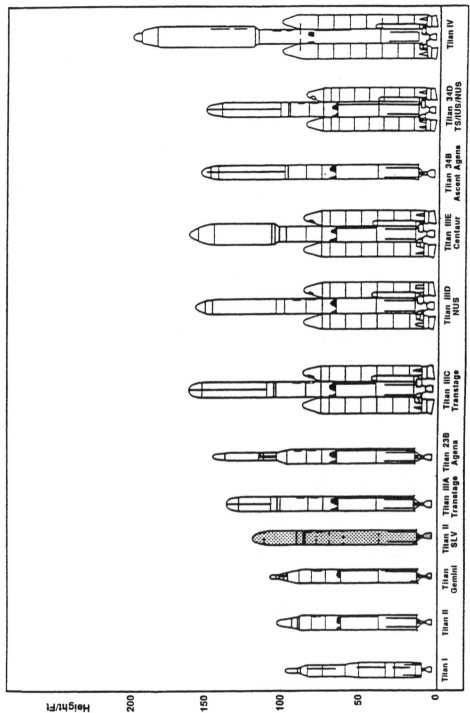

Figure EP.1. Titan launch vehicle family. Titan I was not used as a space launch vehicle. Titan II was refurbished to become Titan II SLV. None of the Titan family from Titan IIIA to Titan IV used Titan II components. Each launch vehicle's core assembly was built specifically for that launch vehicle system. *Courtesy of Lockheed Martin Astronautics, Denver, Colorado.*

the same, but propellant tank sizes varied considerably. Titan II ICBM airframes were not stretched to fit, rather the entire airframes for the Titan IIIC, D, and E series and Titan 34D and Titan IV space launch vehicles were manufactured anew. The solid rocket motors attachment points to the core stage in the Titan IIIC, Titan IIID, Titan IIIE, Titan 34D, and Titan IV necessitated a substantially stronger airframe section due to the immense thrust of the solid rocket motors. Just as airframe reutilization was maximized, so were the Stage I and II engines. The continued testing of the Titan II ICBM engines during the operational life of the program demonstrated that the engines were aging well and could be refurbished for use, saving considerable money.

The selected missile propellant tanks were purged with hot nitrogen gas and disassembled into the major structural elements; corrosion control was used where necessary; welds were x-rayed and rewelded when needed; and the tanks were pressure tested. The tanks were reassembled into complete Stage I and II airframes and stored at Denver or Vandenberg AFB until needed. Figure E.2 illustrates the components used from the Titan II ICBM and the Titan III family of space launch vehicles in the configuration of the Titan II SLV. Titan III vehicle instrumentation and control systems were adapted for use with Titan II SLV; tank level and pressure sensors were added; a new Stage II instrumentation truss and altitude control system were added; new electrical harnesses were fabricated; and the Stage I and II engines were sent to Aerojet-General Corporation for refurbishment. The engines remained rated at 430,000 pounds of thrust for Stage I (sea level) and 100,000 pounds of thrust for Stage II (vacuum).

The major structural modification of the Titan II ICBM airframe for SLV use was to the top of the Stage II oxidizer dome. As an ICBM, the adapter for mounting the Mark 6 reentry vehicle was welded to the Stage II oxidizer dome. This structural member had to be removed and modified to accept both the payload bus, the mounting structure for the satellites or space probe, and the payload fairing structure. The payload fairing now in use measures 10 feet in diameter and 20 to 30 feet in height depending on payload requirements.

Not only the engines and airframe were salvaged for use in the Titan II SLV program. In 1978 the original AC Spark Plug guidance system had been replaced by a more modern and easily maintained Universal Space Guidance System adapted from the Titan III program. The 55 guidance packages were shipped to Delco Electronics where they were inspected, and 16 were selected for modification to provide both boost phase guidance control and payload insertion guidance capability. The main change in the guidance system was to restore the rate of inertial guidance system platform rotation from the one-quarter revolution per minute used for the ICBMs back to the one revolution per minute used in the Titan III Universal Space Guidance System application. The missile guidance computer also was modified to provide telemetry signal capability for use during launch and tracking.[4]

The first launch of a Titan II SLV took place on 5 September 1988 from Space Launch Complex 4 West at Vandenberg AFB. Titan II SLV 23G-1 lofted a classified payload into low Earth orbit. The 23G-1 airframe was originally Stage I of missile B-56 and Stage II of missile B-98. Missile B-56 was originally shipped to Davis-Monthan AFB in October 1963 and stayed on strategic alert for 15 years without a propellant download for Stage I. Missile B-98 was shipped to Davis-Monthan AFB in December 1965. Its last propellant download prior to

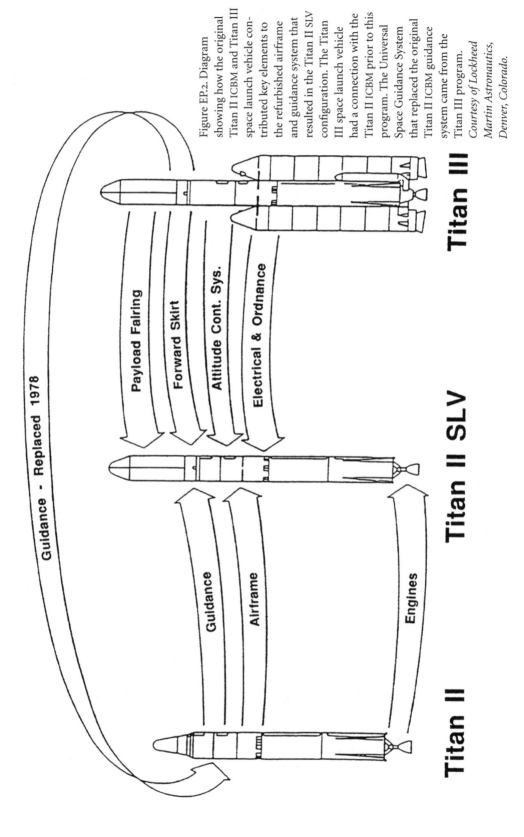

Figure EP.2. Diagram showing how the original Titan II ICBM and Titan III space launch vehicle contributed key elements to the refurbished airframe and guidance system that resulted in the Titan II SLV configuration. The Titan III space launch vehicle had a connection with the Titan II ICBM prior to this program. The Universal Space Guidance System that replaced the original Titan II ICBM guidance system came from the Titan III program.
Courtesy of Lockheed Martin Astronautics, Denver, Colorado.

Titan II

Titan II SLV

Titan III

Guidance - Replaced 1978

Payload Fairing

Forward Skirt

Attitude Cont. Sys.

Electrical & Ordnance

Guidance

Airframe

Engines

Figure EP.3. The first Titan II SLV, 23G-1, is readied for launch of a classified payload on 5 September 1988. *Courtesy of Lockheed Martin Astronautics, Denver, Colorado.*

Figure EP.4. Aerial view of 39 Titan II ICBM Stage I and II airframes in storage at the Aerospace Maintenance and Restoration Center, Davis-Monthan Air Force Base, Arizona. Several of the Stage I airframes look shorter than the rest; this is because the Stage I engines had been removed on these missiles prior to storage. *Courtesy of Ken Sutherland.*

TABLE E.1
TITAN II SLV AIRFRAME DESIGNATIONS AND LAUNCH DATES, 1988–99

STAGE I, STAGE II ICBM	SLV DESIGNATOR	LAUNCH DATE	PAYLOAD
B-56, B-98	23G-1	5 Sep 1988	classified
B-99, B-75	23G-2	5 Sep 1989	classified
B-102 complete	23G-3	25 Apr 1992	classified
B-105 complete	23G-4	not yet launched	
B-65 complete	23G-5	5 Oct 1993	LandSat
B-106 complete	23G-6	4 Apr 1997	Defense Meterological Satellite Program S-14
B-75, B-99	23G-7	19 Jun 1999	QuikSat
B-44, B-94	23G-8	not yet launched	Defense Meterological Satellite Program
B-107 complete	23G-9	not yet launched	
Stage I B-108 fuel tank, B-80 oxidizer; Stage II B-108	23G-10	not yet launched	
B-67, B-89	23G-11	25 Jan 1994	Deep Space Program Clementine Lunar Probe
Stage I B-80 fuel, B-72 oxidizer; Stage II B-84	23G-12	13 May 1998	NOAA-K
B-39, B96	23G-13	not yet launched	
Stage I B-72 fuel, B-92 oxidizer; Stage II B-71	23G-14	not yet launched	

deactivation was in August 1974. Table E.1 lists the airframe sources for the Titan II SLV assemblies. During eight launches from 5 September 1988 to 19 June 1999, the Titan II SLV program has achieved a 100 percent success rate.[5]

Titan Missile Museum National Historic Landmark, Sahuarita, Arizona

Often when a weapons system is retired, representative examples can be seen on pedestals in front of an air base or parked on the grounds of a museum. Rarely, perhaps with ships as an exception, can an example of a weapons system be perserved in operational context. Fortunately one of those rare examples is the Titan Missile Museum National Historic Landmark, composed of the grounds and facilities of Titan II ICBM Launch Complex 571-7. Launch Complex 571-7 was once part of the 571st Strategic Missile Squadron, 390th Strategic Missile Wing, Davis-Monthan AFB, Arizona. This museum is the nation's only surviving operational base Titan II launch complex open on a regular basis for visitation by the public.

The process that resulted in the creation of the Titan Missile Museum National Historic Landmark began on 14 February 1983. Col Paul C. Comeaux, commander of the 390th SMW, contacted Charles T. Niblett, president of the Tucson Air Museum Foundation of Pima County, concerning the foundation's interest in creating a museum out of one of the soon-to-be-deactivated Titan II missile launch complexes. Between the time of this initial contact and the next board meeting when the decision might be made, Niblett contacted Col Hugh Matheson, USAF (Ret.), who had been deputy commander for maintenance at the 390th SMW and LtCol Orville Doughty, USAF (Ret.), who had been commander, 390th Missile Maintenance Squadron, in the 1970s. Many questions had to be answered, both within the Air Force and in the Tucson community, and these three men took on this enormous task. Little did they know that it would take nearly two years and many, many meetings before their goal was reached.

From September 1985 to 21 May 1986 the actual physical work to convert Launch Complex 571-7 took place as well as developing a tour program and recruiting volunteers as tour guides. Matheson and Doughty were instrumental in making this dream a reality. A tour program had to be created, volunteers recruited, and the missile displayed on the surface for satellite observation to confirm it was no longer operational (holes had been cut in each propellant tank and the Mark 6 reentry vehicle had a hole cut in the heat shield). The newly formed Air Force Association, Chapter 106, Green Valley, Arizona, stepped forward to provide an initial cadre of volunteers. Frank Nugent, Frank Knepper, and John Stephenson spear headed this effort. Before the 390th SMW deactivation, officers and enlisted personnel had helped restore the silo and launch control center to its operational configuration, minus a few pieces of equipment that had been sent to the two remaining Titan II wings.

On 8 May 1986 BrigGen John Soper, commander, Strategic Air Command's 47th Air Division, Fairchild AFB, Washington, representing the Air Force, turned over Launch Complex 571-7 to Pima County Supervisor Reg Morrison. Morrison, in turn, gave the ceremonial key to Niblett, representing the Tucson Air Museum Foundation of Pima County.

Figure EP.5. View of the launch duct observation window at Launch Complex 571-7, the Titan Missile Museum National Historic Landmark, Sahuarita, Arizona. The exhaust ducts have a covering of concrete to facilitate access to the viewing area. *Courtesy of the Titan Missile Museum National Historic Landmark Archives, Sahuarita, Arizona.*

Other distinguished guests that had been instrumental in the development of the museum were James Greenwood, vice president of the Tucson Air Museum Foundation of Pima County; BrigGen Lester Brown, 836th Air Division commander; Col John Chambers, the last 390th SMW commander; Ned Robinson, executive director, Pima Air Museum; Col Hugh Matheson, USAF (Ret.); and LtCol Orville Doughty, USAF (Ret.).[6] The Titan Missile Museum was opened for visitors on 21 May 1986 and on 4 April 1994 was designated as a National Historic Landmark. To date the museum has seen over 700,000 visitors from all over the globe.

After a 5-minute briefing, visitors are given a 55-minute guided tour of the above-ground and below-ground features of the site. Included above ground are the Stage I and II engines, a Mark 6 reentry vehicle, fuel and oxidizer conditioning trailers, and a view down into the launch duct. Below ground the visitors tour the blast lock area, Level 2 of the launch control center where a simulated countdown is enacted, and walk down the cableways to Level 2 of the silo where two viewing windows have been cut into the launch duct to permit inspection of Titan II missile N-10. Out of sight from the museum visitors are the remaining levels of the silo equipment area where memorabilia are stored for future

Figure EP.6. Titan II ICBM N-10 (61-2733) is installed in the launch duct at Launch Complex 571-7. Since N-10 was used as a training missile at Sheppard AFB, Texas, no propellants were loaded, and it is safe to display. Phototgraph by author. *Courtesy of the Titan Missile Museum National Historic Landmark Archives, Sahuarita, Arizona.*

Figure EP.7. Both the Stage I and Stage II engine sets are on the surface of the Titan Missile Museum for easy access by visitors. *Courtesy of the Titan Missile Museum National Historic Landmark Archives, Sahuarita, Arizona.*

display. A complete technical order library as well as original construction documents and drawings are also preserved below ground.

The Titan II ICBM program is in a sense a reflection of the best and the worst of what mankind can achieve. The best in that incredibly complex equipment can be designed, tested, and placed in successful operation, protecting our country during a time of considerable peril. The worst in that all the money spent on such ideas is seen by many to have been wasted since we never had to launch an ICBM in anger. Why not just see this story for what it really is? For 24 years, a small but highly important part of our nation's defense literally rested on the shoulders of Titans . . . and did so successfully.

APPENDIX I

TITAN I LAUNCH RECORD; TITAN II AIR FRAME SERIAL NUMBERS, FATE; LAUNCH CREW LIST

TABLE AI.1
TITAN I PROGRAM FLIGHT RECORD SUMMARY

Key:
Lot A—booster flight, dummy second stage
Lot B—two-stage separation, second-stage ignition
Lot C—two-stage performance over limited range
Lot G—two-stage performance over extended range
Lot J—operational prototype—Titan I
Lot M—test bed for Titan II inertial guidance system
Lot V—assembled for OSTF/SLTF and translation rocket tests
Lot SM—operational configuration—Titan I

DASO—demonstration and shakedown operation
OSTF—operational suitability test facility
SLTF—silo launch test facility

All flights were from Cape Canaveral, Florida, except as noted by VAFB (Vandenberg AFB, California). Pickle-barrel refers to the launch being used in determining the impact accuracy of the warhead.

NO.	MISSILE	DATE	OUTCOME	REMARKS
1.	A-3	6 Feb 59	Successful	Stage I operation only, the second stage was filled with water and not equipped with an engine. All objectives were met, structural integrity was demonstrated and Titan I became the first missile program to have a successful flight on its first launch.
2.	A-5	25 Feb 59	Successful	Dummy Stage II
3.	A-4	4 Apr 59	Successful	Dummy Stage II
4.	A-6	4 May 59	Successful	Stage I–II separation test was completed successfully with a water-filled second stage without an engine.
5.	B-5	14 Aug 59	Failure	Premature lift off, automatic destruct
6.	C-3	11 Dec 59	Failure	Accidentally destroyed on pad by destruct system
7.	B-7A	2 Feb 60	Successful	First attempt at complete staging at high altitude was successful. Guidance system was fully operational on this medium-range flight.
8.	C-4	5 Feb 60	Partial Success	Nose cone fairing fell away due to structural failure 50 sec into flight.
9.	G-4	24 Feb 60	Successful	First attempt to separate the Mark IV nose cone was successful. This was the first long-range flight, reaching nearly 5,000 nautical miles.

NO.	MISSILE	DATE	OUTCOME	REMARKS
10.	C-1	8 Mar 60	Successful	
11.	G-5	22 Mar 60	Successful	Capsule recovered
12.	C-5	8 Apr 60	Successful	
13.	G-6	21 Apr 60	Successful	Capsule recovered
14.	C-6	28 Apr 60	Successful	Final limited-range shot
15.	G-7	13 May 60	Successful	Capsule recovered
16.	G-9	27 May 60	Successful	Capsule not recovered due to high seas
17.	G-10	24 Jun 60	Successful	Capsule recovered
18.	J-2	1 Jul 60	Failure	Lost hydraulic power in Stage I, destroyed 11 sec into flight
19.	J-4	28 Jul 60	Partial Success	Stage I premature shutdown
20.	J-7	10 Aug 60	Successful	Capsule not recovered, 5,000 nm flight
21.	J-5	30 Aug 60	Successful	Capsule not recovered, 5,000 nm flight
22.	J-8	28 Sep 60	Successful	Capsule recovered
23.	G-8	28 Sep 60	Successful	6,000 nm flight
24.	J-3	7 Oct 60	Successful	Capsule recovered
25.	J-6	24 Oct 60	Successful	Capsule recovered, 6,100 nm
25A.	V-2	3 Dec 60	No test	VAFB OSTF destroyed during dual propellant loading operation
26.	J-9	20 Dec 60	Partial Success	No ignition Stage II
27.	J-10	20 Jan 61	Partial Success	No ignition Stage II
28.	J-11	9 Feb 61	Successful	5,000 nm flight
29.	J-13	9 Feb 61	Successful	5,000 nm flight
30.	J-12	2 Mar 61	Partial Success	Premature shutdown Stage II
31.	J-14	28 Mar 61	Successful	5,000 nm flight
32.	J-15	31 Mar 61	Partial Success	Premature shutdown Stage II
33.	VS-1	3 May 61	Successful	VAFB SLTF launch
34.	J-16	23 May 61	Successful	5,000 nm flight
35.	M-1	23 Jun 61	Partial Success	Premature shutdown Stage II, inertial guidance system worked well
36.	J-18	20 Jun 61	Successful	5,000 nm flight
37.	M-2	25 Jul 61	Successful	5,000 nm flight
38.	J-19	3 Aug 61	Successful	5,000 nm flight
39.	J-17	5 Sep 61	Successful	6,100 nm, capsule recovered
40.	M-3	7 Sep 61	Successful	4,500 nm flight
41.	SM-2	23 Sep 61	Successful	5,300 nm, launched from VAFB, 395A-1
42.	J-2D	28 Sep 61	Successful	4,500 nm flight
43.	M-4	6 Oct 61	Successful	5,000 nm flight
44.	J-21	24 Oct 61	Successful	6,100 nm flight
45.	J-22	22 Nov 61	Successful	6,000 nm flight
46.	M-5	29 Nov 61	Successful	5,000 nm flight
47.	J-23	13 Dec 61	Successful	5,000 nm flight
48.	M-6	15 Dec 61	Successful	5,000 nm flight
49.	SM-4	20 Jan 62	Partial success	No Stage II ignition, launched from VAFB, 395A-3
50.	M-7	29 Jan 62	Successful	5,000 nm
51.	SM-18	23 Feb 62	Partial success	No Stage II ignition, launched from VAFB, 395A-1
52.	SM-34	4 May 62	Successful	Guidance tape error, VAFB, 395A—1

NO.	MISSILE	DATE	OUTCOME	REMARKS
53.	SM-35	6 Oct 62	Successful	"Pickle-Barrel," launched from VAFB, 395A-1
54.	SM-11	5 Dec 62	Successful	Launched from VAFB, 395A-1
55.	SM-8	29 Jan 63	Successful	"Pickle-Barrel," launched from VAFB 395A-1
56.	SM-3	30 Mar 63	Successful	"Pickle-Barrel," launched from VAFB 395A-2, SAC-DASO
57.	V-1	4 Apr 63	Successful	"Pickel-Barrel," launched from VAFB 395A-1
58.	SM-1	13 Apr 63	Successful	"Pickel-Barrel," launched from VAFB 395A-3, SAC-DASO
59.	V-4	1 May 63	Failure	5 seconds of flight, destructed on pad, launched from VAFB, 395A-1
60.	SM-24	16 Jul 63	Partial success	No Stage II ignition, launched from VAFB, 395A-2
61.	SM-7	15 Aug 63	Successful	"Pickle-Barrel," launched from VAFB, 395A-1
62.	SM-56	30 Aug 63	Partial success	gas generator shutdown, launched from VAFB, 395A-3
63.	SM-83	17 Sep 63	Successful	"Pickel-Barrel," launched from VAFB, 395A-2, SAC
64.	SM-68	14 Nov 63	Successful	"Pickel-Barrel," launched from VAFB, 395A-1, SAC
65.	SM-85	8 Dec 64	Partial success	Premature shutdown Stage I, launched from VAFB, 395A-1, SAC
66.	SM-33	14 Jan 65	Partial success	No Stage II ignition, launched from VAFB, 395A-3, SAC
67.	SM-80	5 Mar 65	Partial success	Propellant depletion, launched from VAFB, 395A-2, SAC

Source: Adapted from the Titan Ballistic Missile Development Plan, 30 April 1960, Air Force Historical Research Agency, Maxwell AFB, Alabama.

TABLE AI.2
TITAN II RESEARCH AND DEVELOPMENT MISSILE SERIAL NUMBERS AND SHIP DATES

Contract AF04(647-576)

MARTIN NUMBER	AF SERIAL NUMBER	DATE SHIPPED
N-1	60-6808	18 Mar 62
N-2	60-6809	27 Jan 62
N-3	60-6810	26 May 62
N-4	60-6811	10 May 62
N-5	60-6812	3 Aug 62
N-6	61-2729	16 Jun 62
N-7	61-2730	21 Aug 62
N-8	61-2731	27 Jan 63
N-9	61-2732	22 Aug 62
N-10	61-2733	2 Apr 62
N-11	61-2734	31 Oct 62
N-12	61-2735	2 Oct 62
N-13	61-2736	8 Nov 62
N-14	61-2737	23 Feb 63
N-15	61-2738	5 Dec 62
N-16	61-2739	18 Dec 62
N-17	61-2740	20 Jan 63
N-18	61-2741	5 Mar 63
N-19	61-2742	12 Feb 63
N-20	61-2743	7 May 63
N-21	61-2744	27 Mar 63
N-22	61-2745	26 Apr 63
N-23	61-2746	14 May 63
N-24	61-2747	28 May 63
N-25	61-2748	22 Aug 63
N-26	61-2749	17 Aug 63
N-27	61-2750	27 Aug 63
N-28	61-2751	10 Sep 63
N-29	61-2752	7 Nov 63
N-30	61-2753	18 Jan 64
N-31	61-2754	21 Dec 63
N-32	62-1867	17 Jan 64
N-33	62-1868	5 Feb 64
N-34	62-1869	not built
N-35	62-1870	not built
N-36	62-1871	not built

Source: Adapted from Titan Master Schedule, Ballistic Systems Division, 3 Jul 63, Section II, pages 2–15. Air Force Historical Research Agency, Maxwell AFB, Alabama.

TITAN II OPERATIONAL MISSILE AIRFRAME SERIAL NUMBERS AND SHIP DATES

Contract AF04(694)-213

MARTIN NUMBER	AF SERIAL NUMBER	DATE SHIPPED	MARTIN NUMBER	AF SERIAL NUMBER	DATE SHIPPED
SM68B-1	61-2755	16 Oct 62	SM68B-36	62-017	6 Jun 63
SM68B-2	61-2756	25 Jul 62	SM68B-37	62-018	17 Sep 63
SM68B-3	61-2757	22 Dec 62	SM68B-38	62-019	8 Jun 63
SM68B-4	61-2758	28 Oct 62	SM68B-39	62-020	2 Jul 63
SM68B-5	61-2759	27 Nov 62	SM68B-40	62-021	5 Jul 63
SM68B-6	61-2760	2 Jan 63	SM68B-41	62-022	22 Sep 63
SM68B-7	61-2761	26 Jan 63	SM68B-42	62-023	21 Aug 63
SM68B-8	61-2762	6 Feb 63	SM68B-43	62-024	27 Sep 63
SM68B-9	61-2763	12 Apr 63	SM68B-44	62-025	20 Sep 63
SM68B-10	61-2764	24 May 63	SM68B-45	62-026	6 Oct 63
SM68B-11	61-2765	29 May 63	SM68B-46	62-027	21 Sep 63
SM68B-12	61-2766	4 Oct 63	SM68B-47	62-028	26 Sep 63
SM68B-13	61-2767	4 Jan 63	SM68B-48	62-029	5 Oct 63
SM68B-14	61-2768	9 Jan 63	B-49	62-12292	7 Oct 63
SM68B-15	61-2769	23 Apr 63	B-50	62-12293	11 Oct 63
SM68B-16	61-2770	7 Aug 63	B-51	62-12294	22 Oct 63
SM68B-17	61-2771	30 Sep 63	B-52	62-12295	16 Oct 63
SM68B-18	61-2772	25 Sep 63	B-53	62-12296	13 Oct 63
SM68B-19	61-2773	15 Feb 63	B-54	62-12297	19 Oct 63
SM68B-20	61-2774	22 Feb 63	B-55	62-12298	27 Oct 63
SM68B-21	62-002	1 Mar 63	B-56	62-12299	24 Oct 63
SM68B-22	62-003	9 Mar 63	B-57	62-12300	31 Oct 63
SM68B-23	62-004	11 Mar 63	B-58	62-12301	4 Nov 63
SM68B-24	62-005	25 Mar 63	B-59	63-7714	14 Nov 63
SM68B-25	62-006	28 Mar 63	B-60	63-7715	28 Oct 63
SM68B-26	62-007	5 Apr 63	B-61	63-7716	31 Oct 63
SM68B-27	62-008	25 May 63	B-62	63-7717	6 Nov 63
SM68B-28	62-009	13 Apr 63	B-63	63-7718	15 Nov 63
SM68B-29	62-010	24 Apr 63	B-64	63-7719	22 Nov 63
SM68B-30	62-011	23 Apr 63	B-65	63-7720	22 Dec 63
SM68B-31	62-012	6 May 63	B-66	63-7721	27 Jan 64
SM68B-32	62-013	9 May 63	B-67	63-7722	15 Jan 64
SM68B-33	62-014	13 May 63	B-68	63-7723	30 Jan 64
SM68B-34	62-015	27 May 63	B-69	63-7724	22 May 64
SM68B-35	62-016	1 Jun 63	B-70	63-7725	24 Feb 64

TITAN II OPERATIONAL MISSILE AIRFRAME SERIAL NUMBERS AND SHIP DATES

Contract AF04(694)-213

MARTIN NUMBER	AF SERIAL NUMBER	DATE SHIPPED
B-71	63-7726	30 Jun 64
B-72	63-7727	27 Jun 64
B-73	63-7728	17 Jul 64
B-74	63-7729	24 Jul 64
B-75	63-7730	10 Aug 64
B-76	63-7731	23 May 64
B-77	63-7732	22 May 64
B-78	63-7733	13 Jul 64
B-79	63-7734	5 Dec 64
B-80	63-7735	28 Jun 64
B-81	63-7736	12 Jul 64
B-82	63-7737	14 Jul 64
B-83	63-7738	31 Jul 64
B-84	63-7739	17 Aug 64
B-85	64-449	3 Mar 65
B-86	64-450	1 Dec 64
B-87	64-451	22 Dec 64
B-88	64-452	19 Jan 65
B-89	64-453	17 Feb 65
B-90	64-454	2 Apr 65
B-91	64-455	3 May 65
B-92	64-456	6 May 65
B-93	64-457	12 May 65
B-94	64-458	24 May 65
B-95	64-459	6 Jul 65
B-96	64-18021	19 Jul 65
B-97	65-10641	24 Aug 65
B-98	65-10642	19 Oct 65
B-99	65-10643	14 Dec 65
B-100	65-10644	29 Jan 66
B-101	65-10645	18 Apr 66
B-102	65-10646	6 Jun 66

Contract AF04(647)-785

MARTIN NUMBER	AF SERIAL NUMBER	DATE SHIPPED
B-103	66-4314	29 Aug 66
B-104	66-4315	27 Oct 66
B-105	66-4316	19 Oct 66
B-106	66-4317	6 Feb 67
B-107	66-4318	18 Apr 67
B-108	66-4319	13 Jun 67

Source: Titan II Operational Missiles, undated operational directive, Martin Marietta Company; Titan Master Schedule, Ballistics System Division, 3 Jul 1963, Section II, pages 2-15. Air Force Historical Research Agency, Maxwell AFB, Alabama.

TABLE AI.4
FATE OF TITAN II MISSILE AIRFRAMES

RESEARCH AND DEVELOPMENT

Number built	**33**
Number flown at Cape Canaveral	23
Number flown at Vandenberg AFB	9
N-10 sent to Sheppard AFB as training missile, presently at Titan Missile Museum, Sahuarita, Arizona	1

OPERATIONAL

Number built	**108**
Number flown from Vandenberg AFB, California	49[a]
Number destroyed in accidents	2
Aircraft Maintenance and Restoration Center, Davis-Monthan AFB, Arizona	39.5[b]
Number selected for refurbishment as Titan II Space Launch Vehicles	14.5[c]
Air Force Museum, Wright-Patterson AFB, Ohio, B-5 (61-2759)	1
National Atomic Museum, Kirtland AFB, New Mexico	1
Huntsville Space Center, Alabama, B-2 (61-2756)	1
TOTAL	141

Note: 141 Titan II airframes were manufactured for the research and development and operational ICBM programs.

[a] *Forty-eight missiles launched in operational training, one launched as a qualification flight during the R&D launch series.*
[b] *The AMARC facility stores 34 assembled airframes and 11 Stage I or Stage II airframes for a total of 39.5 "missiles."*
[c] *Fourteen assembled airframes and one Stage II airframe for a total of 14.5 "missiles."*

TABLE AI.5
TITAN II STRATEGIC AIR COMMAND LAUNCH CREWS

LAUNCH	DATE	SITE	MISSILE	UNIT	S/F/A	IE (NM)	CREW
1	30 Jul 64	395-D	B-28	395th SMS	S	classified	MCCC: Capt Harry K. Hamilton DMCCC: 1Lt Candido J. Corrada BMAT: TSgt David K. Taczek MFT: SSgt Benjamin A. Williams
2	11 Aug 64	395-C	B-9	395th SMS	S	classified	MCCC: Maj Edward Supe DMCCC: Capt Russell A. Cecala BMAT: TSgt James R. Hicks MFT: SSgt Robert E. Oleachea
3	13 Aug 64	395-B	B-7	395th SMS	S	classified	MCCC: Capt Gary E. Marsh DMCCC: 2Lt Steven P. Walker BMAT: SSgt Gerald R. Andrews MFT: SSgt Charles H. Moore
4	2 Oct 64	395-C	B-1	308th SMW	S	< 1	MCCC: Maj Kenneth C. Wine DMCCC: Capt Ted Brown BMAT: MSgt Herbert A. Hancock MFT: SSgt Frank P. Sledge
5	4 Nov 64	395-D	B-32	381st SMW	S	.974	MCCC: Capt Thomas A. Grant DMCCC: Capt William Reinken BMAT: A1C J. G. Montgomery MFT: MSgt Irwin L. Glore
6	25 Mar 65	395-B	B-60	390th SMW	S	2.66	MCCC: Maj D. Charles Jones DMCCC: Capt William R. Dorn BMAT: SSgt Donald J. Horton MFT: A1C Tony M. Walters
7	16 Apr 65	395-C	B-45	390th SMW	S	0.97	MCCC: Maj Kenneth W. Breeding DMCCC: 1Lt George J. Casparro BMAT: MSgt Jerome L. Hood MFT: A1C Bruce M. Bedford
8	30 Apr 65	395-D	B-54	390th SMW	F		MCCC: Maj Norton M. Hewitt DMCCC: 1Lt Joel J. Lefave BMAT: MSgt Miguel L. Dezarraga MFT: A1C Ronald V. Szabat
9	21 May 65	395-B	B-51	381st SMW	S	0.567	MCCC: Capt Robert E. Harrington DMCCC: 1Lt Francis G. Ortiz BMAT: SSgt Joseph S. Avena MFT: SSgt William L. Lowe
10	14 Jun 65	395-C	B-22	381st SMW	F		MCCC: Maj Henry A. Curtin DMCCC: Capt Robert S. Luckett BMAT: SSgt Eugene R. Myer MFT: SSgt Bill J. Lamb
11	30 Jun 65	395-D	B-30	381st SMW	S	0.164	MCCC: Maj Phil R. Young DMCCC: 1Lt W. Alan Perrill BMAT: SSgt Allen T. Brown MFT: A1C James D. Blom

LAUNCH	DATE	SITE	MISSILE	UNIT	S/F/A	IE (NM)	CREW
12	21 Jul 65	395-B	B-62	390th SMW	S	1.14	MCCC: Maj Wendell O. Scott DMCCC: Capt Cal D. Payne BMAT: MSgt William I. Kemp MFT: SSgt Henry Moseley
13	16 Aug 65	395-C	B-6	308th SMW	S	0.324	MCCC: Maj Ronald C. Feavel DMCCC: 1Lt Willam A. Doorley BMAT: MSgt James E. Busby MFT: SSgt Sidney F. Koerlin
14	26 Aug 65	395-D	B-19	308th SMW	S	0.523	MCCC: Capt Richard D. Castleton DMCCC: 1Lt Richard W. Kalishek BMAT: SSgt Jerome E. Patosnak MFT: A1C Doyle V. Smith
15	21 Sep 65	395-B	B-58	381st SMW	F		MCCC: Capt Thomas E. Macomber DMCCC: 1Lt Robert W. Bloodworth Jr. BMAT: SSgt Lester T. Pratt MFT: A1C Jackson G. Shelton III
16	20 Oct 65	395-C	B-33	308th SMW	S	0.582	MCCC: Maj William J. McGee DMCCC: Capt Billy R. Mantooth BMAT: SSgt Ted L. Cook MFT: A1C Henry K. Andrews
17	27 Nov 65	395-D	B-20	308th SMW	S	0.666	MCCC: Maj Alan B. Myler DMCCC: Capt Charles E. Roberson BMAT: TSgt Gener R. Myers MFT: TSgt Jeneral L. Doss
18	30 Nov 65	395-B	B-4	381st SMW	F		MCCC: Capt Milo L. Carlson DMCCC: 2Lt James G. Ellis BMAT: TSgt James A. Collins MFT: A1C Gerald A. Humes
19	22 Dec 65	395-C	B-73	308th SMW	F		MCCC: Maj John Radizietta DMCCC: 1Lt Thomas D. Weaver BMAT: SSgt Richard T. Frenier MFT: A1C Samuel Cirelli
20	3 Feb 66	395-D	B-87	390th SMW	S	1.2	MCCC: Capt James H. Admas DMCCC: 1Lt William S. Powell BMAT: TSgt Maurice E. Edler MFT: A1C Bruce M. Bedford
21	17 Feb 66	395-B	B-61	381st SMW	S	0.940	MCCC: Capt Robert H. Ballinger DMCCC: 2Lt Donald L. McMinds BMAT: MSgt Lavern. A. MacRunnels MFT: A1C Fred S. Hill

LAUNCH	DATE	SITE	MISSILE	UNIT	S/F/A	IE (NM)	CREW
22	25 Mar 66	395-C	B-16	390th SMW	S	0.820	MCCC: LtCol Frank R. Davis DMCCC: Capt William B. Sandmann BMAT: SSgt Arnold B. Evenson MFT: A1C George C. Faulkner Jr.
23	5 Apr 66	395-D	B-50	308th SMW	S	0.358	MCCC: Capt Randolf J. Scheel DMCCC: 1Lt Richard S. Bellas BMAT: SSgt Paul Love MFT: SSgt James L. Tort
24	20 Apr 66	395-B	B-55	381st SMW	S	0.362	MCCC: Capt Robert A. Marshall DMCCC: 1Lt Daniel L. Drew BMAT: TSgt Robert Claycomb MFT: A1C Joe N. Sirratt
25	24 May 66	395-C	B-91	381st SMW	F		MCCC: Capt Marion C. Fasler DMCCC: 1Lt James L. Stapp BMAT: SSgt David Struthers MFT: TSgt Thomas Bloominger
26	22 Jul 66	395-B	B-95	308th SMW	S	classified	MCCC: Capt John Womack Jr. DMCCC: 1Lt James W. Harshbarger BMAT: SMSgt Walter Kundis MFT: MSgt James W. Meddress Jr.
27	16 Sep 66	395-C	B-40	390th SMW	S	classified	MCCC: Capt Charles W. Fowler DMCCC: 1Lt Lester L. Walker BMAT: SSgt Douglas R. Trular MFT: A2C Carl P. Depillo
28	24 Nov 66	395-B	B-68	390th SMW	S	classified	MCCC: Capt Charles R. McQuitty DMCCC: 1Lt James H. Hulse BMAT: SSgt Wilbur J. Walston MFT: A2C Charles Niemchich
29	17 Mar 67	395-C	B-76	381st SMW	S	0.594	MCCC: Capt Thaddeus E. Hughes DMCCC: 2Lt James B. Ward BMAT: TSgt Norman Soloman MFT: A1C Edwin L. Humbarger
30	12 Apr 67	395-B	B-81	308th SMW	F		MCCC: Capt Jacob L. Van Pelt DMCCC: 2Lt James N. Posey BMAT: MSgt William C. Tack Jr. MFT: TSgt Hubert L. Armstrong
31	23 Jun 67	395-B	B-70	381st SMW	S	classified	MCCC: Maj Norbert E. Scheitler DMCCC: 1Lt James A. Trammel BMAT: TSgt Phillip J. Marino MFT: TSgt Bulter J. Kincaid
32	11 Sep 67	395-B	B-21	381st SMW	S	classified	MCCC: Capt Harry C. Williams Jr. DMCCC: 1Lt Hohn H. Bunch Jr. BMAT: SMSgt Rober C. Larsen MFT: SMSgt Richard Herring

LAUNCH	DATE	SITE	MISSILE	UNIT	S/F/A	IE (NM)	CREW
33	30 Nov 67	395-B	B-69	308th SMW	A		MCCC: Maj Douglas C. Cameron DMCCC: 1Lt James C. Swayze BMAT: SSgt Lyle R. Groth MFT: TSgt Robert L. Turner
34	28 Feb 68	395-B	B-88	381st SMW	S	classified	MCCC: Capt Charles E. Dean DMCCC: 1Lt Gary L. George BMAT: TSgt Murley J. Juneau Jr. MFT: SSSgt Gary L. Hartman
35	2 Apr 68	395-C	B-36	395th SMS	S	classified	MCCC: unavailable DMCCC: " BMAT: " MFT: "
36	12 Jun 68	395-C	B-82	390th SMW	S	classified	MCCC: Capt Robert H. Smith DMCCC: 1Lt Edward R. Drews BMAT: SSgt Ronald J. Almasy MFT: SSgt Alfred Ravgiala Jr.
37	21 Aug 68	395-C	B-53	381st SMW	S	1.77	MCCC: Capt Weldon C. Thompson DMCCC: 1Lt Edward R. Lipinski BMAT: TSgt James D. Buck MFT: SSgt Robert D. Richard
38	19 Nov 68	395-C	B-3	390th SMW	S	classified	MCCC: 1Lt Schrade F. Radtke DMCCC: 2Lt Gregory P. Marsh BMAT: SSgt James G. Taylor MFT: Sgt Maxwell G. Evanchak
39	20 May 69	395-C	B-83	395th SMS	S	classified	MCCC: Capt Ted Suchecki DMCCC: 1Lt Jim Buyck BMAT: TSgt Jerry Mayers MFT: Sgt Leroy Cubiciotti
40	26 May 71	395-C	B-69	390th SMW	A		MCCC: 1Lt Ed Moran DMCCC: 1Lt Bruce J. Stensvad BMAT: SSgt Ray Hersey MFT: Sgt William C. Shaff
41	20 Jun 71	395-C	B-12	390th SMW	S	classified	MCCC: Capt John R. Ransome DMCCC: 1st Lt. Charles W. Schubert Jr. BMAT: TSgt George V. Fetchik MFT: SSgt Johnson Randolph
42	27 Aug 71	395-C	B-100	308th SMW	S	classified	MCCC: Capt Billy Van Horn DMCCC: 1Lt Ronald C. Hoff BMAT: SSgt David R. Palmer MFT: SSgt Bernd H. Stayneslie
43	24 May 72	395-C	B-46	381st SMW	S	classified	MCCC: Capt Joe Simonson DMCCC: 1Lt Ken Carter BMAT: Sgt David Beam MFT: A1C Mark Sarnecki

LAUNCH	DATE	SITE	MISSILE	UNIT	S/F/A	IE (NM)	CREW
44	11 Oct 72	395-C	B-78	390th SMW	S	classified	MCCC: Capt Bruce G. Luna DMCCC: 1Lt Joesph E. Snook BMAT: SSgt Ray Hersey MFT: Sgt George F. Rounds
45	6 Oct 73	395-C	B-69	308th SMW	S	classified	MCCC: Capt William J. Phillips DMCCC: 1Lt Gary D. Smith BMAT: SSgt Roger A. Smiley MFT: SSgt Robert L. Sullivan
46	1 Mar 74	395-C	B-85	381st SMW	S	classified	MCCC: 1Lt Terry F. Cook DMCCC: 1Lt C. M. Weber BMAT: A1C Daniel P. Tullio MFT: Sgt Chuck N. Garrlson
47	20 Jun 74	395-C	B-41	390th SMW			AMCCC: Capt G. Nick Emmanuel DMCCC: 1Lt Hank J. Laughlin BMAT: Sgt R. W. Meisenberg MFT: Sgt T. J. Ostrum
48	10 Jan 75	395-C	B-27	308th SMW	S	classified	MCCC: Capt Michael W. Sayer DMCCC: 1Lt M. D. Wilderman BMAT: Sgt Robert B. Ribertone MFT: A1C David W. Fiyak
49	8 Aug 75	395-C	B-52	381st SMW	S	classified	MCCC: Capt William D. Butler DMCCC: 1Lt Mark L. Rogers BMAT: SSgt Howard E. Dumke MFT: SSgt Robert D. Foote
50	4 Dec 75	395-C	B-41(B-18)	390th SMW	S	classified	MCCC: Capt William J. Harrison Jr. DMCCC: 2Lt Gill S. Paszek BMAT: Sgt Madison Smith III MFT: A1C Tobert R. Johnson
51	27 Jun 76	395-C	B-17	308th SMW	S	0.415	MCCC: Capt Roger B. Graves DMCCC: 1Lt Greogry M. Gillum BMAT: SSgt David W. Boehm MFT: SSgt R. Savage

Note: All launch dates are local time. Two launches, 27 August 1971 and 27 July 1976 are listed in other records as 28 August 1971 and 28 July 1976, which are the Greenwich Mean Time launch dates. All other dates are the same, local or Greenwich; B-41(B-18) indicates that the first stage of the missile was B-18, the second stage was B-41; S/F/A = success/failure/abort; IE=impact error, no formula given.

APPENDIX II
SOVIET UNION COUNTERPART TO TITAN II

What was the Soviet Union's counterpart to Titan II? The Soviet Strategic Missile Forces deployed five storeable propellant ICBMs in numbers for the most part greater than Titan II: SS-7 (197 deployed), SS-9 (313), SS-11(970), SS-17 (150), and SS-18 (284). Two can be considered contemporary to Titan II, the SS-7, SS-8, and SS-9. The threat imposed by the SS-18 and SS-19 were the driving force behind the Reagan administration's Strategic Force Modernization Program. A brief description is given for each system. The name after the numerical designator is the NATO code name.

BEFORE SALT I
SS-7 Saddler

The chronological equivalent to Titan II was the SS-7, initially deployed in 1962 in above-ground coffin-style launchers and then in 1964 in 23 three-silo complexes. Final deployment was reached in 1968 with a total of 197 missiles. Unlike Titan II, the SS-7 could store its hypergolic propellants for only two days due to the corrosive nature of the red fuming nitric acid oxidizer. Presumably the fuel, hydrazine, was stored onboard the missile in both the silo-based and coffin-style launchers. Launch preparation time was estimated to be 30 minutes.

The SS-7 was a two-stage missile with a maximum diameter of 10 feet, a length of 100 feet, and a single warhead with a 3-megaton yield (it has been reported that warheads of between 6 and 25 megatons were also deployed). The SS-7 had a payload capacity of 3,000 to 4,400 pounds. The CEP of the Mod-3 version was 1.32 nautical miles at a range of 5,600 nautical miles.

SS-9 Scarp

The SS-9 is probably most identified as the Soviet equivalent to Titan II in both the missile and basing aspect. The SS-9 was a two-stage missile, 118 feet in length, and 10 feet 2 inches in diameter with inertial guidance. Capable of carrying a 9,000- to 11,000-pound payload, the SS-9 is generally considered to have been initially deployed with a 20-megaton warhead with later versions capable of carrying a 25-megaton warhead. The SS-9 had a range of 6,300 nautical miles and a CEP of 0.8 nautical miles in the Mod-2 version reentry vehicle.[2] Unlike the SS-7 silos, hardened to 28 pounds per square inch, the SS-9 was deployed in single missile silos hardened to 280 pounds per square inch. The red fuming nitric acid oxidizer was replaced with nitrogen tetroxide, enabling propellant storage onboard the missile for approximately six months. Deployment was delayed three years, until 1969, due to the complexity of the new silo design. By the time the SALT I agreement was signed in May 1972, 288 SS-9 missiles were deployed. In 1973 the SS-9 silos began conversion to house the SS-18.[3]

SS-11 Sego

The SS-11 was 62 feet 4 inches long, 8 feet in diameter, with two stages and inertial guidance. The SS-11 initially carried a single reentry vehicle estimated to have a yield of 1 megaton, a CEP of 1.0 to 1.5 nautical miles and a range of 6,500 nautical miles. Deployment had reached 970 missiles in silos by the time of the SALT I Agreement in 1972. By 1972 the SS-11 had been flight tested with three reentry vehicles estimated to be carrying 0.5-megaton warheads. The SS-11 was the first Soviet ICBM to carry multiple reentry vehicles.[4]

AFTER SALT I

After the SALT I Agreement, the Soviets continued to deploy storeable propellant missiles to replace those already in their arsenal. Armed with multiple independently targetable reentry vehicles (MIRVs) with accuracies enabling nearly direct hits on Minuteman and Titan II silos, the SS-17, SS-18, and SS-19 presented a serious threat to all of the United States' ICBM forces. With deployment of the SS-18, the Reagan Strategic Force Modernization Program was an attempt to maintain parity with the growing Soviet threat.

SS-17 Spanker

The SS-17 began flight testing in 1974. Developed as a replacement to the SS-11, the SS-17 was a two-stage missile, 75 feet in length with a maximum diameter of 8 feet 6 inches. The SS-17 had a range of 6,500 nautical miles, was deployed in SS-11 silos, and carried four MIRVs of 1 megaton each with a CEP of 0.17 nautical miles.[5]

SS-18 Satan

Test flights of the SS-18 in 1977–78 indicated that a new and serious threat to the Minuteman II and III silos, as well as Titan II, was ready for deployment. Like the SS-9, the SS-18 was 121 feet 4 inches in length and 10 feet 6 inches in diameter. With a range of 6,500 nautical miles, a payload of one 25- to 50-megaton warhead or up to eight MIRVs of 1 to 2 megatons, and the ability to deploy up to 310 missiles within the SALT I Agreement, the SS-18 was a formidable and worrisome weapon.[6]

NOTES

ABBREVIATIONS

AAS (Azimuth Alignment Set)

AFB (Air Force Base)

AFHRA (Air Force Historical Research Agency)

AFLC (Air Force Logistics Command)

ALC (Air Logistics Center)

APS (accessory power supply)

ARDC (Air Research and Development Command)

BMAT (ballistic missile analyst technician)

BMDTTP (Ballistic Missile Development Test Target Program)

BSD (Ballistic Systems Division)

BVL (Butterfly Valve Lock)

CEP (circular error probable)

CSS (Coded Switch System)

CST (Combined Systems Test)

DASO (Demonstration and Shakedown Operations)

DEV (Daily Entry Verification)

DGZ (Designated Ground Zero)

DMCCC (deputy missile combat crew commander)

DSV (Daily Shift Verification)

FBI (Federal Bureau of Investigation)

FOT (Follow-on Operational Test)

GEMSIP (Gemini Stability Improvement Program)

HQ (Headquarters)

ICBM (intercontinental ballistic missile)

IRBM (intermediate range ballistic missile)

JSTPS (Joint Strategic Target Planning Staff)

LCCFC (Launch Control Complex Facilities Console)

LES (Launch Enable System)

LTRE (Long-Term Readiness Evaluation)

MCCC (missile combat crew commander)

MFT (missile facilities technician)

MIMS (Missile Maintenance Squadron)

MIRV (multiple independently targetable reentry vehicle)

MPHT (Missile Potential Hazard Team)

MSAT (Missile Systems Analyst Technician)

NASA (National Aeronautics and Space Administration)

ORT (Operational Readiness Training)

OSTF (Operational Suitability Test Facility)
OT (Operational Test)
PTS (Propellant Transfer System), 184
QMT (Operations and Maintenance Missile Trainer)
RASP (Reliability and Aging Surveillance Program)
RFHCO (rocket fuel handler's clothing outfit)
RP-1 (Rocket Propellant-1)
RTMN (Radio-Type Maintenance Net)
SAC (Strategic Air Command)
SALT I, II (Strategic Arms Limitation Talks)
SATAF (Site Activation Task Force)
SIOP (Single Integrated Operations Plan)
SLAP (Service Life Analysis Program)
SLTF (Silo Launch Test Facility)
SLV (Space Launch Vehicle)
SMS (Strategic Missile Squadron)
SMW (Strategic Missile Wing)
SOFT (Signature of Fragmented Tanks)
SSTTP (Safeguard System Test Target Program)
USAF (United States Air Force)
USGS (Universal Space Guidance System)
VAFB (Vandenberg Air Force Base)
VDAP (Vapor Detector Annunciator Panel)
WDD (Western Development Division)
WSEG (Weapons Systems Evaluation Group)

CHAPTER I: THE AIR FORCE STRATEGIC MISSILE PROGRAM

1. While technically still the Army Air Force at this time, Air Force is used for the sake of continuity.
2. J. L. Chapman, *Atlas: The Story of a Missile* (New York: Harper and Brothers, 1960), 27.
3. R. E. Martin, "The Atlas and Centaur 'Steel Balloon' Tanks, A Legacy of Karel Bossart," 40th International Astronautical Congress, IAA-89-738 (Málaga, Spain, 1989).
4. J. Neufeld, *Ballistic Missiles in the United States Air Force* (Washington, D.C.: U.S. Government Printing Office, 1989), 104–6; Chapters 2, 3, 4. Neufeld's book has an excellent discussion of the entire process. Also see D. Ball, *Politics and Force Levels: The Strategic Missile Program of the Kennedy Administration* (Berkeley and Los Angeles: University of California Press, 1980).
5. Other committee members were Professors Clark B. Millikan, Charles C. Lauritsen, and Dr. Louis G. Dunn, California Institute of Technology; Dr. Hendrik W. Bode, Bell Telephone Laboratories; Dr. Allen E. Puckett, Hughes Aircraft Company; Dr. George B. Kistiakowsky, Harvard University; Professor J. B. Wiesner; Massachusetts Institute of Technology; Mr. Lawrence A. Hyland, Bendix Aviation Corporation; and Drs. Simon Ramo and Dean Wooldridge, Ramo-Wooldridge Corporation. The military advisor was LtGen Bernard A. Schriever.
6. T. E. Caywood et al., "Guidelines for the Practice of Operational Research," *Operations Research* 19 (September 1971): 1187.
7. Neufeld, 261.
8. Neufeld, 106.

9. Neufeld, 107–8.

10. R. L. Perry, *The Ballistic Missile Decision* (RAND Corporation, 1960), 14. R28 P-2686.

11. The Gaither Committee was officially titled "The Security Resources Panel of the Science Advisory Committee of the Federal Civilian Defense Administration." Appointed by President Eisenhower in April 1957, the committee's report was submitted on 7 November 1957. Its recommendations were largely ignored by Eisenhower and served to fuel the missile gap debate during the 1960 presidential campaign.

12. M. Trachtenberg, ed., *The Development of American Strategic Thought: Basic Documents from the Eisenhower and Kennedy Periods, Including the Basic National Security Policy Papers from 1953–1959* (New York: Garland Publishing, 1988), 531.

13. E. G. Schwiebert, *A History of the U.S. Air Force Ballistic Missiles* (New York: Praeger, 1964), 222.

14. Ball, 44–45.

15. Ibid., 34–35.

16. Ibid., 36.

17. Ibid., 47; Schwiebert, 226.

18. Neufield, 213.

19. Ball, 44–45, 114, 122.

20. The nine Vandenberg Altas sites are not counted as alert missiles since by this time they were training sites and not maintained in a strategic alert status.

21. *SAC Missile Chronology: 1939–1988* (May 1990), 47. Office of the Historian, Headquarters Strategic Air Command, Offutt AFB, Nebraska.

22. Neufeld, 237.

23. Neufeld, 234–36; F. Ruggerio, "Missileers' Heritage," Report 2065-81 (Air Command and Staff College, Air University, Maxwell AFB, Alabama, 1981), 34–70.

24. J. M. Collins, *U.S.-Soviet Military Balance: Concepts and Capabilities, 1960–1980* (Washington, D.C.: McGraw Hill, 1980), 443.

25. L. Freedman, *U.S. Intelligence and the Soviet Strategic Threat* (Boulder, Colo.: Westview Press, 1977), 101.

26. Collins, 443.

CHAPTER II: THE FIRST GENERATION: ATLAS AND TITAN I

1. C. Hansen, *U.S. Nuclear Weapons, The Secret History* (Orion Books, 1988), 61–68; T. B. Cochran et al., *Nuclear Weapons Data Book, Volume II: U.S. Nuclear Warhead Production* (Arlington, Tex.: Ballinger Publishing Company, 1987), 154.

2. Neufeld, 104–6.

3. R. A. Smith and H. M. Minami, "History of the Atlas Engines," in *History of Liquid Rocket Engine Development in the United States 1955–1980*, ed. S. E. Doyle (AAS History Series, San Diego: American Astronautical Society, 1992), Vol. 13, pp. 54–56, 63.

4. R. E. Martin, "The Atlas and Centaur 'Steel Balloon' Tanks, A Legacy of Karel Bossart," 40th International Astronautical Congress Paper IAA-89-738, Málaga, Spain, October 7–13, 1989.

5. F. X. Ruggiero, "Missileers' Heritage, Report 2065-81," pp. 19–23, (Air Command and Staff College, Air University, Maxwell AFB, Alabama, 1981); Convair press release for 20 May 1960 Atlas D launch, courtesy of Richard E. Martin; Hansen, 107.

6. Neufeld, 179.

7. *SAC Missile Chronology 1939–1988*, p. 23, (Office of the Historian, Strategic Air Command, 1980, Offutt Air Force Base, Nebraska).

8. Ruggiero, 19.

9. Ibid., 20–21; Neufeld, 195.

10. Neufeld, 21–22. The nine launchers at Vandenberg AFB are not counted as operational

launchers since they did not routinely house missiles or operational crews. Likewise, Vandenberg AFB is not counted as an operational base.

11. W. E. Greene, *The Development of the SM-68 Titan,* AFSC Historical Publications Series, 62-23-1, Air Force Material Center Collection (Wright-Patterson AFB, Ohio, 1962), 11.

12. B. W. Augenstein, "A Revised Development Program for Ballistic Missiles of Intercontinental Range," Special Memorandum 21 (Santa Monica, Calif.: RAND Corporation, 1954), 1–38.

13. Greene, 11–12.

14. Ibid.

15. Ibid., 12.

16. Ibid., 13–14.

17. Neufeld, 114–16, 120, 129.

18. "Semi-Staff Study on Titan ICBM, 30 July 1960," IRIS # 2055891, Air Force Historical Research Agency (AFHRA), Maxwell AFB, Alabama. The document has no page numbers, the data can be found in the General Program Plan section.

19. Greene, 18.

20. "Semi-Staff Study on Titan ICBM," 18; Greene, vii.

21. H. T. Simmons, "Martin's Titan Project," *Missiles and Rockets*, October 1956, p. 20. W. B Harwood, *Raise Heaven and Earth* (Simon and Schuster, 1993), 300–306.

22. Vernon Selby, personal interview and correspondence, June 1996.

23. Ibid. One area of concern early in the fabrication process for Titan I Stage I was that the longeron structures that served as the point of attachment of the missile to the launch mount were bolted onto the Stage I fuel tank and then sealed. Leaks were common in this area.

24. M. J. Chuclick et al., "History of the Titan Liquid Rocket Engines," in *History of Liquid Rocket Engine Development in the United States 1955–1980.*

25. Greene, 106; L. C. Meland and F. C. Thompson, "History of Titan Liquid Rocket Engines, AIAA-89-2389," (AIAA/ASME/SAE/ASEE 25th Joint Propulsion Conference, Monterey, California, July 10–12, 1989) p.2.

26. Greene, 114. For an excellent book on the development of inertial guidance, the reader is directed to D. MacKenzie, *Inventing Accuracy: An Historical Sociology of Nuclear Missile Guidance* (Cambridge, Mass.: MIT Press).

27. "Semi-Staff . . . ," pp. 1-4-7 to 1-4-15.

28. Greene, 91.

29. Greene, 92. Further information on the Titan I launches can be found in the semi-annual reports of the Air Force Missile Test Center, 45th Space Wing History Office, Patrick AFB, Florida (referred to subsequently as 45SWHO).

30. "Atlantic Missile Range Test Activity Report, August 1959," p. 10, 45SWHO, Patrick AFB, Florida.

31. Jim Greichen, personal interview, June 1996.

32. Greene, 129.

33. Greene, 96–97.

34. "Chronological Titan Launch Summary, Test Conductor Summaries and Titan Data Hold Analysis, Updated 31 March 1989," Personal papers of Don Kundich.

35. Titan-Vandenberg, 9 September 1963. Two-page program summary from Martin Company, Author's Collection.

36. "Incident Report of OSTF Site (Titan I) Accident, 3 December 1960," 12–13, provided courtesy of Fred Epler Collection.

37. Personal interview with Robert Rhodus, June 1996.

38. Personal interview with Augie Chiarenza, October 1996.

39. Personal correspondence and interviews, John Carlson Sr., January–February 1997.

40. Rhodus, June 1996.

41. Personal interview with John Adamoli, June 1996.

42. "Incident Report, OSTF Site (Titan I) Vandenberg AFB, 3 December 1960," page 95.

43. "History of the 1st Missile Division, Strategic Air Command, 1 January to 30 June 1961," K-DIV-0001-HI-Vol. 1, p. 78. AFHRA, Maxwell AFB. This document is classified as SECRET. The information used is unclassified.

44. Ibid., 78.

45. "Chronological Titan Launch Summary," Kundich Collection.

46. Personal correspondence with Barrie Ricketson, one of the Blue Streak design engineers for Dehavilland, July 1999.

47. P. Morton, *Fire across the Desert: Woomera and the Anglo-Australian Joint Project, 1946–1980* (Australian Government Printing Service, 1989), 434–43.

48. Memorandum, MajGen B. A. Schriever, Cmdr, AFBMD, to Colonel Terhune and Colonel Leonard, 17 November 1958, subj.: Silo Launchers for TITAN Program; Volume II Supporting Documents for the Development of SM-68 Titan, W. Greene, 1962. (Air Force Material Center Collection, Wright-Patterson AFB, Ohio).

49. Memorandum, C. P. Benedict, Act. Asst. Sec. Def. to Sec. AF, 17 August 1959, subj.: In-Silo Launch for the Titan; Volume II Supporting Documents for the Development of SM-68 Titan, W. Greene, 1962. (Air Force Material Center Collection, Wright-Patterson AFB, Ohio).

50. Personal correspondence, Knisely (K) Dreher, program manager, Silo Launch Test Facility, Ralph M. Parsons Company, April 1997.

51. Dreher, May 1997.

52. Personal interview with John Adamoli, June 1996.

53. "History of the 1st Missile Division, Strategic Air Command, 1 January–30 June 1960," K-DIV-0001-HI-Vol. 1, pp. 78–109. AFHRA, Maxwell AFB, Alabama. This document is classified as SECRET. The information used is unclassified.

54. "Ballistic Systems Division Management Data System Titan Master Schedule, 31 July 1963, Definitions and Planning Factors," IRIS # 2055849, pp. 1–11. AFHRA, Maxwell AFB, Alabama. This document is classified as SECRET. The information used is unclassified.

55. *SAC Missile Chronology, 1939–1988*" 1990, (Office of the Historian, Headquarters Strategic Air Command, Offutt AFB, Nebraska) pp. 40 and 44.

CHAPTER III: THE SECOND GENERATION: TITAN II

1. Neufeld, 194.

2. Personal correspondence with Robert Bolles, June 1997.

3. J. D. Clark, *Ignition! An Informal History of Liquid Rocket Propellants* (New Brunswick, N.J.: Rutgers University Press, 1972), 40.

4. Ibid., 42.

5. Ibid., 43.

6. Ibid., 57–59.

7. Greene, 68.

8. Ibid., 90.

9. Personal correspondence, BrigGen William E. Leonhard, USAF (Ret.), February 1997.

10. Greene, 86.

11. Greene, 52–53.

12. Greene, 55; personal interview, Gen Bernard Schreiver, USAF (Ret.).

13. Greene, 59–60.

14. The final decisions for the Titan program force structure were made on 28 March 1961 when Secretary of the Air Force E. M. Zuckert indicated that the 1962 budget removed the thirteenth and fourteenth Titan squadrons, leaving 12, six Titan I and six Titan II.

15. "Scientific Advisory Committee Meeting Minute Extracts" pertaining to Titan storable propellants, Western Development Division, 26 Feb 60, 1960 Titan, IRIS # 2054117, AFHRA, Maxwell AFB, Alabama.

16. Greene, 74; "WS 107C, Titan II Weapon System Final Report, January 1965," (TRW Space Technology Laboratories) page 3; History Office, Air Force Space Command, Peterson AFB, Colorado. This document is classified as SECRET. The information used is unclassified. Hereafter referred to as STL Final Report.

17. Greene, 74. "Titan Ballistic Missile Development Plan, 30 April 1960," pp. II-4 to II-7, Headquarters Air Research and Development Command. AFHRA, Maxwell AFB, Alabama. This document is classified as SECRET. The information used is unclassified; STL Final Report, page 3. History Office, Air Force Space Command, Peterson AFB, Colorado. This document is classified as SECRET. The information used is unclassified.

18. W. B. Harwood, *Raise Heaven and Earth* (New York: Simon and Schuster, 1993), 320. Seven additional contracts were signed with Martin Company for the Titan II program: AF03(647)-471, 28 November 1960; AF04(647)-695, AF04(647)-577 and AF04(647)-616, 16 January 1961; AF04(647)-847, 14 April 1961; AF04(694)-108, 29 January 1962 and AF04(694)-123, 30 January 1962. Personal papers of Dr. Fred Shaw, Chief, Reference Division, AFHRA, Maxwell AFB, Alabama.

19. "Semi-Staff Study on Titan ICBM, 30 July 1960," IRIS # 2055891, AFHRA, Maxwell AFB, Alabama, Tab M.

20. Greene, 76.

21. Memo: Conversion of the XLR87-AJ-3 and XLR91-AJ-3 to Storable Propellants, R. C. Stiff Jr., Aerojet, 12 October 1959. Howard Smith Collection.

22. "Liquid Rocket Engines," Aerojet Liquid Rocket Company, March 1975, pp. 8 and 22.

23. Louis D. Wilson, personal conversation and correspondence, September 1996; Wally Dinsmore, personal conversation and correspondence, July 1996.

24. Roy Jones, personal interview, August 1996.

25. Jones, August 1996; Wilson, September 1996.

26. Ken Collins and Norman Laux, personal interviews, August 1996.

27. STL Final Report, 8.

28. Jones, August 1996.

29. Ibid.

30. Rollo Pickford, personal communcations, May 1999.

31. B. C. Hacker and J. M. Grimwood, *On the Shoulders of Titans: A History of Project Gemini* (Scientific and Technical Information Office, National Aeronautics and Space Administration, 1977), 140. Hereafter referred to as OTSOT.

32. Jones, August 1996.

33. OTSOT, 168.

34. Jones, August 1996.

35. Air pressure at 70,000 feet is approximately 5 percent that at sea level; air pressure at 250,000 feet is approximately 0.002 percent that at sea level.

36. Collins, August 1996; Wilson, August 1996.

37. Collins, 1996; Jones, 1996; Dinsmore, 1996; Wilson, 1996.

38. "Models Aided Titan II Silo Shot," *Missiles and Rockets,* 19 June 1961, 27–30.

39. Pickford, May 1998.

40. The scaling factor was 36 for a 1/6th scale model so the Nike-Ajax engines were excellent matches for this program.

41. "Models Aided Titan II Silo Shot," 27–30.

42. R. Loya et al., "Scale-Model Test of Silo-Type Ducted Launcher, Volume V-Scale Tests Simulating Titan II (Phase III)," May 1961, Aerojet-General Corporation, p. 20.

43. A 2-inch thick acoustic liner proved as effective as a 6-inch liner used in the earlier work. This meant that a 12-inch thick liner, rather than a 36-inch liner could be used in the full-scale silo.

44. Loya et al., 3.

45. "Models Aided Titan II Silo Shot," 27–30.

46. W. P. Rader and H. N. McGregor, "Engineering Practices in the Solution of Acoustical Fatigue Problems for the Titan," in *Acoustical Fatigue in Aerospace Structures.* Proceedings of the Second International Conference, Dayton, Ohio, April 29 to May 1, 1964, (Syracuse University Press) pp. 707–19.

47. Ibid., 710.

48. Veron Selby, personal interview, June 1996; "The Evolution of the Titan Rocket-Titan I to Titan II," L. J. Adams, 41st Congress of the International Astronautical Federation, IAA-90-631, 1990; "Titan II Will Get More Range and Payload in Production Line Change," *Missiles and Rockets,* 5 September 1960, 7. This is an excellent description of the fabrication process for the Titan I and II airframes.

49. Selby, June 1996.

50. Personal interview, Dale Thompson, Martin Company structural engineeer, June 1996. "GLV POGO Study, Structural Configuration Changes of Titan II R&D Missiles, N-1 thru N-28)," A. E. Bees, December 1963; Addendum February 1964, pp. 1-10. Lockheed Martin Astronautics Research Library, Document 125961.

51. N. Nieberlein, "Titan II Operation Wrap Up and Its Effect upon the Gemini Launch Vehicle," October 1963, LV-301, pp. 3–6. Lockheed Martin Astronautics Research Library Document Number 125960.

52. Martin Company merged with American-Marietta Company on 10 October 1961, creating the Martin Marietta Corporation.

53. Nieberlein, 6.

54. "Assignment of R&D, Operational and CTL Missile Serial Numbers, Operational Directive," Martin Marietta Corporation, 20 April 1965. Courtesy of Hakanson Collection.

55. Contract AF04(694)-314, Order No. 04-964-64-136, (Amendment #1). Author's Collection.

56. *Aerojet: The Creative Company,* (Aerojet History Group), 1995. pp. III-79 and III-118.

57. "Titan Improvement Program, First Quarterly Progress Report," 1964. AFHRA, Maxwell AFB, pp. viii–47; Improved Titan Concepts, 6 January 64; Aerojet-General. AFHRA Maxwell AFB, slide presentation materials.

58. T. Von Kármán, "Aerodynamic Heating—The Temperature Barrier in Aeronautics," pp. 140–42. Proceedings of the Symposium on High Temperature—A Tool for the Future, Berkeley, California, June 1956.

59. Dr. William T. Barry, personal interview and correspondence, February 1997. Barry was a materials scientist consultant at the General Electric Space Sciences Laboratory in Philadelphia during the development of the Mark 6 reentry vehicle.

60. Mark Morton, *Progress in Reentry Recovery Vehicle Development,* Missile and Space Vehicle Department, (General Electric, Philadelphia), Pennsylvania, 2 January 61, pp. 6–9.

61. Donald Schmidt, personal communication, April 1997.

62. I. J. Gruntfest and L. H. Shenker, "Behavior of Reinforced Plastics at Very High Temperatures," *Modern Plastics* 35 (June 1958): 155-63. The stabilized arc jets were not used for Titan II reentry vehicle development.

63. Barry, February 1997; Schmidt, February 1997.

64. "Technical Developments of the Missile and Space Division," (General Electric, no date given) pp. 3–8. Provided from the Donald Schmidt collection.

65. The loss of weight, as ablated material sloughs off the vehicle, as well as changes in the coefficient of drag due to surface roughening, caused changes to the ballistic coefficient.

66. "Thor-Able and Atlas Reentry Recovery Programs," (Missile and Space Vehicle Department, General Electric Defense Electronics Division, no date); M. Morton, "Progress in Reentry Recovery Development," Missile and Space Vehicle Department, (General Electric Defense Electronics Division), 2 January 1961. Personal collection of Donald Schmidt.

67. "Detail Design Specifications for Model SM-68B Missile, Including Addendum for XSM-68B," 1 September 1961, p. A-7. BSM-TII-C20000, Lockheed Martin Astronautics Research Library, Denver, Colorado. Hereafter referred to as DDS.

68. DDS, p. A-8. The Mark 4 reentry vehicle, including the W-49 warhead, weighed 4,000 pounds.

69. While all of the components were critical, the suggestion of n-butylphosphoric acid by I. J. Gruntfest was a particularly key point. Barry, February 1997.

70. Both dimensions are from a cross section drawing of the reentry vehicle created by Martin Marietta Company for use as a satellite fairing. Author's Collection.

71. Barry, February 1997. This mechanism was developed by Barry in conjunction with the formulation of the Series 100 plastic, U.S. Patent 3,177,175 (1965).

72. Ibid.

73. "Operational Test Report, LONG LIGHT, Titan II, LGM-25C, AFSN 62-12298 (B-55), 20 April 1966," Microfilm N0568, p. 23. AFHRA, Maxwell AFB, Alabama. All material on this reel was unclassified.

74. DDS, p. 16.

75. Collins, 446. "Hearings before the Committee on Armed Services, United States Senate, Part 7, FY 1983," p. 4172. Testimony before the committee indicated that the B-53 Y1 gravity bomb had a yield of 9 megatons.

76. DDS, p. 8. The Model 1037J decoy was a modified version of the optical reentry decoy 1033.

77. T. Greenwood, *Making the MIRV: A Study of Defense Decision Making* (Cambridge, Mass.: Ballinger Publishing, 1975), 4.

78. Ibid., 42.

79. R. Smoke, *National Security and the Nuclear Dilemma* (New York: Random House, 1987), 161.

80. Collins, 446.

81. "Titan II System Improvements," slide presentation materials from Ballistic Systems Division presentation, 1965, AFHRC, Maxwell AFB.

82. G. Spinardi, *From Polaris to Trident: The Development of the US Fleet Ballistic Missile Technology* (New York: Cambridge University Press, 1994), 106.

83. Greenwood, 43.

84. Greene, 114.

85. A total of seven Titan I Lot M missiles were used to test the prototype inertial measurement units and missile guidance computers for Titan II.

86. STL Final Report, p. 177; "Illustrated Parts Breakdown, Inertial Measurement Unit MX-6363/DW-11E," Technical Order 11G22-4-2-4, pp. 1–9, Titan Missile Museum National Historic Landmark Archives, Sahuarita, Arizona.

87. "History of Strategic Air Command, 1 July 1975 to 31 December 1976, Historical Study No. 161, Vol. XI, Chapter V, Exhibit 40-Operations Plan Number 83-76: Titan II Launches in Support of Universal Space Guidance System," K416.01-161, p. ii. This document is classified SECRET. The information used is unclassified. AFHRA, Maxwell AFB, Alabama.

88. *Universal Space Guidance System—System Technical Description,* December 1971. (Delco Electronics, GMC, Milwaukee, Wisconsin); Titan II RIVET HAWK briefing documents, Delco Electronics, 10 April 1978. Personal collection of Robert Popp, Delco engineer who was a member of the USGS team for the Titan II program.

89. Robert Popp, personal interviews, October 1996.

90. Popp, October 1996; F. Charlie Radaz, October 1996.

91. Robert Popp, May 1998. Popp was the Delco field service supervisor who scheduled all of the installation of the USGS modification. He credits the success of the USGS modification program during RIVET HAWK to six key Delco staff members: Marion Sanders, engineering test conductor; Hal Gebhardt, field service senior engineer; Jack Hoeft, Titan electronic technician; John Coutley, senior telemetry engineer; Kamal Odeh, senior technical writer; and Don Bueschel, quality control engineer.

92. Radaz, October 1996.

93. Popp, May 1998.

94. Jack Cozzens, personal interview, October 1996.

95. Originally the Stage I oxidizer had to be offloaded before the BVL could be removed for maintenance. With special tools, the second model of the BVL could be removed without the need for propellant transfer. Technical Order 21M-LGM25C-2-26, pp. 1-1 to 1-15; Titan Missile Museum National Historic Landmark Archives, Sahuarita, Arizona.

96. "History of the 381st SMW, McConnell AFB, July–September 1973," K-WG-381-HI-Vol. 1, p. 62. AFHRA, Maxwell AFB, Alabama. This document is classified as SECRET. The information used is unclassified.

97. Cozzens, 1996.

98. Ron Hakanson, personal interview, July 1996.

99. Missile recycle logs courtesy of Ron Hakanson.

CHAPTER IV: TITAN II RESEARCH AND DEVELOPMENT FLIGHT TEST HISTORY

1. The Titan II production line lot number assignments continued directly from the Titan I Lot M missiles which were flown to test the Titan II inertial guidance system. Thus Lot N was the designation for the research and development missiles. "WS 107C, Titan II Weapon System Final Report, January 1965," p. 32. STL Final Report, History Office, Air Force Space Command, Peterson AFB, Colorado.

2. The missiles were grouped into two categories: (a) Category I Design Verification, Eastern Test Range; N-1, N-2, N-4, N-5, N-6, N-9, N-11, N-12, N-13 and N-14; (b) Category I Design Verification, Western Test Range; N-7, N-8, N-19, N-22, N-26, N-27, N-30; Category I Weapon System Effectiveness Verification, Eastern Test Range; N-15, N-16, N-17, N-18, N-20, N-21, N-24, N-25, N-29, N-31, N-32, N-33, and N-3A; (d) Category II, Western Test Range; N-23, N-28 and B-15. Ibid., p. 392; personal communication with Harlan Weissenborn, AC Spark Plug engineer, June 1996.

3. "History of the WS-107C Weapons Branch (Titan), 1 January to 30 June 1962," 45SWHO, chap. 3. Unless otherwise noted, all flight tests utilized the Mark 6 Mod 4 General Electric reentry vehicle.

4. W. E. Greene, "The Development of the SM-68 Titan, Volume I, Narrative," 1962, Historical Office, Deputy Commander for Aerospace Systems, Air Force Systems Command. (AFSC Historical Publications Series, 62-23-1), p. 77. History Office, Air Materiel Command, Wright-Patterson AFB, Ohio.

5. OTSOT, 105.

6. Historical Report BSTRA 1 July 62-31 December 1962. p. 3, Author's collection. What BSTRA stands for is unknown.

7. Personal correspondence with Al Schaefle, Titan II program manager for Martin Company at this time, September 1997.

8. Historical Report, BSTRA, p. 4; "History of WS-107C Weapons Division (Titan): 1 July to 31 December 1962," chap. 3, pp. 3–4. 45SWHO.

9. Historical Report, BSTRA, pp. 4–5; "History of WS-107C Weapons Division (Titan): 1 July to 31 December 1962," chap. 3, p. 4. 45SWHO.

10. Ibid.

11. Historical Report, BSTRA, p. 6. "History of WS-107C Weapons Division (Titan): 1 July to 31 December 1962," chap. 3, pp. 4–5. 45SWHO.

12. Schaefle, 1997.

13. OTSOT, 126; Historical Report, BSTRA, p. 7. "History of WS-107C Weapons Division (Titan): 1 July to 31 December 1962," chap. 3, p. 5. 45SWHO.

14. OTSOT, 126; Historical Report, BSTRA, 8.

15. OTSOT, 126; "History of WS-107C Weapons Division (Titan): 1 January to 30 June 1963," chap. 3, p. 1. 45SWHO.

16. While N-16 was rated as a successful flight, the coolant used in the inertial measurement unit developed bubbles and the reentry vehicle impact was considerably off target. STL Final Report, p. 177, History Office, Air Force Space Command, Peterson AFB, Colorado.

17. OTSOT, 134, 140.

18. OTSOT, 135.

19. OTSOT, 137. This failure was followed by a partially successful launch, N-22 (61-2745) from Vandenberg Air Force Base on 20 June 1963.

20. OTSOT, 168–70.

21. STL Final Report, 534–36.

22. "N-7 Launch Program Report of 15 March 1963," p. 4. Lockheed Martin Astronautics Collection, Vandenberg Flight Operations, courtesy of Jim Touts; Personal communications with Harlan Weissenborn, June 1997.

23. Personal interview with Andy Hall, October 1996.

24. Hall, October 1996; Personal correspondence, Elmer Dunn, April 1997.

25. Dunn, April 1997.

26. Personal interview, George Sansone, July 1996.

27. Personal interviews with Don Kundich, October 1996, and John Adamoli, June 1996.

28. "Special Report N-7 Flight Analysis, March 1963," Lockheed Martin Astronautics Research Library #56269.

29. Robert Popp, personal interview, October 1996.

30. "N-7 Launch Progress Report," 15 March 1963, 10–11.

31. Kundich, October 1996; Adamoli, June 1996.

32. "N-7 Launch Progress Report," 15 March 1963, 11.

33. Personal interview, Ed Carson, October 1996.

34. Personal correspondence, Joe Melendez, tracking team leader, 23 July 1997.

35. Compiled from interview material supplied by Leo Blickley and Sid Kuphal, July 1997. Diving was a volunteer sideline for engineers and technicians in the Sea Operations Department. During 1963, 17 divers made 453 working dives and 192 training dives.

36. STL Final Report, pp. 177–78. "History of the 1st Strategic Aerospace Division, 1 January 1963–30 June 1963," pp. 64–67. K-DIV-0001-HI, Vol 1. AFHRA, Maxwell AFB, Alabama. This document is classified as SECRET. The information used is unclassified.

37. Titan II Flight Test Record, courtesy Don Kundich.

CHAPTER V: TITAN II LAUNCH COMPLEX DESIGN AND CONSTRUCTION

1. J. J. O'Sullivan, ed., "The Proceedings of the Second Protective Construction Symposium (Deep Underground Construction)," Vol. 1, Project RAND, R-341, March 24–26, 1959. This volume provides an interesting look at the early design considerations for hardened ballistic missile facilities utilizing mine construction techniques and mine shafts as possible missile sites.

2. W. R. Large Jr., "Ballistic Missile Hardening Study," 10 July 1958, p. 23, Air Force Material Center History Office, Wright-Patterson AFB, Ohio. BrigGen W. R. Large Jr. was Assistant Commander in Chief, Strategic Air Command.

3. "Ballistic Missile Hardening Study," 25–27.

4. "Missile Site Separation, October 1959," Advanced Planning Office, Deputy Commander Ballistic Missiles, Air Force Ballistic Missile Division, p. ii; Air Force Material Center, Wright-Patterson AFB, Ohio.

5. "Missile Site Separation," p. 53; S. Glasstone and P. J. Dolan, eds., *The Effects of Nuclear Weapons*, 3d edition (Department of Defense and Energy Research and Development Administration, 1977). The nuclear bomb effects computer/circular sliderule was used to make this computation.

6. "Missile Site Separation," p. 3.

7. Ibid., 6. Glasstone and Dolan, 253–56.

8. Design Analysis WS 107 A-2 Titan II Operational Base Facilities, Phase II Construction, Volume II, 5 January 1961, p. 2. Titan Missile Museum National Historic Landmark Archives, Sahuarita, Arizona.

9. Ralph M. Parsons, Titan II, First Construction Phase Structural Calculations, Standard Design (1961), p. 304. Titan Missile Museum National Historic Landmark Archives, Sahuarita, Arizona.

10. Glasstone and Dolan, *The Effects of Nuclear Weapons,* 233–34.

11. Design Analysis, 1.

12. The author wishes to thank Dr. Ted Krauthammer of Penn State University's Department of Civil and Environmental Engineering and John McCarney, vice president (retired), Ralph M. Parsons Company, for reviewing this section.

13. "Missile Systems Analyst Specialist, Volume 2," pp. 1–14, Extension Course Institute, Air Training Command. "Missile Systems Analyst Specialist, Volume 3," p. 78, Extension Course Institute, Air Training Command. Titan Missile Museum National Historic Landmark Archives, Sahuarita, Arizona.

14. Stan Goldhaber, personal communication, July 1997.

15. "Final Report Silo Closure Door Test Program, WS 107 A-2 Titan II," 22 June 1962, pp. 1–4, Ralph M. Parsons Company. Titan Missile Museum National Historic Landmark Archives.

16. Dates for the various phases of work at each base are given in the wing histories. Three training sites at Vandenberg Air Force Base were also constructed at the same time.

17. A general description of the construction process is given here. See the individual wing histories for wing-specific differences.

18. Don Boomhower, personal interview, September 1997.

19. The installation and welding of the reinforcing steel included precisely locating approximately 400 inserts for use in the subsequent construction.

20. "Titan II Structural Calculations, First Design Phase," Silo Wall Construction, Section 5-2, p. 36. Titan Missile Museum National Historic Landmark Archives, Sahuarita, Arizona.

21. "History of Davis-Monthan Air Force Base, Arizona, October 1960–January 1964," Corps of Engineers Ballistic Missile Construction Office, p. VIII-5. Office of History, Headquarters, U.S. Army Corps of Engineers, Alexandria, Virginia.

22. "History of Davis-Monthan Air Force Base," p. VIII-6.

23. "Titan II, Phase I Construction Structural Calculations, Standard Design," Ralph M. Parsons Company, 1961, pp. 327–36. Titan Missile Museum National Historic Landmark Archives.

24. U.S. Army Corps of Engineers Ballistic Missile Construction Office Series Drawings: Blast Lock Structure, Blast Doors, Plan Elevation and Construction, 105/89.0; Author's calculations based on dimensioned steel in these as-built blueprints; "Titan II Phase I Structural Calculations," p. 187. Titan Missile Museum National Historic Landmark Archives

25. "History of Davis-Monthan Air Force Base," p. VIII-9. Office of History, HQ U.S. Army Corps of Engineers, Alexandria, Virginia.

26. "U.S. Army Corps of Engineers Ballistic Missile Construction Office, History of Little Rock Area Office, 5 October 1960 to 31 July 1963," pp. 21–22. Office of History, HQ U.S. Army Corps of Engineers, Alexandria, Virginia; Launch Silo Launch Duct Steel Upper Section, 118/199. Titan Missile Museum National Historic Landmark Archives.

27 "U.S. Army Corps of Engineers Ballistic Missile Construction Office Summary Reports for McConnell AFB, 3 Oct 60 to 30 Sep 63; Davis-Monthan AFB, Oct 60 to Jan 64; Little Rock AFB, 5 Oct 60 to 31 July 63. Office of History, HQ U.S. Army Corps of Engineers, Alexandria, Virginia; "Titan II Phase II, All Bases, Final Design Calculations, Structural, Vol. I," Ralph M. Parsons Company, no date, Author's Collection; "Titan II Phase II, All Bases, Final Design Calculations, Structural, Vol. II," Ralph M. Parsons Company, no date, Section 6A, Launch Silo Shock Isolation System, Sheet 79; As-Built Drawings: Shock Isolation Platfoms, Level 3, Sheet 119/21. Titan Missile Museum Archives.

28. "U.S. Army Corps of Engineers Ballistic Missile Construction Office, History of Little Rock

Area Office," 5 October 1960–31 July 1963, pp. 22–23; McConnell AFB, pp. 38–40. Office of History, HQ U.S. Army Corps of Engineers, Alexandria, Virginia.

29. "Titan II Phase II Structural Calculations," Ralph M. Parsons Company, 1961, pp. 2, 103. Titan Missile Museum Archives, Sahuarita, Arizona. The author measured the silo closure door upper and lower surfaces at the Titan Missile Museum National Historic Landmark.

30. "Final Design Calculations, Phase II, All Bases," Ralph M. Parsons Company 1960, Sheet 103, pp. 108–10. The 3.5-inch top and bottom plates were used in this calculation. Titan Missile Museum National Historic Landmark Archives, Sahuarita, Arizona.

31. "WS 107 A-2 Titan II Operational Base Facilities Design Analysis, Phase II Construction," Ralph M. Parsons Company, 6 January 1961, p. 12. Titan Missile Museum Archives, Sahuarita, Arizona.

32. "WS 107 A-2 Titan II," p. 42. Calculated by author from as-built drawing, Launch Silo Surface Closure Door, Plan and Section 119/76, Sheet 76. Titan Missile Museum Archives.

33. Boomhower, September 1997 and November 1998.

34. "Titan II Phase II Structural Calculations," Ralph M. Parsons Company, 1961, Sheet 110. Author's collection.

35. Boomhower, September 1997, and Harry Christman, July 1999.

36. "U.S. Army Corps of Engineers Ballistic Missile Construction Office, History of McConnell Area Office," 6 October 1960–30 September 1963, pp. 42–44. Office of History, HQ U.S. Army Corps of Engineers, Alexandria, Virginia.

37. Christman, July 1999.

38. "History of the 1st Missile Division, Strategic Air Command, 1 Jan–30 Jun 1961, Volume I," Narrative, pp. 100–105; K-DIV-0001-HI-Vol. 1, AFHRA, Maxwell AFB, Alabama.

39. While technically called Site 11 by the Corps of Engineers, I have used the SAC Launch Complex number designation throughout the history to prevent confusion. A pairing of site and complex numbers for each operational base can be found in figures 5.17, 5.18, and 5.19.

40. Selection of the individual site locations took into consideration proximity to the air base while maintaining a 7 nautical mile separation between sites, accessibility off of paved roads and geological factors such as depth of water table and type of soil. The sites ranged from 2,040 to 4,183 feet in elevation and from 24.4 to 54.5 miles from Davis-Monthan. "History of Davis-Monthan Air Force Base, October 1960 to January 1964," U.S. Army Corps of Engineers Ballistic Missile Construction Office, pp. II-3 and II-4. Office of History, Headquarters U.S. Army Corps of Engineers, Alexandria, Virginia.

41. The 390th Bombardment Group (Heavy), the predecessor to the 390th SMW, had four squadrons assigned, two of which were the 570th and 571st. These two unit designations were assigned as strategic missile squadrons. "History of the 390th Strategic Missile Wing, January 1962," K-WG-390-HI-Vol. 1, pp. 1–2. AFHRA, Maxwell AFB, Alabama. This document is classified as SECRET. The information used is unclassified. "A Brief History of the 390th SMW, 15 January 1943 to 25 July 1968," pp. 5–10. Titan Missile Museum National Historic Landmark Archives, Sahuarita, Arizona.

42. "A Brief History of the 390th SMW, 15 January 1943 to 25 July 1968," pp. 17–18. Titan Missile Museum National Historic Landmark Archives, Sahuarita, Arizona.

43. The missile is listed as B-4 but with the B-5 serial number. Data from the Titan Master Schedule shows that B-5 was the missile shipped. "History of the 390th SMW, November to December 1962," K-WG-390-HI-Vol. 1, pp. 61–62. "BSD Management Data System, Titan Master Schedule, 3 July 1963," pp. 2–16. AFHRA, Maxwell AFB, Alabama. These documents are classified as SECRET. The information used is unclassified.

44. "History of the 390th SMW, April 1963," K-WG-390-HI-Vol. 1, p. 25. AFHRA, Maxwell AFB, Alabama. This document is classified as SECRET. The information used is unclassified.

45. "History of the 390th SMW, April 1963," K-WG-390-HI-Vol. 1, p. 18. AFHRA, Maxwell AFB, Alabama. This document is classified as SECRET. The information used is unclassified.

46. "History of the 390th SMW, May 1963," K-WG-390-HI-Vol. 1, pp. 32–34. AFHRA, Maxwell AFB, Alabama. This document is classified as SECRET. The information used is unclassified.

47. "History of the 1st Strategic Aerospace Division, 1 January to 30 June 1964," Microfilm P0367, p. 1682. AFHRA, Maxwell AFB, Alabama. This film roll is unclassified.

48. "History of the 390th SMW, June 1963," K-WG-390-HI-Vol. 1, p. 16. AFHRA, Maxwell AFB, Alabama. This document is classified as SECRET. The information used is unclassified; "History of the Site Activation Task Force, Davis-Mothan Air Force Base, Tucson, Arizona, 1 January 1963 to 30 June 1963), Titan Missile Museum National Historic Landmark Archives. This document lists the dates of launch complex turnover at ECC level, then an operational readinesss and launch sequench demonstration followed by acceptance by SAC. Wing histories indicate that the sites were now declared on alert.

49. "History of the 390th SMW, July–September 1963," K-WG-390-HI-Vol. 1, pp. 1, 9–12. AFHRA, Maxwell AFB, Alabama. This document is classified as SECRET. The information used is unclassified; "Historical Report, Administrative History of Ballistic Systems Division, 1 July to 31 December 1963." K243.012 Vol. 1, pp. 116–18. AFHRA, Maxwell AFB, Alabama. This document is classified SECRET. The information used is unclassified. Additional data from missile and engine history documents provided by Ron Hakanson, July 1997.

50. "History of the 390th SMW, October–December 1963," K-WG-390-HI-Vol. 1, p. 8. AFHRA, Maxwell AFB, Alabama. This document is classified as SECRET. The information used is unclassified.

51. "History of McConnell Area U.S. Army Corps of Engineers Ballistic Missile Construction Office, McConnell AFB, Kansas, 6 October 1960 to 30 September 1963," p. 27. Office of History, Headquarters, U.S. Army Corps of Engineers, Alexandria, Virginia.

52. Ibid., Appendix C–H; "BSD Management Data System Titan Master Schedule, 31 March 1964," pp. 8.1–8.43, IRIS 2055856, AFHRA, Maxwell AFB, Alabama.

53. "History of the 381st SMW, McConnell AFB, 1 August to 31 October 1962," K-WG-381-HI-Vol. 1, pp. 1–2. AFHRA, Maxwell AFB, Alabama. This document is classified as SECRET. The information used is unclassified; "The Missile Years, 1961–1981," p. 1. A partial copy of a McConnell AFB history, undated. Titan Missile Museum National Historic Landmark Archives, Sahuarita, Arizona.

54. Ibid.

55. History of the 381st SMW, McConnell AFB, December 1962," K-WG-381-HI-Vol. 1, Document 1 and 3. Available as microfilm, reel # N0565. AFHRA, Maxwell AFB, Alabama. This document is classified as SECRET. The information used is unclassified.

56. Missile airframe numbers are not available. Delivery schedules for Titan II operational airframes do not list shipment of missiles to the 381st SMW until April 1963.

57. "History of the 381st SMW, McConnell AFB, January 1963," K-WG-381-HI-Vol. 1, pp. 12–13. Available on microfilm, reel # N0565. AFHRA, Maxwell AFB, Alabama. This document is classified as SECRET. The information used is unclassified.

58. "History of the 381st SMW, McConnell AFB, February 1963," K-WG-381-HI-Vol. 1, Document 3. Available on microfilm, reel # N0565. AFHRA, Maxwell AFB, Alabama. This document is classified as SECRET. The information used is unclassified.

59. "BSD Management Data System, Titan Master Schedule, Change Notice, 31 March 1964," IRIS 205586, Section II, p. 1. AFHRA, Maxwell AFB, Alabama. This document is classified as SECRET. The information used is unclassified.

60. "Administrative History of Ballistic Systems Division, 1 July–31 December 1963," K243.012, p. 118. AFHRA, Maxwell AFB, Alabama. This document is classified as SECRET. The information used is unclassified.

61. "History of the Little Rock Area Office, Corps of Engineers Ballistic Missile Construction Office, 5 October 1960–31 July 1963," Appendix 4. Office of History, Headquarters, U.S. Army Corps of Engineers, Alexandria, Virginia."The 308th, From the Past into the Unknown," undated, p. 43. Suchecki Collection.

62. Ibid., 16–17.

63. Personal communications, MSgt Don Rawlings, USAF (Ret.), April 1998.

64. "History of the 308th Strategic Missile Wing, February 1963," K-WG-308-HI-Vol. I, pp. 13–19. AFHRA, Maxwell AFB, Alabama. This document is classified as SECRET. The information used is unclassified. "Ballistic Systems Division Management Data System Titan Master Schedule, 31 July 1963." pp. 2–16. AFHRA, Maxwell AFB, Alabama. There is confusion in the February 1963 report that implies missile B-8 was installed first. According to the Titan II Master Schedule, B-20 was installed first. To further confuse the issue, "The 308th, From the Past into the Unknown," p. 45, states that missile B-68 was the first to arrive. This appears to be an error in nomenclature, i.e., the writer mistakenly took SM-68B as a B number. The missile serial number given was for B-20, supporting the data from the Titan II Master Schedule.

65. CMSgt Walter Kundis, USAF (Ret.), personal communications, and Frank Ainsworth, May 1998.

66. See Chapter 3 for details on "Operation Wrap Up."

67. "History of the 308th Strategic Missile Wing, June 1963," K-WG-308-HI Vol. I, p. 14. AFHRA, Maxwell AFB, Alabama. This document is classified as SECRET. The information used is unclassified.

68. "Administrative History of Ballistic Systems Division, 1 July–31 December 1963," K243.012, Vol. 1, p. 118. AFHRA, Maxwell AFB, Alabama. This document is classified as SECRET. The information used is unclassified.

69. "History of the 308th Strategic Missile Wing, October to December 1963," K-WG-308-HI Vol. I, p. 7. AFHRA, Maxwell AFB, Alabama. This document is classified as SECRET. The information used is unclassified.

70. "Headquarters SAC Operations Plan, Titan II Weapons System, "Green Jug" Launch Duct Dehumidification and Miscellaneous Work," 17 March 1964, no page given. Titan Missile Museum National Historic Landmark Archives, Sahuarita, Arizona.

71. "History of the 381st SMW, McConnell AFB, 1 October to 31 December 1964," K-WG-381-HI-Vol. 1, pp. 28–29. AFHRA, Maxwell AFB, Alabama. This document is classified as SECRET. The information used is unclassified.

72. "Detailed Design Specification for Model SM-68B Missile (Including Addendum for XSM-68B, 1 September 1961," p. 30, Lockheed Martin Astronautics Research Library, Denver, Colorado.

73. "Titan II (LGM-25C) Long-Term Readiness Evaluation Management Plan, Fiscal Year 1967," p. 1. Titan Missile Museum National Historic Landmark Archives, Sahuarita, Arizona.

74. "Titan II (LGM-25C) Long-Term Readiness Evaluation Management Plan, Final Report Phase II, February 1967," pp. 4–6. Titan Missile Museum National Historic Landmark Archives, Sahuarita, Arizona.

75. "History of the 390th SMW, October–December 1967," K-WG-390-HI-Vol. 1, pp. 21–22. AFHRA, Maxwell AFB, Alabama. This document is classified as SECRET. The information used is unclassified.

76. "History of the 390th SMW, April–June 1977," K-WG-390-HI-Vol. 1, p. 96. AFHRA, Maxwell AFB, Alabama. This document is classified as SECRET. The information used is unclassified.

77. Ibid., 96–98. Summary reports of U.S. Army Corps of Engineers for each operational wing: Vandenberg AFB facilities costs were estimated from: "Sundt Corporation and Subsidiaries Space, Missile and Astronomy Facilities," courtesy Sundt Corporation, Tucson, Arizona; "Parsons Space Launch and Test Facilities," courtesy Parsons Infrastructure and Technology Group, Inc., Pasadena, California.

CHAPTER VI: TITAN II IN CONTEXT

1. Collins, 130.

2. Ruggiero, 55, 65, 67; Neufeld, 234–37.

3. S. Sagan, "SIOP-62: The Nuclear War Plan Briefing to President Kennedy," *International Security* 12:1 (Summer 1987): 35.

4. Sagan, 24.

5. D. Ball and J. Richelson, eds., *Strategic Nuclear Targeting* (Ithaca, N.Y.: Cornell University Press, 1986), 55.

6. Sagan, 22, 29, 44–50.

7. Ibid., 37–38.

8. Ball and Richelson, 62–63.

9. C. S. Gray, "The Future of Land-Based Missile Forces," Adelphi Papers 140 (London: International Institute for Strategic Studies, 1978), 32; Sagan, 24n8.

10. Hearings before the Committee on Armed Services, United States Senate, Ninety-Seventh Congress, Second Session, on S.2248, Department of Defense Authorization for Appropriations for Fiscal Year 1983, p. 4325.

11. Robert S. McNamara, Annual Defense Department Report, Fiscal Year 1965 (Washington, D.C.: U.S. Government Printing Office, 1967), 12.

12. Ball and Richelson, 74–75; Terry Terriff, *The Nixon Administration and the Making of U.S. Nuclear Strategy* (Ithaca, N.Y.: Cornell University Press, 1995), 207–8.

13. Collins, 452–53.

14. *Modernizing U.S. Strategic Offensive Forces: The Administration's Program and Alternatives,* Congressional Budget Office, 1983, p. 111.

15. M. Wilrich and J. B. Rhinelander, eds., *SALT, the Moscow Agreements and Beyond,* (New York: Free Press, 1974), 297.

16. R. P. Labrie, ed., *SALT Handbook: Key Documents and Issues, 1972–1979* (Washington, D.C.: American Enterprise Institute for Public Policy Research, 1979), 42. *SALT II and the Costs of Modernizing U.S. Strategic Forces,* Staff Working Paper, September 1979, Congressional Budget Office, Washington, D.C., p. 10.

17. Labrie, *SALT Handbook,* Article IV, 631.

CHAPTER VII: MANNING TITAN II

1. BMAT is used for continuity sake. Technical Order 21M-LGM25C-1, Section 7, pp. 7–1 to 7–5.

2. MSgt Bill Shaff, USAF (Ret.), personal correspondence, April 1999. Shaff was both an MFT and a BMAT in the Titan II program.

3. CMSgt Miguel DeZarraga, personal interview. June 1997; personal communications, January 1998.

4. Earl Blackaby, personal correspondence, March 1998.

5. "A History of the 533rd Strategic Missile Squadron, 26 February 1962–1 November 1985," p. 3, courtesy Tom Veldtman. This was an informal history of the squadron.

6. Personal interview with Bill Kelley, February 1998.

7. Personal interview and communications, Maj Thomas Herring, USAF (Ret.), July 1996.

8. Personal communications with SMSgt William C. Shaff, USAF (Ret.), and Capt William D. Leslie Jr., April 1998.

9. Leslie, April 1998.

10. Veltman, July 1998.

11. LtCol John Haley III, USAF (Ret.), and LtGen Jay Kelley, USAF (Ret.), October 1997.

12. Maj Kenneth Grunewald, USAF (Ret.), personal interview, August 1996.

13. LtCol John K. Powers (Ret.), personal interview, April 1998.

14. "History of the 390th SMW, 1 July 1973–30 September 1973," K-WF-390-HI-Vol. 2, Exhibit 27 "Report of Collateral Investigation of Missile/Explosives Accident LGM-25C, Missile Serial # 62-0012, Launch Complex 570-4 at Davis-Monthan AFB, Arizona, on 24 July 1973," pp. 6–7; Exhibit 27-D, pp. 1–10. AFHRA, Maxwell AFB, Alabama. This document is classified as SECRET. The information used is unclassified.

15. LtCol Bob Buzan, USAF (Ret.), and Capt Robert Bohnsak, USAF (Ret.), personal correspondance, February 1997.

16. "History of the 308th Strategic Missile Wing, January–March 1978," K-WG-308-HI-Vol. 1, pp. 23–26, Exhibits 45, 46. AFHRA, Maxwell AFB, Alabama. This document is classified as SECRET. The information used is unclassified

17. "History of the 381st SMW, McConnell AFB, October to December 1981," K-WG-381-HI-Vol. 1, p. 14. AFHRA, Maxwell AFB, Alabama. This document is classified as SECRET. The information used is unclassified; personal communications with SSgt Mark Hess, USAF (Ret.), February–March 1998. Hess was a member of the public relations team that dealt with the press during Cooke's stay at McConnell AFB; personal communications with Col Kenneth L. Hollinga, USAF (Ret.), April 1998.

18. Hess, February–March 1998.

CHAPTER VIII: TITAN II OPERATIONAL FLIGHT AND EVALUATION PROGRAMS HISTORY

1. "Titan II Missile Synopsis, Strategic Air Command," Attachments 1–5, 15 August 1985, Titan Missile Museum National Historic Landmark Archives.

2. "Detail Design Specifications for Model SM-68B Missile Including Addendum for XSM-68B," pp. 21–23. Research Library, Lockheed Martin Astronautics, Denver, Colorado.

3. The launch of B-1 was rescheduled due to inoperative equipment on the thrust mount damper, part of the fire and blast protective coverings found only in Launch Complexes 395-B, -C, and -D and peculiar to VAFB. Since these were not installed at the operational complexes, this hold was not counted against the program.

4. Martin Marietta Company Titan DASO booklet, F. Charles Radaz Collection.

5. R. R. Gunter, *An Introduction to Ballistic Missiles. Volume IV: Guidance Techniques* (Air Force Ballistic Missile Division and Space Technology Laboratories, 1960), 207.

6. "Titan II System Status," a briefing package, unclassified, that contains a description of the DASO results and upcoming OT flight program goals. Author's Collection.

7. "Operational Test Final Report for the LGM-25C Weapon System (Titan II), 1st Strategic Aerospace Division, Vandenberg AFB, California, 31 May 1966," K-WG-381-HI-Vol. 1, pp. 6, 44–45; microfilm reel N0568, AFHRA, Maxwell AFB, Alabama. This document is classified as SECRET. The information used is unclassified.

8. C. Hanson, *U.S. Nuclear Weapons, the Secret History* (Tex.: Aerofax, Inc., 1988), 84–85.

9. "Operational Test Final Report," 34–35.

10. "History of the 381st Strategic Missile Wing, McConnell AFB, Kansas, 1 April to 30 June 1966," K-WG-381-HI-Vol. 1, Exhibit 9, pp. 7–9; Microfilm reel N0568, AFHRA, Maxwell AFB, Alabama. This document is classified as SECRET. The information used is unclassified.

11. "History of the 1st Strategic Aerospace Division, Vandenberg AFB, California, January–June 1966," K-DIV-0001-HI-Vol. 2, pp. 40–55. Microfilm reel P0370, AFHRA, Maxwell AFB, Alabama. This document is classified as SECRET. The information used is unclassified.

12. "Operational Test Final Report," 6. The CEP value is the author's calculation using data from the OT report. Exercise GOLD FISH had a vernier engine failure, resulting in an impact that was short by 7 nautical miles. If this value is removed from the CEP calculation the result is 0.78 nautical miles. The Air Force deemed the removal of the GOLD FISH data acceptable since this was clearly an anomaly significantly outside the impact errors of the other flights.

13. "History of the 381st Strategic Missile Wing, McConnell AFB, Kansas, 1 April to 30 June 1966," K-WG-381-HI-Vol. 1, Exhibit 15, pp. 18–19; Microfilm reel N0568, AFHRA, Maxwell AFB, Alabama. This document is classified as SECRET. The information used is unclassified.

14. "Titan II Missile Synopsis, Strategic Air Command," Attachment 4, 15 August 1985, Titan Missile Museum National Historic Landmark Archives.

15. Ibid., Attachment 5.

16. "SOFT Vandenberg Operations Program Plan/Installation Plan," 1 March 1973, p. ii, Martin Marietta Company. Personal collection of F. Charles Radaz.

17. Personal communications with Robert Popp, RIVET HAWK program Delco engineer, December 1997.

18. "Detail Design Specification for Model SM-68B Missile Including Addenum for XSM-68B," p. 53a. Research Library, Lockheed Martin Astronautics, Denver, Colorado.

19. Gray, 32; R. L. Leggett, "Two Legs Do Not a Centipede Make," *Armed Forces Journal International,* pp. 20–32; MacKenzie, Appendix A, Table A.1; Collins, 446.

20. "History of the 381st Strategic Missile Wing, Kansas, McConnell AFB, July 66 to September 66," Exhibit #12, "Operational Test Final Report for the LGM-25C Weapon System, 1st Strategic Aerospace Division, Vandenberg AFB," pp. 34–35. Microfilm N0568. AFHRA, Maxwell AFB, Alabama. This document is classified as SECRET. The information used is unclassified.

21. Formulas for calculating the CEP proved hard to find. The formula is given on page 63 of "Final Report for LGM30B Minuteman Weapon System Demonstration and Shakedown Operation Program," Microfilm 2405. AFHRA, Maxwell AFB, Alabama. This document is classified as SECRET. The information used is unclassified. T. E. Caywood et al., "Guidelines for the Practice of Operational Research," *Operations Research* 19 (September 1971): 1187n; "History of the 381st Strategic Missile Wing, Kansas, McConnell AFB, July 66 to September 66," Exhibit #12, "Operational Test Final Report for the LGM-25C Weapon System, 1st Strategic Aerospace Division, Vandenberg AFB, California," p. 25. Microfilm N0568. AFHRA, Maxwell AFB, Alabama. This document is classified as SECRET. The information used is unclassified; MacKenzie, 1990. MacKenzie has an excellent discussion of the determination of CEP and arguements over its meaning. Gunter has a detailed mathematical discussion of the errors associated with CEP determinations.

22. S. Talbott, *Endgame, the Inside Story of SALT II* (New York: Harper and Row, 1979), 25, is but one example that perpetuates the idea that Titan II was a crude weapon.

23. Gunter, 207.

24. MacKenzie, Appendix A, Table A.1

25. "Titan II Service Life Analysis Program, Testing of Engines, 22–24 January 1985," p. 2. Aerojet Techsystems Company, Sacramento, California. Eugene Scoular Collection.

26. Ibid.

27. "History of the Strategic Air Command, January to June 1967, Offutt AFB, Nebraska" p. 371, Exhibit 95. AFHRA, Maxwell AFB, Alabama.This document is classified as TOP SECRET/RESTRICTED DATA. The information used is unclassified.

28. "History of Strategic Air Command FY-1969, Historical Study 116, Offutt AFB, Nebraska," Vol. XIX, chap. 4, Exhibit #7. AFHRA, Maxwell AFB, Alabama. This document is classified SECRET. The information used is unclassified.

29. "Titan II Reliability and Aging Surveillance Program Management Plan, 1981," Titan Missile Museum National Historic Landmark Archives, Sahuarita, Arizona.

30. "Assessment Report," Appendix F, pp. 12–14, 28–29; missile recycle records are courtesy of Ron Hakanson, 1997.

31. "Titan II Reliability and Aging Surveillance Program Management Engineering Test Report # 31, 6 September 1985," Titan Missile Museum National Historic Landmark Archives, Sahuarita, Arizona.

32. "Missileers' Heritage," Maj F. X. Ruggiero, USAF, 1981, p. 37. Student Research Report 2065-81, Air Command and Staff College, Air University, Maxwell AFB, Alabama.

33. "History of the 1st Strategic Aerospace Division, Vandenberg AFB, California, January–June 1965," K-DIV-0001-HI-Vol. 1, p. 127, AFHRA, Maxwell AFB, Alabama; "History of the 1st Strategic Aerospace Division, July–December 1967," K-DIV-0001-HI-Vol. 1, pp. 180–83, AFHRA, Maxwell AFB, Alabama. This document is classified as SECRET. The information used is unclassified.

34. "History of the 1st Strategic Aerospace Division, Vandenberg AFB, California, January–June 1965," K-DIV-0001-HI-Vol. 1, p. 39, AFHRA, Maxwell AFB, Alabama. This document is classified as SECRET. The information used is unclassified.

35. "History of the 1st Strategic Aerospace Division, Vandenberg AFB, California, July–December 1966," K-DIV-0001-HI-Vol. 1, pp. 10–11, 91–92, AFHRA, Maxwell AFB, Alabama. This document is classified as SECRET. The information used is unclassified.

36. "History of the 1st Strategic Aerospace Division, Vandenberg AFB, California, January–June 1967," pp. 15–16, AFHRA, Maxwell AFB, Alabama. This document is classified as SECRET. The information used is unclassified.

37. "History of the 1st Strategic Aerospace Division, Vandenberg AFB, California, July 1969–June 1970," K-DIV-0001-HI-Vol. 1, pp. 178–81; Vol. 2, Exhibit 2, AFHRA, Maxwell AFB, Alabama. This document is classified as SECRET. The information used is unclassified.

38. "Missile Systems Analyst Specialist, Volume 3: Launch Site Operations," 1973, p. 56, Titan Missile Museum National Historic Landmark Archives, Sahuarita, Arizona.

39. Personal correspondence and communication with CMSgt Carl Duggan, USAF (Ret.), July 1996.

40. Personal correspondence with LtCol Ted Suchecki, USAF (Ret.), September 1997.

41. Personal communication and interview with Randy Welch, September 1996.

42. Personal communication with Al Howton, April 1997.

43. The Titan II tank is 120 inches at the outside diameter, or outside skin line. The Titan III tank is 120 inches on the inside diameter, or inside skin line. Radaz, October 1996.

44. "History of the 308th Strategic Missile Wing, Little Rock AFB, Arkansas, 1 October to 31 December 1964," K-WG-308-HI-Vol. 1, Appendix 24, "DASO Narrative Report," pp. 4–5. AFHRA, Maxwell AFB, Alabama. This document is classified as SECRET. The information used is unclassified.

45. "History of the 308th Strategic Missile Wing, Little Rock AFB, Arkansas, 1 October to 31 December 1964," K-WG-308-HI-Vol. 1, "Performance of Operations Personnel, Project Black Widow, 13 Oct 64," Attachment 10. Appendix 24, "DASO Narrative Report," p. 12. AFHRA, Maxwell AFB, Alabama. This document is classified as SECRET. The information used is unclassified.

46. Personal communications, LtCol Ted Brown, USAF (Ret.), May 1998.

47. "DASO Narrative Report," pp. 12–13; Attachment 10. "Titan II Status," undated briefing papers. Author's Collection.

48. "History of the 308th Strategic Missile Wing, Little Rock AFB, Arkansas, 1 July to 30 September 1965," Microfilm N0297A, p. 24. Personal correspondence with LtCol Richard W. Kalishek, USAF (Ret.), March 1998. "History of the 381st Strategic Missile Wing, Kansas, McConnell AFB, 1 April to 30 June 1966," K-WG-381-HI-Vol. 1, Exhibit 12, "Operational Test Final Report for the LGM-25C Weapon System (Titan II), 31 May 1966, Headquarters, 1st Strategic Aerospace Division, Vandenberg, AFB, Californina," Microfilm N0568, pp. 8–40. AFHRA, Maxwell AFB, Alabama.

49. "History of the 308th Strategic Missile Wing, Little Rock AFB, Arkansas, 1 October to 31 December 1965," K-WG-308-HI- Vol. 1, p. 19. AFHRA, Maxwell AFB, Alabama. This document is classified as SECRET. The information used is unclassified; "History of the 381st Strategic Missile Wing, McConnell AFB, 1 April to 30 June 1966," K-WG-381-HI-Vol. 1, Exhibit 12, "Operational Test Final Report for the LGM-25C Weapon System (Titan II), 31 May 1966, Headquarters, 1st Strategic Aerospace Division, Vandenberg, AFB, California, " Microfilm N0568, pp. 8–40. AFHRA, Maxwell AFB, Alabama.

50. "History of the 381st Strategic Missile Wing, Kansas, McConnell AFB, 1 April to 30 June 1966," K-WG-381-HI-Vol. 1, Exhibit 12, "Operational Test Final Report for the LGM-25C Weapon System (Titan II), 31 May 1966, Headquarters, 1st Strategic Aerospace Division, Vandenberg, AFB, California, " Microfilm N0568, pp. 8–40. AFHRA, Maxwell AFB, Alabama.

51. "History of the 308th Strategic Missile Wing, Little Rock AFB, Arkansas, October–December 1967," K-WG-308-HI-Vol. 1, p. 37. AFHRA, Maxwell AFB, Alabama. This document is classified as SECRET. The information used is unclassified. "History of the 308th Strategic Missile Wing, Little Rock

AFB, Arkansas, October–December 1973," K-WG-308-HI-Vol. 1, p. 75. AFHRA, Maxwell AFB, Alabama. This document is classified as SECRET. The information used is unclassified.

52. F. Charlie Radaz, personal interview, October 1996.

53. *25 Years of Excellence: 308th Strategic Missile Wing 1962–1987, Little Rock AFB, Arkansas.* Booklet from the inactivation ceremony, p. 51. Author's Collection.

54. Radaz, October 1996

55. *25 Years of Excellence,* 51.

56. "History of the Strategic Air Command, Offutt AFB, Nebraska, 1 July 1975 to 31 December 1976, Historical Study 161, Volume XI, Chapter V, Exhibits 1-50," K416-01-161. Exhibits 40, 42, 43. AFHRA, Maxwell AFB, Alabama. This document is classified as SECRET. The information used is unclassified. The error distances are given as 5,777 feet long and 1,449 feet cross-range, accounting for approximately 60–70 percent of the impact error. Assuming 65 percent error in both distances yields 1.46 and 0.366 nautical miles, respectively, as the original impact values. Corrected miss distances are therefore 0.511 and 0.128 nautical miles, respectively.

57. The author wishes to thank Robert Popp, Ed Stapp, and John Hanna, all Delco guidance system engineers, for patiently explaining this problem and solution.

58. "Operations Plan Number 83-76," Exhibit 44.

59. Col Charles Simpson, USAF (Ret.), personal interview and correspondence, July 1996 and June 1997.

60. Dick Rector and CMSgt Paul Seashore, USAF (Ret.), personal interviews, October 1996.

61. "History of the 381st Strategic Missile Wing, McConnell AFB, Kansas, 1 October to 31 December 1964," K-WG-381-HI-Vol. 1, p. 39. AFHRA, Maxwell AFB, Alabama. This document is classified as SECRET. The information used is unclassified. Maj. Robert Arnold, USAF (Ret.), personal interview February 1998.

62. Arnold, February 1998.

63. "History of the 381st Strategic Missile Wing, McConnell AFB, Kansas, 1 October to 31 December 1964," K-WG-381-HI-Vol. 1, pp. 39–40. AFHRA, Maxwell AFB, Alabama. This document is classified as SECRET. The information used is unclassified; Rector, October 1996.

64. "Ordnance Study Report for Titan II (SM-68B) Missiles, DSR S11299A, Rev. 1," December 1964; p. 1, Courtesy of F. Charles Radaz.

65. Radaz, October 1996.

66. "History of the 381st Strategic Missile Wing, McConnell AFB, Kansas, 1 October to 31 December 1964," K-WG-381-HI-Vol. 1, pp. 39–40. AFHRA, Maxwell AFB, Alabama. This document is classified as SECRET. The information used is unclassified.

67. "Titan DASO 100%," p. 6. DASO booklet published by Martin Company. Author's Collection. Three of the five DASO reentry vehicle impacts were closer than one nautical mile of the intended impact point. The remaining two were just outside the one nautical mile target ring. The figure used for this information was presented in an undated but declassified Titan II Present Status briefing package, Author's Collection.

68. "History of the 381st Strategic Missile Wing, McConnell AFB, Kansas, 1 April to 30 June 1965," K-WG-381-HI-Vol. 1, Exhibit 20, p. 13. AFHRA, Maxwell AFB, Alabama. This document is classified as SECRET. The information used is unclassified.

69. Ibid., 16–18; Exhibit 20, p. 13.

70. "History of the 381st Strategic Missile Wing, McConnell AFB, Kansas, 1 July to 30 September 1965," K-WG-381-HI-Vol. 1, Exhibit 12, pp. 3–14. Microfilm 23793. AFHRA, Maxwell AFB, Alabama.

71. "History of the 381st Strategic Missile Wing, McConnell AFB, Kansas, 1 April to 30 June 1966," K-WG-381-HI-Vol. 1, Exhibit 12, "Operational Test Final Report for the LGM-25C Weapon System (Titan II), 31 May 1966," Headquarters, 1st Aerospace Division, Vandenberg AFB, California, pp. 39–40. Microfilm N0568. AFHRA, Maxwell AFB, Alabama.

72. "History of the 381st Strategic Missile Wing, McConnell AFB, Kansas, 1 October to 31 December 1965," K-WG-381-HI-Vol. 1, Exhibit 14, pp. 14–16. Microfilm N0568. AFHRA, Maxwell AFB, Alabama.

73. "History of the 381st Strategic Missile Wing, McConnell AFB, Kansas, 1 April to 30 June 1966," K-WG-381-HI-Vol. 1, Exhibit 12, "Operational Test Final Report for the LGM-25C Weapon System (Titan II), 31 May 1966," Headquarters, 1st Strategic Aerospace Division, Vandenberg, AFB, California, pp. 39–40. AFHRA, Maxwell AFB, Alabama. This document is classified as SECRET. The information used is unclassified.

74. The date listed for the launch is local time, the actual date of launch based on Zulu time would be 25 May 1966.

75. "History of the 381st Strategic Missile Wing, McConnell AFB, Kansas, 1 April to 30 June 1966," K-WG-381-HI-Vol. 1, Exhibit 15, "Follow-on Operational Test Program Report for the LGM-25C Weapon System (Titan II), Silver Bullet," Headquarters, 1st Strategic Aerospace Division, Vandenberg, AFB, California, pp. 1-29. AFHRA, Maxwell AFB, Alabama. This document is classified as SECRET. The information used is unclassified.

76. "A Brief History of the 390th Strategic Missile Wing, Davis-Monthan AFB, Arizona, 15 January 1943 to 31 December 1970," p. 26. Titan Missile Museum National Historic Landmark Archives, Sahuarita, Arizona.

77. "History of the 381st Strategic Missile Wing, McConnell AFB, Kansas, July 66 to September 66," Exhibit #12. "Operational Test Final Report for the LGM-25C Weapon System, 1st Strategic Aerospace Division, Vandenberg AFB, California" pp. 1–46. Microfilm N0568. AFHRA, Maxwell AFB, Alabama.

78. Radaz, October 1996, and Suchecki, September 1997.

79. Suchecki, September 1997

80. Ed Moran and SMSgt Ray Hersey, USAF (Ret.), personal interview, October 1996; SMSgt William C. Shaff, USAF (Ret.), written correspondence, December 1996; Radaz, October 1996.

81. Moran and Hersey, October 1996.

82. MSgt Jim Purkey, USAF (Retired), personal interview, October 1996; "History of the 390th Strategic Missile Wing, Davis-Monthan AFB, April–June 1971," K-WG-390-HI-Vol. 2, Exhibit 52, pp. 1–10 (p. 4 is missing). AFHRA, Maxwell AFB, Alabama. This document is classified as SECRET. The information used is unclassified.

83. "History of the 390th Strategic Missile Wing, Davis Monthan AFB, April–June 1971," K-WG-390-HI-Vol. 2, Exhibit 52, p. 5, AFHRA, Maxwell AFB, Alabama. This document is classified as SECRET. The information used is unclassified.

84. Ed Ryan, personal correspondence, January 1997.

85. "Hangar Queen Flies Like a Champion," pp. 8, 9, *Desert Airman,* Davis-Monthan AFB newspaper, 26 November 1971.

86. Ibid.

87. On 5 October 1973, B-69 was finally launched successfully by a crew from the 308th SMW, Mission M2-27 SSTTP. No equipment modifications were made in the operational fleet due to this incident.

CHAPTER IX: FATAL ACCIDENTS IN THE TITAN II PROGRAM

1. "Air Force Instruction 51-503: Aircraft, Missile, Nuclear and Space Accident Investigations," Department of the Air Force, 1 December 1998, Chap. 1, p. 1.

2. Ibid.

3. "Report of USAF Aerospace Safety Missile Accident Investigation Board, Missile Accident, LGM-25C-62-006, Site 373-4, Little Rock AFB, 9 August 1965," Air University Library, Maxwell AFB, Alabama, p. W-1-5.

4. Ibid., W-1-7; Narrative Description, p. 3.

5. Ibid., Investigation and Analysis, W-4, 1–7.

6. Ibid., Investigation and Analysis, Findings and Recommendations, p. W-3-3.

7. Ibid.

8. Keith Wanklyn, personal interview, July 1997. Wanklyn was the Martin Marietta Company engineer assigned the task of investigating replacement hydraulic fluids. He found a likely candidate but it was judged too expensive. Additional precautions during welding operations near hydraulic fluid sources were used instead of replacing the fluid.

9. "History of the 308th Strategic Missile Wing, Little Rock AFB, Arkansas, October–December 1965," K-WG-308-HI-Vol. 1, p. 1. AFHRA, Maxwell AFB, Alabama. This document is classified as SECRET. The information used is unclassified.

10. "History of the 308th Strategic Missile Wing, Little Rock AFB, Arkansas, January–March 1966," K-WG-308-HI-Vol. 1, p. 13. AFHRA, Maxwell AFB, Alabama. This document is classified as SECRET. The information used is unclassified.

11. "History of the 308th Strategic Missile Wing, Little Rock AFB, Arkansas, July–September 1966," K-WG-308-HI-Vol. 1, p. iii. AFHRA, Maxwell AFB, Alabama. This document is classified as SECRET. The information used is unclassified.

12. "History of the 308th Strategic Missile Wing, Little Rock AFB, Arkansas, April to September 1986," K-WG-308-HI-Vol. 1, p. 48. AFHRA, Maxwell AFB, Alabama. This document is classified as SECRET. The information used is unclassified.

13. Col Dan Jacobowitz, USAF (Ret.), and Col N. Hartman, USAF (Ret.), personal correspondence, May 1998.

14. "History of the 308th Strategic Missile Wing, Little Rock AFB, Arkansas, October–December 1976," K-WG-308-HI-Vol. 1, Exhibit 11; "AF Sergeant, 24, Died from Exposure to Freon-113 Vapors," *Arkansas Gazette,* 1 April 1977.

15. Capt Bill Howard, USAF (Ret.), personal communications, April 1998.

16. Jimmy McFadden, personal communications, April 1998.

17. Howard and McFadden, April 1998.

18. Air Force Publicity Release, 4 April 1977. Titan Missile Museum National Historic Landmark Archives, Sahuarita, Arizona.

19. "History of 381st Strategic Missile Wing, McConnell AFB, Kansas, 1 October to 31 December 1964," K-WG-381-HI-Vol. 1, pp. 45–47. AFHRA, Maxwell AFB, Alabama. This document is classified as SECRET. The information used is unclassified. Missile history records indicate that missile B-79 was installed in Launch Complex 533-7 in late December 1964, but the site was not returned to operational status until January 1965.

20. "Report of Missile Accident Investigation, Major Missile Accident, Titan II Complex 533-7, 22 September to 10 October 1978," Vol. 1, pp. 1–12. Titan Missile Museum National Historic Landmark Archives.

21. Ibid., and Tabs W-1 through W-3. *Wichita Eagle-Beacon,* 18 January 1979, pp. 1A, 6A. Bill Livingston Collection.

22. Ibid., 9–12.

23. "Assessment Report: Titan II LGM 25C Weapon Condition and Safety, for the Senate Armed Services Committee and the House Armed Services Committee," May 1980, pp. 35–38. Titan Missile Museum National Historic Landmark Archives.

24. "Major Missile Accident," pp. Tab-I-1 to Tab-I-2.

25. Wanklyn, July 1997. Mr. Wanklyn provided a summary of his diary entries for this time, as well as copy summary by John McDonald.

26. LtCol Craig Allen, personal interview, February 1998.

27. Wanklyn, July 1997.

28. Allen, February 1998; Wanklyn, July 1997.

29. "History of the 381st SMW, McConnell AFB, Kansas, October to December 1978," K-WG-381-HI-Vol. 1, p. 24. AFHRA, Maxwell AFB, Alabama. This document is classified as SECRET. The information used is unclassified. Personal interview with LtCol Craig Allen, USAF (Ret.), February 1998.

30. Personal interview with Maj Jerald Bozeman, USAF (Ret.), February 1998.

31. N. Polmar and T. Laur, eds., *Strategic Air Command* (Baltimore, Md.: Nautical & Aviation Publishing Company of America, 1990), 166.

32. "History of the 381st SMW, McConnell AFB, Kansas, January–March 1981," K-WG-381-HI-Vol. 1, p. 44. AFHRA, Maxwell AFB, Alabama. This document is classified as SECRET. The information used is unclassified. "History of the 381st SMW, McConnell AFB, Kansas, April–June, 1981," K-WG-381-HI-Vol. 1, p. 46. AFHRA, Maxwell AFB, Alabama. This document is classified as SECRET. The information used is unclassified.

33. "History of the 381st SMW, McConnell AFB, Kansas, October to December 1981," K-WG-391-HI-Vol. 1, Exhibit 57. AFHRA, Maxwell AFB, Alabama. This document is classified as SECRET. The information used is unclassified.

34. "Assessment Report: Titan II LGM 25C Weapon Condition and Safety, for the Senate Armed Services Committee and the House Armed Services Committee," May 1980. Titan Missile Museum National Historic Landmark Archives.

35. The missile was a composite of Stage I from B-98 (65-10642) and Stage II from B-25 (62-006) The Air Force assigned the Stage II missile number to composite missiles. Stage I was involved in an oxidizer leak at Launch Complex 570-8, Davis-Monthan AFB, in March 1966. A bolt patch was installed. A Phase I Reliability and Aging Surveillance Program evaluation had been conducted in February 1976 with no defects noted. Stage II was in place at Launch Complex 373-4 when the facility fire occurred on 9 August 1965. It was returned to the depot for repair which included installation of two oxidizer tank patches to correct suspected weld flaws. Examination of missile fragments after the explosion revealed no failure of either patch. Personal papers of LtCol Craig Allen, USAF (Ret.).

36. Maj Rodney Holder, USAF, personal correspondence, June 1998.

37. Comprehensive use was made of the "Report of Missile Accident Investigation: Major Missile Accident, 18 and 19 September 1980, of Titan II Complex 374-7," microfilm reel K520.3913, AFHRA, Maxwell AFB, Alabama. A complete copy was provided to the author by Col Lloyd Houchin, the Accident Investigation Board president.

38. Col John Moser, USAF (Ret.), personal interview and correspondence, October 1996.

39. SSSgt Donald V. Green, USAF (Ret.), personal conversation and communications, May 1998.

40. Col Richard A. Sandercock, USAF (Ret.), personal correspondence and conversations, October 1996–April 1997.

41. SAC sent its own Disaster Response Force, with Vice Commander in Chief, SAC, LtGen Lloyd Leavitt Jr. commanding it from the SAC command post initially. He later came to Little Rock.

42. Sandercock, April 1997.

43. BrigGen Ronald Gray, USAF (Ret.), personal interview, June 1996.

44. "Report of Missile Accident Investigation: Major Missile Accident, 18 and 19 September 1980, of Titan II Complex 374-7," microfilm reel K520.3913, pp. 10–12. AFHRA, Maxwell AFB, Alabama.

45. Ibid., "Technical Engineering Evaluation," Tab I-7, "The Effects of Equipment Reconfiguration on the Air Balance and Air Flows at Missile Complex 374-7," pp. 1–7 and Tab I-8, "Investigation of Scenarios and Sequence of Events for the Titan II Complex 374-7 Explosion," pp. 1–4.

46. Ibid., "Summary Statement Regarding Condition and Disposition of Titan II Warhead," Tab I-4, pp. 1–2.

47. Ibid., Tab L, pp. 1–3.

48. "History of the 308th Strategic Missile Wing, Little Rock AFB, Arkansas, October to December 1980," K-WG-308-HI-Vol. 1, pp. 33–34. AFHRA, Maxwell AFB, Alabama. This document is classified as SECRET. The information used is unclassified.

49. "History of the 308th Strategic Missile Wing, Little Rock AFB, Arkansas, March to June 1981," K-WG-308-HI-Vol. 1, p. 4. AFHRA, Maxwell AFB, Alabama. This document is classified as SECRET. The information used is unclassified.

50. "History of the 308th Strategic Missile Wing, Little Rock AFB, Arkansas, October to December 1983," K-WG-308-HI-Vol. 1, p. 72. AFHRA, Maxwell AFB, Alabama. This document is classified as SECRET. The information used is unclassified.

51. "History of the 308th Strategic Missile Wing, Little Rock AFB, Arkansas, October to December 1981," K-WG-308-HI-Vol. 1, pp. 50–51. AFHRA, Maxwell AFB, Alabama. This document is classified as SECRET. The information used is unclassified. Work on the safe and seal of Launch Complex 374-7 was completed in early December 1982.

52. "Titan II Weapon System: Review Group Report," December 1980; "Assessment Report: Titan II LGM 25C Weapon Condition and Safety, for the Senate Armed Services Committee and the House Armed Services Committee," May 1980. Titan Missile Museum National Historic Landmark Archives.

53. "Review Group Report," Executive Summary, p. v.

54. Ibid.

55. Ibid., vii–x.

56. Ibid., x.

CHAPTER X: THE END OF THE TITAN II ERA

1. "Hearings before the Committee on Armed Services, Unites States Senate, 97th Congress, First Session, on S.815, Department of Defense Authorization for Appropriations for Fiscal Year 1982," Part 7 Strategic and Theater Nuclear Forces, Civil Defense, p. 3841.

2. Ibid.

3. *New York Times,* 24 September 1981, 1.

4. Ibid., 3 October 1981, 12.

5. Ibid., 12 November 1981, 18.

6. "Hearings before the Subcommittee on Strategic and Theater Nuclear Forces of the Committee on Armed Services, United States Senate, Ninety-Seventh Congress, First Session" (Washington, D.C.: U.S. Government Printing Office), p. 47.

7. "Hearings before the Committee on Armed Services, Unites States Senate, 97th Congress, 2nd Session, on S.2248, Department of Defense Authorization for Appropriations for Fiscal Year 1983," Part 7 Strategic and Theater Nuclear Forces, 23 February 1982, p. 4325.

8. Ibid., 4232; *Modernizing U.S. Strategic Offensive Forces: The Administration's Program and Alternatives,* Congressional Budget Office, May 1983, p. 84.

9. Ibid., 4285.

10. "Felling the Giants: Titan II Deactivation at Davis-Monthan AFB, Arizona," Special Study 84-2, 25 October 1984, Fifteenth Air Force Office of History, pp. 3–4. Air Force Space Command History Office, microfilm roll 38522. This document is classified as SECRET. The information used is unclassified.

11. "Assessment of the 390th SMW Inactivation," undated, p. 4. Titan Missile Museum National Historic Landmark Archives, Sahuarita, Arizona.

12. "Felling the Giants, pp. 6–7. The author thanks Maj Mark Clark, USAFR, for details pertaining to the routine during deactivation.

13. Ibid.

14. Fred Shaw, "Titan II Working Papers," p. 61, AFHRA, Maxwell AFB, Alabama. This document is classified as SECRET/RESTRICTED DATA. The information used is unclassified.

15. "Detail Design Specification for Model SM-68B Missile (Including Addendum for XSM-68B)," 1 September 1961, p. 37. Lockheed Martin Astronautics Research Library, Denver, Colorado.

16. "History of the 381st Strategic Missile Wing, McConnell AFB, Kansas, 1 April–8 August 1986," K-WG-381-HI-Vol. 1, p. 2, AFHRA, Maxwell AFB, Alabama. This document is classified as SECRET. The information used is unclassified.

17. "History of the 308th Strategic Missile Wing, Little Rock AFB, Arkansas, July–September 1984," K-WG-308-HI-Vol. 1, pp. 46–47, AFHRA, Maxwell AFB, Alabama. This document is classified as SECRET. The information used is unclassified.

18. "History of the 308th Strategic Missile Wing, Little Rock AFB, Arkansas, April–June 1985," K-WG-308-HI-Vol. 1, pp. 50–52. AFHRA, Maxwell AFB, Alabama. This document is classified as SECRET. The information used is unclassified.

19. Maj Neil Couch, USAF, personal interview, December 1997.

20. Col Richard Sandercock, USAF (Ret.), personal communications, 1998.

EPILOGUE

1. Titan II Systems Overview and Considerations—System Summary, 25 October 1972, Martin Marietta Company. Courtesy F. Charlie Radaz.

2. Martin Marietta Public Relations Fact Sheet, Author's Collection.

3. Titan II Manufacturing Concept Data Book, Martin Marietta, Rev A 8-29-88, p. 1-1. Author's Collection.

4. Robert Popp, personal communications, December 1999.

5. Darl D. Kemper, Titan II SLV project engineer, Lockheed Martin Astronautics, personal communications, July 1999. Museum history courtesy of Titan Missile Museum Heritage Program, LtCol Orville Doughty, USAF (Ret.), December 1998.

APPENDIX 2

1. S. J. Zagola, *Target America: The Soviet Union and the Strategic Arms Race, 1945–1964* (Novato, Calif.: Presidio Press, 1993), 258.

2. Russia's Strategic Ballistic Missile Forces CD, ©Tomma X, New Jersey; L. Freedman, *U.S. Intelligence and the Soviet Strategic Threat,* 2d Edition (Princeton, N.J.: Princeton University Press, 1986), 172.

3. *Air Force Magazine,* March 1978, 106.

4. Russia's Strategic Ballistic Missile Forces CD; Freedman, 109–10, 169.

5. Freedman, 172.

6. *Air Force Magazine,* March 1981, page 113; Russia's Strategic Ballistic Missile Forces CD.

INDEX

(An *f* following a page number refers to a figure, a *t* to a table.)

Navaho cruise missile, 1, 2
Niblett, Charles T., 271
Nike-Ajax missile engines. *See* Titan II, silo
    1/6th scale model
nitrogen tetroxide. *See* hypergolic propellants
North American Aviation Company. *See* Atlas
    engines
Nugent, Frank, 271

Offutt AFB, 14t
"On Target" 133
Operational Suitability Test Facility (OSTF), 22–
    25, 25f
"Operation Wrap-Up," 51, 134, 135, 139, 143

Parade Magazine, 176
Parsons Company (Ralph M. Parsons
    Company), 27, 110, 124
Patosnak, Jerome E., 198
Patrick AFB (Cape Canaveral, Fla.). *See* Eastern
    Test Range
Petyrk, Ray, 92
Pickford, Rollo, 44
Plattsburgh AFB, 14t
Plumb, Jeffery L., 235
Pogo phenomenon (pogo stick), 75, 77, 78, 79
Polaris submarine launched ballistic missile,
    4, 63, 181
Polumbo, Dennis, 167
Popp, Robert, 66, 86, 201
Poseidon submarine launched ballistic missile,
    63, 151
Potter, David, 92
Powell, David F., 235, 236, 237
Power, Thomas S., 3, 8
Powers, John K., 169
Pratt, Lester T., 206
Project DEFENDER, 81
Project EXTENDED LIFE, 145
Project GREEN JUG, 143
Project PACER DOWN, 233, 234
Project TOP BANNANA, 144, 162
Project YARD FENCE, 145
propellants, hypergolic. *See* hypergolic
    propellants

Radaz, Charlie, 66, 200, 204, 208, 209
Radizietta, John, 198
Ramo-Wooldridge Corporation, 3, 34

RAND Corporation, 1, 3, 13
Randolph, Johnson, 212
Ransome, John R., 212
RASP. *See* Titan II non-flight test programs
Rawlings, Don, 141
Ray, Daryl L., 170
Reagan administration: Strategic Forces
    Modernization Program, 253
Rector, Dick, 203–4
reentry vehicle. *See also* Atlas; Titan I; Titan II;
    Minuteman I, II, III
    development, 56–63, 59f, 60f, 61f
    improved reentry vehicle for Titan II, 62–63
    Mark 2, 57, 59f
    Mark 3, 58
    Mark 4, 31t, 36, 57, 60, 77, 78
    Mark 6, 36, 49, 60, 61f, 62, 62t, 76, 77, 147, 180
    Mark 12, 62, 63
    Mark 17, 62, 63
Regulus II cruise missile: inertial guidance, 64
Reinken, William, 205
Rhoads, John R., 142
Rhodus, Robert, 23–25
Ribertone, Robert B., 200
Richardson, Tom, 127
Richmond, A. P., 126
Ritland, Osmond J., 36
Riva, Stephen L., 238
RIVET CAP. *See* Titan II deactivation
RIVET HAWK, 66, 201–2. *See also* Titan II,
    inertial guidance system
Roberts, Jimmy, 239, 241, 242, 248
Robinson, Ned, 272
Rock, Kans. *See* fatal accidents, Launch
    Complex 533-37
Rollings, Joseph D., 221
Rolls Royce (of Britain). *See* Blue Streak
Romig, James C., 228
Rossborough, D. G., 248
Runholt, John, 164
Rust, Clayton A., 126, 130
RVX-1, RVX-2. *See* reentry vehicle development
Ryan, Ed, 210, 211

SALT I, SALT II. *See* Strategic Arms Limitation
    Talks
Sandaker, James R., 238
Sandercock, Richard A., 242, 243, 244
Sanders, Hurbet A., 216
Sanders, Robert A., 228
Sansone, George, 85

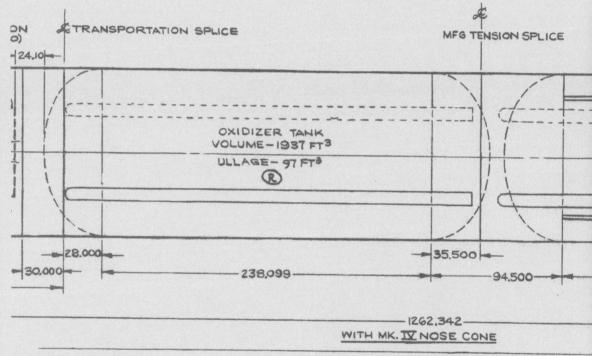

TRANSPORTATION SPLICE

MFG TENSION SPLICE

24.10

OXIDIZER TANK
VOLUME—1937 FT³
ULLAGE— 97 FT³
®

28.000

30.000

238.099

35.500

94.500

1262.342

WITH MK. IV NOSE CONE

NOSE CONE

Titan II outboard configuration.
6 July 1960. The Martin Co., Denver.